The Accuracy of Spatial Databases

The Accuracy of Spatial Databases

Edited by

Michael Goodchild and Sucharita Gopal

National Center for Geographic Information & Analysis
University of California
Santa Barbara, CA 93106

CRC Press is an imprint of the
Taylor & Francis Group, an **informa** business
A TAYLOR & FRANCIS BOOK

First published 1989 by Taylor & Francis Ltd.

Published 2020 by CRC Press
Taylor & Francis Group
6000 Broken Sound Parkway NW, Suite 300
Boca Raton, FL 33487-2742

First issued in paperback 2020

ISBN 13: 978-0-367-58019-3 (pbk)
ISBN 13: 978-0-85066-847-6 (hbk)

Visit the Taylor & Francis Web site at
http://www.taylorandfrancis.com

and the CRC Press Web site at
http://www.crcpress.com

British Library Cataloguing in Publication Data

Goodchild, Michael F
The accuracy of spatial databases.
1. Geography. Information systems. Machine-readable files
I. Title
910'.28'5574
ISBN 0-85066-847-6

Library of Congress Cataloguing in Publication Data is available

Contents

List of figures

List of tables

Preface

In December, 1988 a group of over 50 people drawn from universities, the GIS industry and data dissemination agencies met at La Casa de Maria in Montecito, California to discuss accuracy problems in spatial data. The meeting was convened by the new National Center for Geographic Information and Analysis (NCGIA) to lay out a research agenda in this area for the Center for the following two years. This volume is an edited collection of the position papers presented at that meeting.

The NCGIA was announced by the National Science Foundation on 19th August 1988, and awarded to a consortium of universities led by the University of California at Santa Barbara, and including the State University of New York at Buffalo and the University of Maine at Orono. Its published research plan (NCGIA, 1989) outlines a program aimed at the systematic removal of perceived impediments to the adoption and use of GIS technology, particularly in the social sciences. The Center's research program consists of a series of initiatives in certain key problem areas, 12 of which are planned to begin in the first three years of the Center's operation. Each initiative begins with a meeting of specialists to lay out the detailed agenda, and proceeds with research distributed over the three sites of the Center. The Montecito meeting was the first of these specialist meetings, to lay out the agenda for the first initiative, on the Accuracy of Spatial Data.

As Ronald Abler noted in his description of the processes leading to the establishment of the NCGIA,

> "Two GIS capabilities that excite enthusiasm among potential users are the ability to change map scales and the ability to overlay maps at random. Both capabilities are indeed exceedingly useful; they constitute much of the comparative advantage GIS holds over spatial analysis based on analogue maps. Both capabilities may also mislead decision makers who are unaware of the imprecision inherent in all cartography and who are untutored in the ways errors compound when map scales are changed or when maps are merged." (Abler, 1987 p.305).

Peter Burrough introduces his chapter "Data Quality, Errors and Natural Variation" by identifying:

> "a false lure (in) the attractive, high quality cartographic products that cartographers, and now computer graphics specialists, provide for their colleagues in environmental survey and resource analysis...Many soil scientists and geographers know from field experience that carefully drawn boundaries and contour lines on maps are elegant misrepresentations of changes that are often gradual, vague or fuzzy. People have been so conditioned to seeing the variation of the earth's surface portrayed either by the stepped functions of choropleth maps or by smoothly varying mathematical surfaces that they find it difficult to conceive that reality is otherwise." (Burrough, 1986 p.103).

If the burgeoning GIS industry is indeed being driven by false perceptions of data accuracy, then the truth will be devastating: even the simplest products will be suspect.

The best insurance at this point is surely to sensitize the GIS user community to the accuracy issue, and to develop tools which allow spatial data handling systems to be used in ways which are sensitive to error.

Map accuracy is a relatively minor issue in cartography, and map users are rarely aware of the problem. So why does digital spatial data handling raise issues of error and accuracy, when conventional map use does not? The following seven points establish the basis and objectives of the research initiative.

1. The precision of GIS processing is effectively infinite.

The typical vector-based GIS allocates 8 decimal digits of precision to each of its coordinates, and many allocate 16. The precision of processing is normally dictated by the precision of input, so one would expect line intersection points, for example, to be computed with at least the precision of input coordinates. However such levels of precision are much higher than the accuracy of typical GIS data. On a topographic map of 100cm by 100cm, 8 digits of precision would resolve coordinates to the nearest micron, while the average line on a topographic map is about 0.5mm wide. On the globe, 8 decimal digits would resolve positions to the nearest 10cm, which is far more precise than the typical global dataset.

The consequence of very high precision is often unwanted artifacts. Spurious or sliver polygons result when two versions of the same line, which are consistent to the accuracy of the data, are overlaid with a much higher precision. Goodchild (1978, 1979) has analyzed some of the statistical properties of spurious polygons, while Dougenik (1980) has described an algorithm for overlay which avoids the problem by working to the accuracy of the data rather than to the precision of floating point arithmetic. Raster overlay achieves much the same end by adopting a fixed spatial resolution.

2. All spatial data are of limited accuracy.

Positional data are ultimately based on measurements of distances or angles, and their accuracy depends on the instruments used. Many methods for determining positions on the earth's surface require the user to assume a particular mathematical shape, or figure, for the earth, and are therefore limited by the goodness of fit of the figure. Many geographic concepts are inherently fuzzy, such as "is North of" or "is in the Indian Ocean", so data which make use of such concepts must be correspondingly inaccurate.

More significantly in a GIS context, many of the objects which populate spatial databases are abstractions or generalizations of actual spatial variation. For example, the area labeled "soil type A" on a map of soils is not in reality all type A, and its boundaries are not sharp breaks but transition zones. Similarly, the area labeled "population density 1,000-2,000/sq.km" does not in fact have between 1,000 and 2,000 in every square km, or between 10 and 20 in every hectare, since the spatial distribution of population is punctiform and can only be approximated by a smooth surface.

The accuracy of spatial data is often described in terms of positional and attribute accuracy as if these were separable. For example, a spot elevation at a benchmark has a vertical accuracy which may be quite independent of the accuracy of its position, as the two were likely determined by different instruments. Similarly the error in the attributes of a pixel in a Landsat scene may be quite different from the error in the pixel's position.

3. The precision of GIS processing exceeds the accuracy of the data.

In principle, data should be expressed with a precision which reflects their accuracy. Thus if one measures a room temperature with a thermometer graduated in degrees, it should be quoted to the nearest degree, e.g. 70°F. After conversion to Celsius the result should be quoted as 21°C and not 21.1111..., as the accuracy is still approximately one degree. However this practice is rarely if ever followed in data processing: GIS designers typically carry the maximum precision through GIS operations, and print results to a precision which far exceeds the level justified by the accuracy of the data. In part this may be due to a lack of appropriate methods for describing and tracking accuracy, and in part to a reluctance to throw away digits, however meaningless they may be.

4. In conventional map analysis, precision is usually adapted to accuracy.

The limit to precision on a paper map is the line width, or typically about 0.5mm. Distortion in the paper medium due to changing humidity may often reduce accuracy substantially below this level. However 0.5mm remains the target accuracy in data collection and map compilation for many large series. Because it is difficult to change scale using conventional techniques, it is possible to integrate the entire process of collection, compilation, distribution and analysis around this target level of accuracy, and in this way the map user is often sensitized to the nature of the data.

The tools of conventional map analysis - overlay of transparencies, planimetry, dot counting - are crude, and operate at similar levels of precision. Thus a planimeter will generate an estimate of area with a precision consistent with the accuracy of the source document: 0.5mm accuracy in boundary lines justifies roughly 1 sq.mm accuracy in a planimetered area estimate.

5. The ability to change scale and combine data from various sources and scales in a GIS means that precision is usually not adapted to accuracy.

In the digital world, precision is no longer tied to accuracy. No current GIS warns the user when a map digitized at 1:24,000 is overlaid with one digitized at 1:1,000,000, and the result is plotted at 1:24,000, and no current GIS carries the scale of the source document as an attribute of the dataset. Few even adjust tolerances when scales change. Most vector systems perform operations such as line intersection, overlay or buffer zone generation at the full precision of the coordinates, without attention to their accuracy. As a result inaccuracy often comes as a surprise when the results of GIS analysis are checked against ground truth, or when plans developed using GIS are implemented. An agency proposing a GIS-based plan loses credibility rapidly when its proposals are found to be inconsistent with known geographical truth. Abler (1987) quotes Rhind on GIS: "we can now produce rubbish faster and with more elegance than ever before". Geographic information systems encourage the user to do things which are clearly not justified by the nature of the data involved.

6. We have no adequate means to describe the accuracy of complex spatial objects.

For simple points, positional accuracy can be described by two- dimensional measurement models, and a large literature exists on this subject in geodetic science. For complex objects, lines and areas, we now have methods for describing some of the errors introduced by the digitizing process, but none for modeling the complexity of the relationship between abstract objects and the spatial variation which they represent. Moreover digitizing errors are often relatively minor compared to the errors inherent in the process of cartographic abstraction itself. Given more comprehensive models, we would be in a position to develop methods for describing accuracy within spatial databases, and for tracking its propagation through the complexity of processes in a GIS.

7. The objective should be a measure of uncertainty on every GIS product.

Ideally, the result of describing error in the database and its propagation through GIS processes would be a set of confidence limits on GIS products. In some cases this would be relatively straightforward: an error model of the positional accuracy of a polygon could be used to generate confidence limits on an estimate of the polygon's area, as papers in this volume by Griffith and Chrisman show. In other cases the problem is more difficult, in view of the complex set of rules which are used during overlay in many GIS applications. In severe cases the concept of confidence limits may have to be replaced by a system of warnings or traps to help unwary users, to identify meaningless operations or ones where the results would be hopelessly erroneous.

The subject of the NCGIA research initiative might well be summed up in a single anecdote. GIS technology is now widely used in forestry, often to make estimates of timber yield. A forester was asked how he dealt with the problem of inaccuracy in yield estimates, and answered that the accuracy was known to be 10%. Given the problems of making such estimates based on poorly delimited stands, the natural spatial heterogeneity of forest species, and the dubious accuracy of the conversion tables used to compute board feet from tree counts, we asked for the basis of the accuracy estimate. "It's very simple - we take the estimate from the GIS and subtract 10% to be safe".

The chapters of the book represent a number of different perspectives on the accuracy problem, from efforts to simulate and model error to techniques for accommodating to it. We have chosen to organize the 23 chapters into seven sections, and to provide a brief introduction to each. The first and last sections are general overviews. In between, the papers in Section Two through Four look at error in GIS with increasing technical complexity and generality. Section Five groups together four papers on various aspects of error in the spatial analysis of socio-economic data, and Section Six contains four papers on more fundamental approaches to the particular problems of aggregation and reporting zone effects which underlie much modeling in human geography and related disciplines.

The book raises many problems and issues, and offers relatively few solutions. Although GIS is often seen as little more than a set of technical tools, we believe that the issues of accuracy which emerge when spatial data are subjected to the high-precision, objective processing of GIS are fundamental to geography and the spatial sciences generally. GIS allows cartographic and numerical data to be analyzed simultaneously rather than separately, and alters the role of the map from end product to intermediate data store. So it is inevitable that renewed interest in the analysis and modeling of spatial data would lead to questions about accuracy, reliability and meaning. In the long term, we hope that the result of this interest will be an improved understanding of the world which spatial data tries to represent, because understanding of spatial process is essential to a full understanding of the spatial forms which it generates. In the short term, we hope that this book will lead to greater awareness of accuracy issues among GIS users, and to more extensive research on error handling and modeling within spatial databases.

Acknowledgements

We wish to acknowledge the support of the US National Science Foundation, grant SES 88-10917, and to thank David Lawson for his excellent work on the figures, and Sandi Glendinning, Phil Parent and Karen Kemp for their assistance in organizing the conference.

References

Abler, R. F., 1987, The National Science Foundation National Center for Geographic Information and Analysis. *International Journal of Geographical Information Systems* **1**, 303-26.

Burrough, P. A., 1986, *Principles of Geographical Information Systems for Land Resources Assessment*. (Oxford: Clarendon)

Dougenik, J. A., 1980, WHIRLPOOL: a processor for polygon coverage data. *Proceedings, AutoCarto IV*. ASPRS/ACSM, Falls Church, VA, 304-11.

Goodchild, M. F., 1978, Statistical aspects of the polygon overlay problem. *Harvard Papers on Geographic Information Systems 6*. (Reading, MA: Addison-Wesley)

Goodchild, M. F., 1979, Effects of generalization in geographical data encoding. In *Map Data Processing*, Freeman, H., and Pieroni, G., (eds.), pp. 191-206. (New York: Academic Press)

NCGIA, 1989, The research plan of the National Center for Geographic Information and Analysis. *International Journal of Geographical Information Systems* **3**, 117-36.

Contributors

Carl G. Amrhein	Department of Geography, University of Toronto, Toronto, Ontario, M5S 1A1, Canada.
Giuseppe Arbia	Istituto di Statistica Economica, University of Rome, "La Sapienza", P. le Aldo Moro, 5-00137 Rome, Italy.
Michael Arno	Tydac Technologies Inc., 1600 Carling Avenue, Suite 310, Ottawa, Ontario, Canada, K1Z 8R7.
Rajan Batta	Department of Industrial Engineering, State University of New York, Buffalo, NY 14260.
Wolfgang Bitterlich	Tydac Technologies Inc., 1600 Carling Avenue, Suite 310, Ottawa, Ontario, Canada, K1Z 8R7.
David Brusegard	The Institute for Market and Social Analysis(IMSA), 344 Dupont Street, Suite 401, Toronto, Ontario, M5R 1V9, Canada.
Nicholas R. Chrisman	Department of Geography, University of Washington, Seattle, WA 98195.
Geoffrey Dutton	Prime Computer, Inc., Prime Park, MS-15-70, Natick, MA 01760.
Peter F. Fisher	Department of Geography, Kent State University, Kent, OH 44242-0001.
Robin Flowerdew	Department of Geography, University of Lancaster, Lancaster, England.
A. Stewart Fotheringham	National Center for Geographic Information & Analysis, State University of New York, Buffalo, NY 14260.
Michael F. Goodchild	National Center for Geographic Information & Analysis, University of California, Santa Barbara, CA 93106.
Mick Green	NorthWest Regional Research Laboratory, University of Lancaster, Lancaster, England.
Daniel A. Griffith	Department of Geography, 343, H.B. Crouse Hall, Syracuse University, Syracuse, NY 13244-1160.

Stephen C. Guptill	U.S. Geological Survey, 521 National Center, Reston, VA 22092.
Adrian Herzog	Department of Geography, University of Zurich/Irchel, Winterthurerstrasse 190, 8057 Zurich, Switzerland.
Susan Kennedy	Department of Geography, University of California, Santa Barbara, CA 93106.
Piotr H. Laskowski	Intergraph Corporation, One Madison Industrial Park, Huntsville, AL 35807.
Weldon A. Lodwick	Department of Mathematics, Box 170, 1200 Larimer Street, University of Colorado, Denver, CO 80204.
Guilio Maffini	Tydac Technologies Inc., 1600 Carling Avenue, Suite 310, Ottawa, Ontario, Canada, K1Z 8R7.
Gary Menger	The Institute for Market and Social Analysis(IMSA), 344 Dupont Street, Suite 401, Toronto, Ontario, M5R 1V9, Canada.
Stan Openshaw	NorthEast Regional Research Laboratory, CURDS, University of Newcastle, Newcastle upon Tyne, NE1 7RU, England.
Paul B. Slater	National Center for Geographic Information & Analysis, University of California, Santa Barbara, CA 93106.
David M. Theobald	Department of Geography, University of California, Santa Barbara, CA 93106.
Waldo R. Tobler	Department of Geography, University of California, Santa Barbara, CA 93106.
Howard Veregin	Department of Geography, University of California, Santa Barbara, CA 93106.
Stephen J. Walsh	Department of Geography, University of N. Carolina, Chapel Hill, NC 27599-3220.

Section I

Inaccuracy in spatial databases raises research questions at a number of levels. From the standpoint of pure research, the problems of modeling error in complex spatial data are closely related to much of the agenda of spatial statistics, which deals primarily with statistical processes in two dimensions. At the same time not everyone would agree that the problem is statistical, and fuzzy logic has been identified as a possible avenue for basic research as well. Others would argue that the problem can be tackled by searching for mathematical schemas which are robust under uncertainty.

The main motivation for interest in the accuracy of spatial databases comes from an applied perspective - the problem is real, and we need better methods for addressing it. Besides basic research, we need greater sensitivity to error on the part of GIS users, greater awareness of the kinds of errors which can occur, and techniques for recognizing and perhaps reducing their impact.

Finally, the problem is also technical, in the sense that we need explicit methods of tracking and reporting error in GIS software, and algorithms and data structures which recognize uncertainty and inaccuracy directly: in short, we need greater sensitivity to accuracy issues among GIS designers and developers.

The first section of the book contains a single paper, by Howard Veregin. The author takes a comprehensive view at the subject matter and finds a hierarchy of needs among the GIS user community. The wide-ranging, applied perspective of the paper makes a useful introduction to the sections and chapters which follow.

Chapter 1

Error modeling for the map overlay operation

Howard Veregin

Abstract

This paper critically examines some current approaches to error modeling for GIS operations, with particular emphasis on map overlay. A five-level 'hierarchy of needs' for modeling error in GIS operations is presented. This hierarchy serves to identify how the applicability of error models for map overlay is context-dependent. In applying such models it is important to consider, not only the sources of error present in spatial data, but the methods used to detect and measure these errors. The ability to successfully manage and reduce error in output products is in turn dependent on the nature of the error model applied.

Introduction

Geographic information systems permit a wide range of operations to be applied to spatial data in the production of both tabular and graphic output products. Too frequently, however, these operations are applied with little regard for the types and levels of error that may result. In numerous published articles detailing GIS applications, a critical examination of error sources is conspicuously absent and output products are presented without an associated estimate of their reliability. Unfortunately, in most cases these omissions do not imply that errors are of a sufficiently low magnitude that they may safely be ignored. Moreover, the fact that input data are themselves of relatively high quality is no guarantee that output products will be error-free.

The utility of GIS as a decision-making tool is dependent on the development and dissemination of formal models of error for GIS operations. While the goal of eliminating errors from output products may at present prove to be infeasible, decision-makers should at minimum be provided with a means of assessing the accuracy of the information upon which their decisions are based. Although error models have been developed for certain GIS operations, these models have not been widely adopted in practice. One impediment to their more widespread adoption is that no single error model and indeed no single definition of 'accuracy' is applicable in all instances.

This paper briefly reviews some existing error models for the map overlay operation, with the aim of elucidating how the applicability of such models is context-dependent. In applying such models it is important to consider, not only the sources of error present in the input data, but also the methods used to detect and measure these errors. The ability to successfully manage and reduce error associated with map overlay is in turn dependent on the nature of the error model applied.

The hierarchy of needs concept

Figure 1 presents a 'hierarchy of needs' for modeling error in GIS operations. The hierarchy of needs concept originated in the field of Humanistic Psychology, where it is used to describe the process of human emotional development. Maslow (1954) has shown that basic human needs are organized into a hierarchy of 'relative prepotency.' That is, the attainment of successively higher-order needs is dependent on the satisfaction of the needs at all lower levels in the hierarchy. The needs at a specific level must be met before an individual can progress to higher levels, and if these needs are not met, the attainment of higher levels may be retarded or thwarted

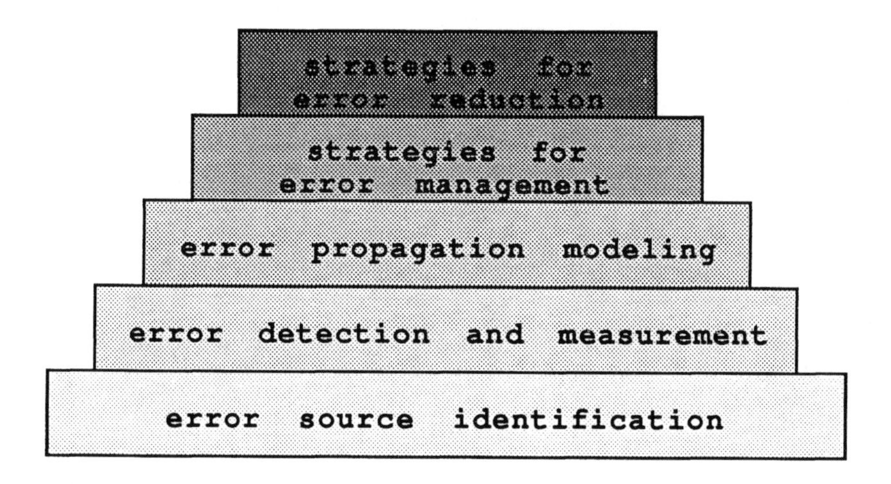

Figure 1 A hierarchy of needs for modeling error in GIS operations.

Level I of the hierarchy shown in Figure 1 (error source identification) is concerned with the basic problem of identifying and classifying sources of error in spatial data. One of the most basic error classifications (well-represented in the papers in this compendium) distinguishes between "cartographic" error, or error in positional features such as points and lines, and "thematic" error, or error in the values of the thematic attribute. Chrisman (1987) refers to these two sources of error as "positional" and "attribute" error, respectively, while Bedard (1987) calls them "locational" and "descriptive" error. Another important distinction is between "measurement" error, or imprecision in the assigned value of the cartographic or thematic features, and "conceptual" error, or error associated with the process of translating real-world features into map objects. Conceptual error is essentially synonymous with fuzziness (Chrisman, 1987). Some authors have also distinguished between "categoric" or "qualitative" error for nominal and ordinal data, and "numeric" or "quantitative" error for interval and ratio data (Bedard, 1987).

Level II (error detection and measurement) focuses on methods of assessing accuracy levels in spatial data. For thematic error, error measurement is frequently performed by constructing a confusion matrix. This approach is common when evaluating the performance of a classification algorithm for remotely sensed data (e.g., Card, 1982). While the confusion matrix is appropriate for categorical data, such as land cover, a different approach would need to be applied to numerical data. A variety of methods have also been proposed for assessing cartographic error (e.g., Chrisman, 1983;

Muller, 1987). Griffith (this volume) examines cartographic error from the standpoint of both measurement error and conceptual error, which are referred to as "digitization" error and "generalization" error, respectively.

Level III (error propagation modeling) is concerned with the consequences of applying GIS operations to spatial data. This level may be subdivided into error propagation *per se* and error "production." The former refers to the process in which errors present in spatial data are passed through a GIS operation and accumulate in output products. In some cases, the resulting errors may bear little resemblance to those in the input data, particularly when a series of operations have been applied. Error production, in contrast, refers to a situation in which errors in output products are attributable mainly to the operation itself. Thus errors may be present in output products where none existed in the data used to construct them.

In keeping with standard nomenclature (Maslow, 1962), levels I to III of the hierarchy of needs may be referred to as "basic" needs. The two highest levels of the hierarchy may be referred to as "metaneeds" since they are of equal importance and are not hierarchical. These two levels go beyond error assessment *per se* and are concerned with the inferences that may be drawn from the results of the error propagation model.

Level IV of the hierarchy (strategies for error management) focuses on methods for coping with errors in output products, or on methods for decision-making in the presence of error. A simple example is the specification of some minimum accuracy standard that the output products must attain if they are to be accepted. Such standards might apply to the thematic attribute or to the cartographic features of the output. More complex examples include methods of inference derived from artificial intelligence and information theory (e.g., Smith & Honeycutt, 1987; Stoms, 1987).

Level V (strategies for error reduction) is concerned with methods for reducing or eliminating errors in output products. Among the simplest strategies are heuristics and decision-trees which can be followed in performing particular operations (e.g., Alonso, 1968; Chrisman, 1982). Other examples include the use of expert knowledge, ancillary data, or logical consistency checks.

Map overlay

The implications of the hierarchy of needs concept for modeling error in GIS operations may be illustrated with reference to the map overlay operation. Map overlay involves the superimposition of two or more input maps, or data layers, with the aim of producing a composite map showing the intersection of the mapping units on the individual data layers. Figure 2 gives a simple example for two categorical data layers with a vector-based structure. Here the mapping units are polygons. Data layer 1 represents land cover and data layer 2 represents agricultural capability. The objective of map overlay in this case might be to delineate areas for future residential development that are non-urban and have low agricultural capability. Thus the Boolean AND operator would be applied.

Figure 3 gives an example of map overlay for two numerical data layers with a raster-based structure. Here the mapping units are cells or pixels. In this example the data layers might represent remotely sensed digital brightness values (DBVs) in two different spectral bands. Map overlay is achieved in the example by means of the arithmetic operator "minus." In remote sensing this is known as band differencing. Other arithmetic operators might also be applied, as in the calculation of band sums, ratios and normalizations. This approach is also common in suitability analysis, where scores on a set of suitability ratings are weighted and then added to derive a composite suitability score.

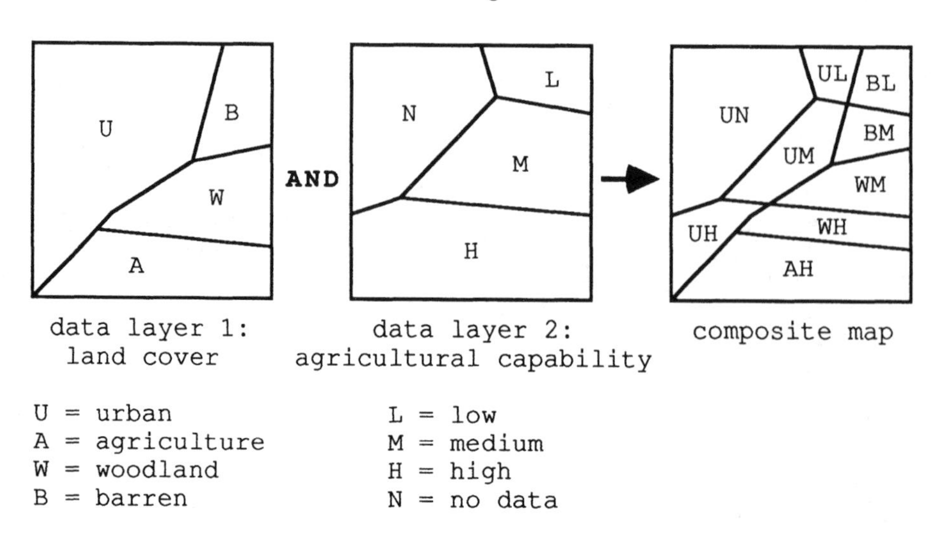

Figure 2 Map overlay for categorical data.

Figures 2 and 3 identify several factors of importance in applying an error propagation model for map overlay. Note that in Figure 2, no data are available for agricultural capability in the north-west corner of data layer 2. Presumably this is because this is part of the urban built-up area (as shown in data layer 1). However, the boundary of the urban area on data layer 1 does not coincide exactly with the area containing no data on data layer 2. There are a number of possible reasons for this discrepancy. The maps may be of different scales, they may be based on different projections, or one of the maps may have been digitized with a higher density of points along the mapping unit boundaries (Dahlberg, 1986). These are examples of cartographic errors as described above under level I of the hierarchy of needs.

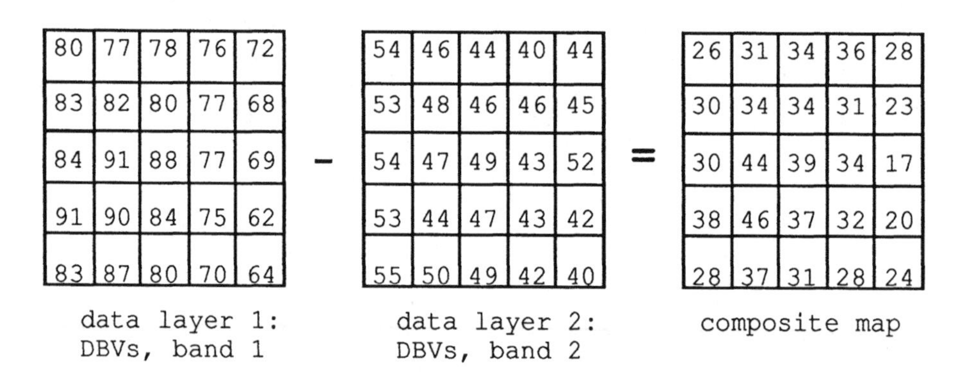

Figure 3 Map overlay for numerical data.

The discrepancy between the two data layers may also be attributed to thematic error. For example, the land cover map in data layer 1 may have been derived from remotely sensed data using a classification algorithm that is inefficient at classifying land cover at the rural-urban fringe. In this case the positions of the mapping unit boundaries are artifacts of the assignment of cover classes.

The type of error model applied in map overlay depends on the degree to which cartographic and thematic errors may be assumed to be significant. In the case of thematic error it is also important to distinguish between models for numerical data and those for categorical data. Although it is also important to distinguish between models for measurement error and those for conceptual error, the models reviewed here typically focus on the former. That is, they are based on an implicit assumption that errors are attributable, not to conceptual fuzziness and inherent uncertainty, but to imprecision in measurements.

Cartographic error

The main consequence of applying map overlay to data layers containing cartographic error is the creation of spurious polygons on the composite map. Spurious polygons are artifacts of the map overlay operation that do not represent features in the real world. They arise from positional discrepancies between representations of the same line on alternate data layers.

The spurious polygon problem has long received attention in the error modeling literature. In an early study, McAlpine and Cook (1971) observed that the number of composite map polygons, m_c, was related to the number of polygons on the individual data layers, m_i, by the equation

$$m_c = \left[\sum_{i=1}^{n} m_i^{1/2} \right]^2 \quad (1)$$

where n is the number of data layers. This equation was found to give a good approximation of the actual number of polygons produced in an empirical test. Equation (1) shows that m_c tends to rise exponentially as n increases. Empirical tests conducted by the authors revealed that small polygons tended to predominate on the composite map and that many of these polygons were indeed spurious.

Since the model is based on the number of polygons on each data layer rather than the presumed accuracy of the digital representations of these polygons, polygon size (or some other surrogate measure of spuriousness) must be employed as a basis for error management or reduction. A potential error reduction strategy, for example, might be to delete or close those polygons below a certain threshold size. This strategy would require further testing to determine an appropriate threshold value. The utility of such a strategy would be limited by the lack of a perfect correlation between polygon size and the real variable of interest. Thus the inferences that might be derived from the error model are circumscribed by reliance on a method of error detection and measurement that is not entirely appropriate to the purpose at hand.

An alternate error model for the spurious polygon problem was proposed by Goodchild (1978). As shown in Figure 4, this model is based on the intersection of a "true" cartographic line and the digitized representations of this line on different data layers. Each point of intersection along the true line is labelled with a binary symbol referring to the data layer containing the digitized representation of the line at that point. Spurious polygons are then identified by certain sequences of binary symbols. In Figure 4, for example, the string ...12121... would be generated. In this example a spurious polygon occurs for every sequence of five symbols for which adjacent symbols are different. Thus one spurious polygon would be detected, although others would also be detected if the sequence were extended.

As the true shape and location of any digitized line is rarely known, this method of detecting spurious polygons is used to derive statistics for runs of symbols under different conditions. These statistics are based on the number of vertices on each data layer, which is potentially calculable in an operational environment.

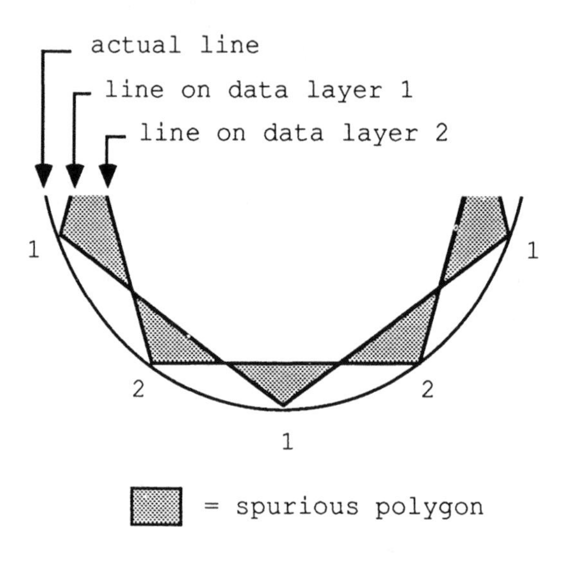

Figure 4 Detection of spurious polygons (source: Goodchild, 1978).

The maximum number of spurious polygons, s_{max}, is given by

$$s_{max} = 2 \min(v_1, v_2) - 4 \quad (2)$$

where v_1 and v_2 are the number of vertices in data layers 1 and 2, respectively. For random sequences of binary symbols, the number of spurious polygons, s, is given by

$$s = \frac{2 v_1 v_2}{v_1 + v_2} - 3 \quad (3)$$

Experimental results showed that equation (3) overestimated the number of spurious polygons to a high degree. The relationship between the number of vertices and the number of spurious polygons also showed considerable variability.

This model, like the previous one, does not suggest a definitive set of strategies for error management or reduction, although equation (3) indicates that a stochastic approach might be feasible. The model facilitates estimation of the number of spurious polygons for worst and random cases, but provides no criteria for determining whether a particular composite map polygon is spurious and should therefore be closed.

A third error model, which has been implemented in some modern commercial GIS packages, employs the epsilon band concept as a measure of positional accuracy. Algorithms now exist that eliminate spurious polygons by using this band to determine whether points on different data layers are essentially identical. An alternate method of error reduction involves the merging of data from different data layers to acquire more accurate cartographic detail. Chrisman (1987) discusses a example in which the land/water boundary on a generalized geological map was replaced with the boundary from a land use map of larger-scale.

Thematic error

The preceding models are applicable if it may be assumed that the errors present in the input data layers are cartographic in origin. As mentioned previously, errors may also be attributed to the thematic component of these data. The presence of thematic error demands a much different approach to error modeling for map overlay. Moreover, such models may be differentiated according to their applicability to numerical and categorical data. For reasons of brevity the following discussion focuses on raster data, as the principles applied in error modeling are not necessarily applicable to vector data.

Numerical data

For numerical data, the accuracy of a given data layer might be defined in terms of the deviations between the actual and estimated values associated with each pixel or cell. One approach to error modeling for map overlay might then involve calculation of the error variance of each data layer and the error covariance of each pair of data layers. The error covariance for data layers i and j, s_{ij}, is defined as

$$s_{ij} = \frac{1}{M} \sum_{m=1}^{M} (z_{mi} - \hat{z}_{mi})\ (z_{mj} - \hat{z}_{mj}) \qquad (4)$$

where M is the number of cells in a data layer, and z_{mi} and $\hat{z}_{mi}$ are the actual and estimated values for cell m in data layer i, respectively (and similarly for data layer j). When i=j the equation defines the error variance of a data layer.

This approach facilitates the calculation of the error variance of the composite map as a function of the arithmetic operator applied in map overlay. For example, when n data layers are added, the error variance of the composite map, s_c^2, is given by

$$s_c^2 = \sum_{i=1}^{n} \sum_{j=1}^{n} s_{ij} \qquad (5)$$

Although the error variance of each data layer is always positive, the error covariance may be either positive or negative. As Chrisman (1987) points out, the presence of negative covariances implies that the error variance of the composite map may be lower than the error variances of the individual data layers. This observation runs contrary to the notion that the accuracy of the composite map can never be higher than the accuracy of the least accurate data layer.

It is also possible to define the error variance of the composite map for other arithmetic operators. In the case of "minus," for example, the error variance of the composite map is defined as

$$s_c^2 = \sum_{i=1}^{n}\sum_{j=1}^{n} s_{ij} - 4\sum_{j=2}^{n} s_{1j} \tag{6}$$

A simple example is shown in Figure 5. Equations can also be derived for more complex situations involving more than one operator. Such situations are common in remote sensing in the calculation of normalizations and other band combinations.

	data layer 1: DBVs, band 1						data layer 2: DBVs, band 2						composite map				
actual	91	76	79	84	77		58	51	52	42	55		33	25	27	42	22
	87	84	80	79	77		57	55	43	49	42		30	29	37	30	35
	88	99	90	90	72	−	59	58	47	57	53	=	29	41	43	33	19
	92	94	88	76	72		60	51	47	47	44		32	43	41	29	28
	84	93	85	79	62		67	63	51	53	51		17	30	34	26	11
estimated	80	77	78	76	72		54	46	44	40	44		26	31	34	36	28
	83	82	80	77	68		53	48	46	46	45		30	34	34	31	23
	84	91	88	77	69	−	54	47	49	43	52	=	30	44	39	34	17
	91	90	84	75	62		53	44	47	43	42		38	46	37	32	20
	83	87	80	70	64		55	50	49	42	40		28	37	31	28	24

$$s_c^2 = \sum_{i=1}^{n}\sum_{j=1}^{n} s_{ij} - 4\sum_{j=2}^{n} s_{1j} = s_{11} + s_{22} + 2s_{12} - 4s_{12}$$

$$= 34.4 + 53.84 + 2(26.08) - 4(26.08) = 36.08$$

Figure 5 An error model for subtraction.

The model described above depends on the ability to measure the error variances and covariances for a set of data layers. This implies that the mapping units on the data layers coincide spatially, which is unlikely to occur for irregular polygons (in contrast to cells or pixels as described above). One solution might be to rasterize the vector data layers prior to map overlay. This would, however, introduce another source of error which would have to be accounted for in the error model (Veregin, in press).

Categorical data

For categorical data, errors in the individual data layers might be measured using the contingency table approach often applied in remote sensing. In this approach the actual and estimated cover classes for a selected sample of cells on a map are cross-tabulated. It is then possible to calculate the proportion of cells that are correctly classified and estimate the accuracy of the entire map based on inferential statistics. A more sophisticated approach utilizes a random-stratified sampling method in which the same number of samples is chosen from each cover class. This has the advantage that minor classes are not under-represented in the sample, which makes it possible to calculate the accuracy of individual classes.

As in the case of numerical data, error models for categorical data depend on the operator applied in the map overlay operation. Newcomer and Szajgin (1984) present a model for the Boolean AND operator in which the accuracy of a given data layer i, $P[\bar{E}_i]$, is defined as the proportion of cells in the data layer that are correctly classified. Given two data layers with accuracies of $P[\bar{E}_1]$ and $P[\bar{E}_2]$, the accuracy of the composite map, $P[\bar{E}_c]$, is given by

$$\begin{aligned} P[\bar{E}_c] &= P[\bar{E}_1 \cap \bar{E}_2] \\ &= P[\bar{E}_1]\, P[\bar{E}_2 | \bar{E}_1] \end{aligned} \qquad (7)$$

The conditional probability $P[\bar{E}_2 | \bar{E}_1]$ is the proportion of correctly classified cells in data layer 1 that are correctly classified in data layer 2. Equation (7) defines composite map accuracy as the intersection of the correctly classified cells on each data layer. The equation can be expanded for situations involving more than two data layers as follows:

$$\begin{aligned} P[\bar{E}_c] &= P[\bar{E}_1 \cap \bar{E}_2 \cap \ldots \cap \bar{E}_n] \\ &= P[\bar{E}_1]\, P[\bar{E}_2 | \bar{E}_1] \prod_{i=3}^{n} P[\bar{E}_i | \Theta(\bar{E}_i)] \end{aligned} \qquad (8)$$

where

$$\Theta(\bar{E}_i) = \bar{E}_1 \cap \bar{E}_2 \cap \ldots \cap \bar{E}_{i-1} \qquad (9)$$

A simple example is shown in Figure 6.

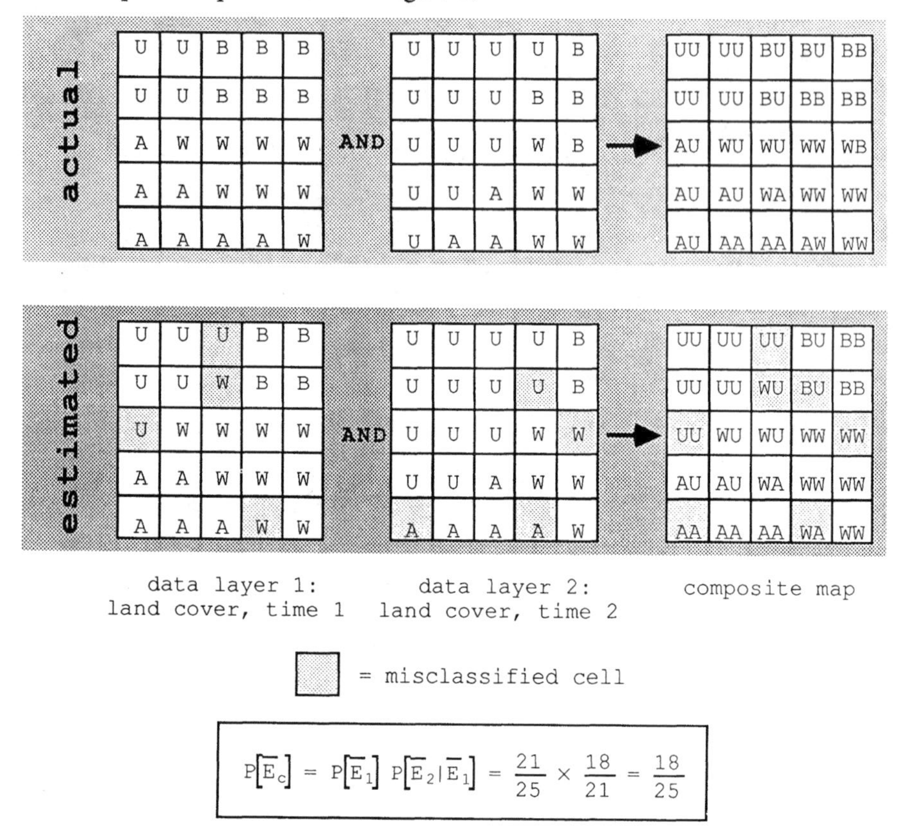

Figure 6 An error model for the AND operator.

In this example the two data layers represent land cover classes at two different dates. The objective of map overlay might be to identify changes in land cover over time.

From the preceding equations the maximum and minimum composite map accuracy can be determined. Maximum accuracy is given by

$$P\left[\bar{E}_c^{max}\right] = \min\left\{P\left[\bar{E}_i\right]\right\} \qquad i=1,2,\ldots,n \tag{10}$$

Minimum accuracy is given by

$$P\left[\bar{E}_c^{min}\right] = \max\left\{0, \left(1 - \sum_{i=1}^{n} P\left[E_i\right]\right)\right\} \tag{11}$$

where $P[E_i]$ is the proportion of cells in data layer i that are misclassified.

These formulae lead to several general conclusions about composite map accuracy for the AND operator. Composite map accuracy will at best be equal to the accuracy of the least accurate data layer. This will occur when the misclassified cells on each data layer coincide spatially with those on the least accurate data layer. At worst composite map accuracy will be equal to one minus the sum of the proportion of misclassified cells on each data layer (or to zero if this value is negative). The decline in accuracy as the number of data layers increases resembles a negative exponential curve, as shown in Figure 7. For the sake of generality, this figure portrays a set of curves for various levels of data layer accuracy (i.e., $P[\overline{E}_i] = 0.9, 0.7, 0.5, 0.3$ and 0.1) where the accuracy of each data layer is the same. These curves also depict a situation in which errors are assumed to be independent. In this case equation (8) becomes

$$P[\overline{E}_c] = \prod_{i=1}^{n} P[\overline{E}_i] \tag{12}$$

This is the form suggested by MacDougall (1975). Deviations from the general trend shown in Figure 7 will result from variations in the degree of spatial coincidence of misclassified cells. Composite map accuracy will tend to rise as the degree of coincidence increases (i.e., as $P[E_j | E_i] \rightarrow 1$).

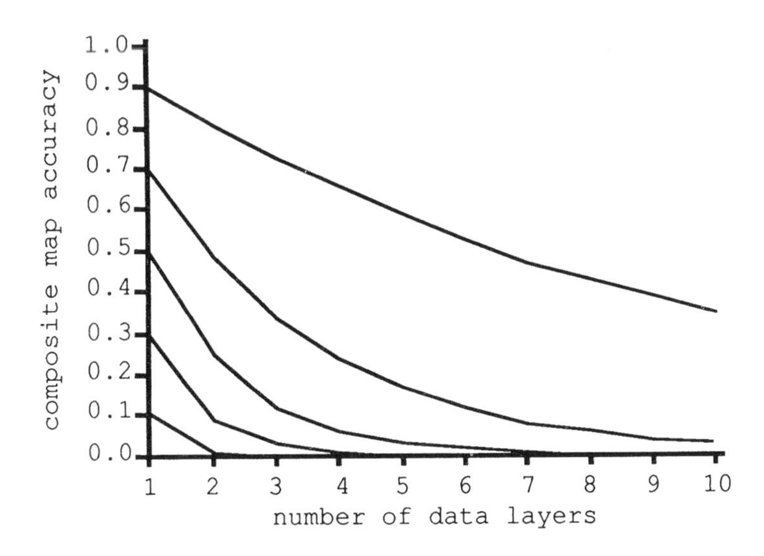

Figure 7 Composite map accuracy for the AND operator.

The preceding equations form an appropriate model for the AND operator, but are not applicable to other operators. For the AND operator, the composite map portrays the presence of a particular cover class from data layer 1 *and* a particular cover class from data layer 2 (in the case of two data layers). Thus for a cell to be accurate on the composite map it must be accurate on *all* individual data layers. For the OR operator, in contrast, the composite map portrays the presence of a particular cover class from data layer 1 *or* a particular cover class from data layer 2. In this case a cell need be accurate on only *one* data layer in order to be accurate on the composite map.

Composite map accuracy for the OR operator may be defined in terms of the intersection of the misclassified cells on each data layer. In the case of two data layers,

$$\begin{aligned} P\left[\overline{E}_c\right] &= 1 - P\left[E_1 \cap E_2\right] \\ &= 1 - P\left[E_1\right] P\left[E_2 \mid E_1\right] \end{aligned} \tag{13}$$

For situations involving more than two data layers, equation (13) may be expanded as follows:

$$\begin{aligned} P\left[\overline{E}_c\right] &= 1 - P\left[E_1 \cap E_2 \cap \ldots \cap E_n\right] \\ &= 1 - P\left[E_1\right] P\left[E_2 \mid E_1\right] \prod_{i=3}^{n} P\left[E_i \mid \Theta\left(E_i\right)\right] \end{aligned} \tag{14}$$

where

$$\Theta\left(E_i\right) = E_1 \cap E_2 \cap \ldots \cap E_{i-1} \tag{15}$$

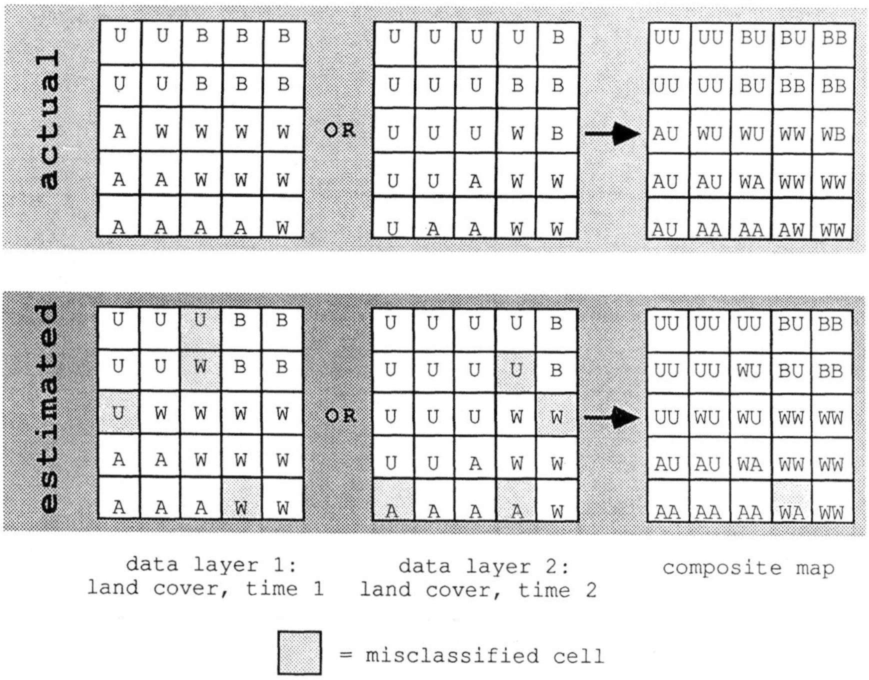

$$P\left[\overline{E}_c\right] = 1 - P\left[E_1\right] P\left[E_2 \mid E_1\right] = 1 - \frac{4}{25} \times \frac{1}{4} = \frac{24}{25}$$

Figure 8 An error model for the OR operator.

A simple example is presented in Figure 8. The objective of map overlay in this example might be to identify cells with a particular cover class on either of the two dates. The data layers depicted in Figure 8 are the same as those in Figure 6, but the composite map is more accurate in Figure 8, where the OR operator is applied.

Maximum and minimum composite map accuracy can be determined from the preceding equations. Maximum accuracy is given by

$$P\left[\bar{E}_c^{\,max}\right] = \min\left\{1,\ \left(\sum_{i=1}^{n} P\left[\bar{E}_i\right]\right)\right\} \qquad (16)$$

Minimum accuracy is given by

$$P\left[\bar{E}_c^{\,min}\right] = \max\left\{P\left[\bar{E}_i\right]\right\} \qquad i=1,2,\ldots,n \qquad (17)$$

The preceding formulae lead to several general conclusions about composite map accuracy for the OR operator. In contrast to the AND operator, composite map accuracy can never fall below the accuracy of the *most* accurate data layer. As the number of data layers increases, composite map accuracy will tend to increase, as shown in Figure 9. Thus application of the OR operator may actually produce a composite map that is *more* accurate than any of the individual data layers on which it is based. This finding runs contrary to the prevailing notion that the accuracy of the composite map can never be higher than the accuracy of the least accurate data layer (e.g., Walsh *et al.*, 1987). Figure 9, like Figure 7, depicts a situation in which errors are assumed to be independent. In this case composite map accuracy is defined as follows:

$$P\left[\bar{E}_c\right] = 1 - \prod_{i=1}^{n} P\left[E_i\right] \qquad (18)$$

The main impediment in implementing this model for categorical data lies in the calculation of the conditional probabilities. The standard technique applied in error detection and measurement for categorical data is the confusion matrix, in which the actual and estimated cover classes for a sample of cells are cross-tabulated. While this technique permits calculation of the percent of cells correctly classified, the actual locations of these cells cannot be incorporated into the structure of the matrix. This means that the conditional probabilities used in the model will normally not be calculable without an additional ground-based accuracy assessment for a set of cells that coincide spatially on each data layer. In this light the more general form of the model, as represented by equations (12) and (18), seems more applicable than the more detailed form represented by equations (8) and (14).

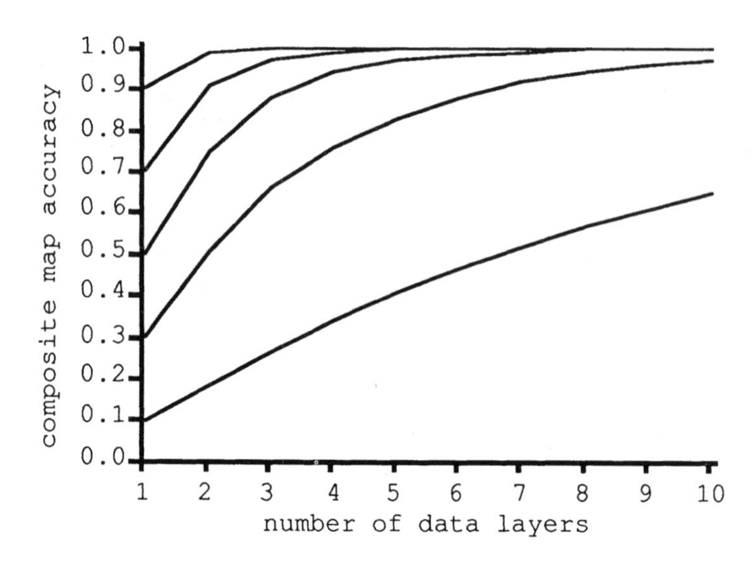

Figure 9 Composite map accuracy for the OR operator.

By de Morgan's law, the NOT of an AND operation is equivalent to the OR of the NOTs, and *vice versa.* The accuracies in these two cases are, however, not identical. For example, one might be interested in identifying cells that are non-urban on the two data layers shown in Figure 6. If the NOT of the AND operation was obtained, then composite map accuracy would be assessed by equation (8). An identical composite map could be produced by applying the OR operator to the NOTs of each data layer, but in this case composite map accuracy would be assessed by equation (14). The discrepancy in accuracy reflects a difference in the interpretation of the two composite maps. The first depicts the NOT of the intersection of the urban cells, while the second depicts the union of the non-urban cells.

This example shows that the inferences that may be derived from an error propagation model cannot be isolated from the error model itself. Even when output products appear to be identical, they may exhibit marked differences in accuracy that are explicable only in light of the types of errors that the model is capable of assessing. This observation suggests that the assessment of output product accuracy based solely on the products themselves (e.g., Estes, 1988) may not prove to be particularly cost-effective. Such assessments may be useful in themselves, but cannot necessarily be incorporated into a more comprehensive error model, which would be desirable, for example, when additional GIS operations were applied.

Conclusion

There are additional issues of importance in error modeling for map overlay that are beyond the scope of this paper. First, the error models discussed here tend to focus on measurement, rather than conceptual, error. Conceptual error is, however, of considerable significance in map overlay. A commonly-cited example is the overlay of data layers with indeterminate or fuzzy boundaries, such as soil types. In this case there exists no accurate standard against which to assess data layer accuracy, because the cartographic features of the data layers do not exist in the real world. Instead they represent abstractions of real-world features, created for the purpose of map representation. The error models applied in this case would be interpretable only with explicit reference to the abstractions that have been made.

Another issue of importance stems from the use of sampling methodology. The models discussed here would need to be extended to account for the probabilistic nature of samples. These models also tend to focus on raster data and would need to be extended to account for irregularities in the shape and size of polygons. Reliance on a rater-based data structure ignores the problem of mapping units which do not coincide on different data layers, which complicates the assessment of composite map accuracy through the introduction of a weighting factor.

Finally, the models reviewed in this paper typically provide no means of assessing the spatial distribution of error in composite maps. Thus one might reject a map as inaccurate when portions of it contain little error, or accept a map as accurate when portions of it contain a great deal of error. Clarke (1985) gives some examples of error surfaces for the operation of vector to raster conversion which exhibit spatial heterogeneity in the distribution of error.

This paper has attempted to elucidate, within the framework of a hierarchy of needs, how the applicability of error models for the map overlay operation is dependent on the methods employed for detecting and measuring errors in spatial data. These methods are in turn dependent on the error sources deemed significant in a given situation. The ability to apply a specific error model cannot be isolated from these more basic questions. Likewise, the inferences that may be drawn from an error propagation model, or more specifically, the strategies that may be developed for error management and reduction, cannot be divorced from the nature of the error model itself.

Acknowledgement

This paper is part of a larger research project entitled *Error modeling for spatial data: The case of the nuclear winter base term*, which is funded by the Institute on Global Conflict and Cooperation, University of California, San Diego.

References

Alonso, W., 1968, Predicting best with imperfect data. *AIP Journal*, 248-255.

Bedard, Y., 1987, Uncertainties in land information systems databases. *Proceedings, Auto-Carto 8*, 175-184.

Card, D. H., 1982, Using known map category marginal frequencies to improve estimates of thematic map accuracy. *Photogrammetric Engineering and Remote Sensing* **48**(3), 431-439.

Chrisman, N. R., 1982, Beyond accuracy assessments: correction of a misclassification. *Proceedings, American Society for Photogrammetry and Remote Sensing*, 123-132.

Chrisman, N. R., 1983, A theory of cartographic error and its measurement in digital data bases. *Proceedings, Auto-Carto 5*, 159-168.

Chrisman, N. R., 1987, The accuracy of map overlays: a reassessment. *Landscape and Urban Planning* **14**, 427-439.

Clarke, K. C., 1985, A comparative analysis of polygon to raster interpolation methods. *Photogrammetric Engineering and Remote Sensing* **51**(5), 575-582.

Dahlberg, R. E., 1986, Combining data from different sources. *Surveying and Mapping* **46**(2), 141-149.

Estes, J., 1988, GIS product accuracy *Paper presented at Specialist Meeting on Accuracy of Spatial Databases*, National Center for Geographic Information and Analysis, Montecito CA, December 1988.

Goodchild, M. F., 1978, Statistical aspects of the polygon overlay problem. *Harvard Papers on Geographic Information Systems*, Vol. **6**. (Reading, MA: Addison-Wesley)

Griffith, D., 1989, Distance calculations and errors in geographic databases, *this volume*.

MacDougall, E. B., 1975, The accuracy of map overlays. *Landscape and Planning* **2**, 23-30.

Maslow, A. H., 1954, *Motivation and Personality*. (New York: Harper & Row)

Maslow, A. H., 1962, *Toward a Psychology of Being*. (New York: Van Nostrand).

McAlpine, J. R., and Cook, B. G., 1971, Data reliability from map overlay. *Proceedings, Australian and New Zealand Association for the Advancement of Science, 43rd Congress*.

Muller, J. C., 1987, The concept of error in cartography. *Cartographica* **24**(2), 1-15.

Newcomer, J. A., and Szajgin, J., 1984, Accumulation of thematic map errors in digital overlay analysis. *The American Cartographer* **11**(1) 58-62.

Smith, G. R., and Honeycutt., D. M., 1987, Geographic data uncertainty, decision making and the value of information. *Proceedings, Second Annual International Conference, Exhibits and Workshops on Geographic Information Systems*, 300-312.

Stoms, D., 1987, Reasoning with uncertainty in intelligent geographic information systems. *Proceedings, Second Annual International Conference, Exhibits and Workshops on Geographic Information Systems*, 693-700.

Veregin, H., in press, A review of error models for vector to raster conversion. *The Operational Geographer*.

Walsh, S. J., Lightfoot, D. R., and Butler, D. R., 1987, Recognition and assessment of error in geographic information systems. *Photogrammetric Engineering and Remote Sensing* **10**,1423-1430.

Section II

The power of GIS to combine data from many sources, using many different scales, projections and data models is one of its major strengths. But the inevitable consequences of combining data sources and changing scales is a loss of sensitivity to each data set's idiosyncracies, particularly its accuracy. The papers in this section look at the nature of errors and propose various approaches to coping with them.

Map overlay is perhaps the key operation in GIS, and Chrisman's paper appropriately looks at overlay from the perspective of data accuracy. The paper proposes a model of error based on separation into attribute and positional components, and looks at some promising directions for further research.

Walsh's paper illustrates the complexity of data in many GIS applications by combining remotely sensed imagery from Landstat, elevation data in the form of digital elevation models (DEMs) derived from photogrammetry, and soil information derived from maps, which in turn were compiled from field surveys. By the time they enter the GIS, such data have been repeatedly processed and stripped of much of the information on which knowledge of uncertainty might be based. It is rare, for example, to find any form of data quality report attached to the typical map product. Thus it is hardly surprising that the complex GIS analysis on the database treats its input as error-free.

Surveyors and cartographers traditionally attempt to compensate for uncertainty by applying expert knowledge in the preparation of maps and map-based data. For example, a cartographer would not allow a stream to cross a slope whose contours indicated convex-upwards curvature, but would modify the contours or move the stream. Fisher proposes to extend this approach by constructing an expert system to detect and possibly to correct errors in soil maps.

Perhaps the best understood and most studied errors in GIS are those introduced during digitizing, particularly by inaccurate placement of the digitizer cross-hairs during point-mode digitizing. Maffini and his colleagues from TYDAC Technologies have looked at the effects of map scale and speed of digitizing on errors, and how such errors later propagate through GIS operations. They suggest ways of ensuring quality control in digitizing, and ways in which the user can accommodate to expected errors.

The previous papers in this section discussed the types and nature of errors in map overlay and digitizing operations. Lodwick's paper, which is the last in this section, develops the mathematical framework for the analysis of error propagation through the overlay operation, given knowledge of errors in the input layers. Two types of suitability analysis are compared although as Veregin's chapter showed earlier, the full set of possibilities in map overlay presents a very extensive agenda for future work.

Chapter 2

Modeling error in overlaid categorical maps

Nicholas R. Chrisman

Abstract

A model for error in geographic information should take high priority in the research agenda to support the recent explosion in use of GIS. While the treatment of surfaces and spatial autocorrelation is more mathematically tractable, much of the GIS layers consist of categorical coverages, analyzed through polygon overlay. This paper provides a basic taxonomy of forms of error in this type of information and provides some questions for future research.

Background

Geographic information systems (GIS) need little introduction. Over the past twenty years, this technology has grown from rare to prevalent. Twenty, or even ten, years ago there were few operational systems. Recently, hundreds of systems have been sold to federal, state and local governments. Utilities, forestry, mineral and other industries are investing millions of dollars in geographically referenced data bases. One market study expects an investment of $90 billion in facility management systems before the turn of the century, while another study projects a solid start in this direction with $500 million in 1989. Despite these developments, the tools of analysis have not expanded at the same rate. A pervasive use of GIS technology simply automates traditional map overlays. In addition, most of the results do not include the likely error associated with the data.

The general topic of error in GIS deserves high priority in research. Spatial information comes in many forms, so it will be necessary to follow parallel lines of investigation. This paper argues that many categorical maps used in GIS overlay analysis require a different error model from the more common ones developed in previous research.

Polygon overlay

A large portion of the data collected and managed by geographic information systems comes in the form of polygons, closed two-dimensional figures with associated attributes. The current wave of GIS software concentrates much effort on the relational structure of these attributes, harking back to the old geographic paradigm of the "Geographical Matrix" presented by Brian Berry (1964). However, GIS involves operations beyond the standard manipulation of attributes provided by packages such as SPSS, SAS or a relational database. The distinctive operations require spatial structure and an explicitly spatial formulation of error.

A number of authors have studied the diverse operations performed in spatial analysis (Chrisman, 1982a; Tomlin, 1983; Guevarra, 1983). Each study has found that co-location, discovered by map overlay, is fundamental to the construction of more complex procedures. Overlay uses positional information (geometry) to construct new zones which share parentage from the separate source layers. Map overlay has been used as a visual, non-quantitative procedure for many years (Sauer, 1918; Steinitz *et al.*, 1976). Fundamentally, overlay applies Venn diagrams (set logic) to geographic questions as described by Warntz (1968). Ten years ago, the research problem was to calculate a map overlay efficiently, and thus turn an inaccurate manual process into a more reliable automated one. Now many software packages can perform the operation and they have begun to flood the market.

The development process is not complete yet. No current software system tempers its results with an error estimate. This failure is not due to devious motives, but rather the lack of appropriate tools. There have been a few cautionary articles describing errors and misapplications of overlay analysis (MacDougall, 1975; Hopkins, 1977), but the subject is not completely explored.

Error in overlaid maps must be studied to ensure proper use of geographic information systems. A first goal is to build a model of error appropriate to the kind of information commonly used. Then the effects of this error must be carried through to construct new analytical procedures. This paper presents some new approaches that may provide the basis for a more comprehensive theory of error. Section 1, below, describes the concept for a theory of map error. Section 2 describes previous work on the analysis of overlaid maps and the analytical procedures that form the basis for this project. Section 3 discusses questions for further research.

Building a model of map error

The most critical task in improving spatial analytical methods for GIS is to develop a generally accepted model of error in the data. Most advances in statistical analysis for geographic data have relied upon models of error developed in other disciplines. The problems of overlaid categorical maps, however, will not respond to a borrowing strategy.

The statistical concept of error is hard to grasp by those who come from the long tradition of manual cartography. As the term is used by traditional cartographers, error is a bad thing, and the profession has the intention to wipe out all error. However, this feeling is based on a different definition of error from the statistical one. In cartographic practice, the phenomenon termed error fits most closely to the concept of *blunder* in photogrammetry. A blunder is a mistake which can be easily detected, and hence should be recognized and removed. There is no quarrel with procedures to remove blunders, but the profession must recognize that there are other, more subtle, forms of error. This form of error represents unavoidable fluctuation, more akin to the sampling errors treated by classical statistics. Errors creep into maps from a variety of sources, and each source may lead to a different type of effect. Although it may be possible to reduce these fluctuations with higher cost technology, the effort may not be worth the investment. Once errors are understood they form the basis for probabilistic statements. The analytical strengths of many sciences are based on stochastic models, not deterministic structures with zero error.

Definition of categorical coverage

Eventually, there is a need for a model of error treating all forms of spatial information. For the purposes of a manageable project, it is necessary to restrict attention to a specific type of information. This paper limits its attention to one kind of two-dimensional distribution, termed a *categorical coverage.*

A categorical coverage is a specific type of polygon map used quite frequently for GIS applications. To arrive at a detailed definition, it is important to distinguish this form of polygon data from a similar, better studied case. Much of the development of spatial statistics has been fueled by the availability of census (or similar) information for spatial collection units. In particular, spatial autocorrelation was developed to treat census-like zones, often termed "modifiable units". Although spatial autocorrelation is a powerful tool and provides a useful beginning, it is based on arbitrary collection units (Figure 1).

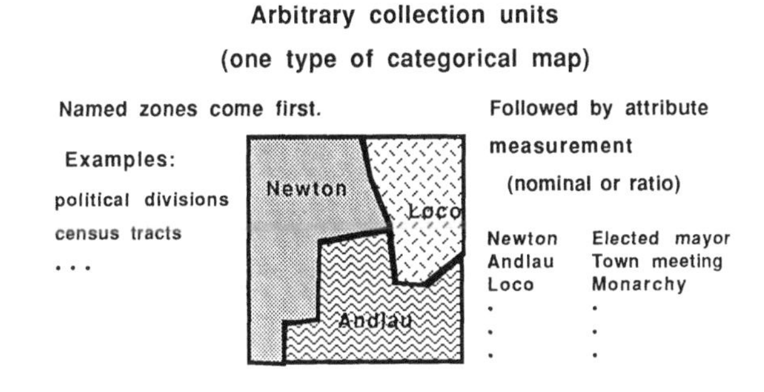

Figure 1 Arbitrary collection units (one type of categorical map).

The important consideration is which component, the spatial description or the attribute, takes logical precedence. In the pure case (administrative units such as municipalities), the positional description of the object precedes any attributes assigned. These maps are *choropleth* maps in the purest sense, because the places exist, then they are filled. [Choropleth has now come to refer to categorical maps derived from classed continuous distributions, but that does not alter the etymology.] Collection zones are usually arbitrary; so it is important to develop a statistical procedure, such as spatial autocorrelation, that attempts to remove the bias introduced by accidents of zone boundaries. This procedure assumes that some underlying spatial distribution has been obscured by the use of collection zones. The crux of the procedure is the weight matrix, which usually is based on some topological or geometric relations between the collection units. In effect, the weight matrix encapsulates a specific model of the likely error in using attribute information attached to the collection zones.

A substantial body of spatial analysis is applied to arbitrary collection zones, particularly in fields such as social and economic geography. Often the unit of analysis should be the person, firm or household, but spatially aggregated statistics are the only sources available. Substantial progress has been made on these spatial models, however, other forms of spatial data require a different model.

Many of the users of GIS software do not rely upon collection unit sources. The layers fed into a GIS are more likely to be soil maps, vegetation maps, ownership parcels, and many more. Although the distinction is not absolute, these maps derive from a different approach (Figure 2). These maps may be displayed as choropleth, but similarity of graphic display obscures fundamental differences. Some system of classification (the soil taxonomy, the vegetation classes, and even the list of taxable parcels) logically precedes the map. The map results from assigning each portion of the area into one class or another. Issues of positional accuracy, scale and other cartographic concerns become much more prominent than they are in the collection zone case. The model of error implicit in spatial autocorrelation relies on an underlying continuous distribution, aggregated into discrete and arbitrary spatial units. A model of error for categorical coverages reverses the logic. Spatial units are adjusted on a continuous space to reflect the categorical distinctions. An error model for categorical coverages has to recognize the constraints of cartographic production processes.

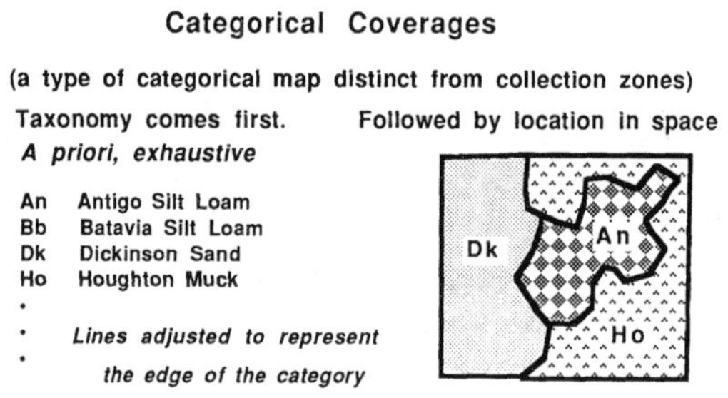

Categorical Coverages
(a type of categorical map distinct from collection zones)
Taxonomy comes first.
A priori, exhaustive
An Antigo Silt Loam
Bb Batavia Silt Loam
Dk Dickinson Sand
Ho Houghton Muck
Lines adjusted to represent the edge of the category
Followed by location in space
Dk
An
Ho

Figure 2 Categorical coverages (a type of categorical map distinct from collection units).

Just as categorical data analysis of social surveys requires error models that differ from classical regression error models, categorical coverages will require different error models from continuous spatial distributions.

Previous work on models of spatial error

A large number of different, incompatible models of spatial error have been developed between the different disciplines that treat spatial information. As a crude characterization, there are inductive and deductive schools of thought, usually divided on disciplinary lines. The inductive disciplines, such as survey engineers, have developed meticulous tools to study certain problems of error that can be carefully measured. They are cautious about sweeping generalizations. The deductive approach, based on stochastic models, is more likely to cover the breadth of the problem. However, stochastic modelling is limited by the range of tractable models available. Most models currently available do not apply directly to the problems of categorical coverages. A new model can be developed by reconciling different approaches to error and the numerical models that derive from them.

Some of the most developed models have been created for the different positioning systems used in geodetic science, survey engineering and photogrammetry. These related disciplines rely on a general model of redundant measurement, resulting in rather massive systems of simultaneous equations. Like most least-squares models, an adjustment of survey data can estimate an error ellipse for each point. In geodetic applications, this model is inescapable. In some applications, such as parcel maps, the location of boundary corners may also follow these rules. The error ellipse for points is not the final product, since many analytical applications use area as the primary description of frequency in spatial data. A model for variance in area is required. An estimate of variance in area for polygons described by well-defined points was presented by Neumyvakin and Panfilovich (1982). Omissions and mathematical flaws were corrected by Chrisman and Yandell (1988), but the results have not been adopted for practical use.

The data reduction power of surveying models is great, but the information model does not apply particularly closely to most categorical coverages. A surveying error model ascribes error to points, while a categorical coverage is not constructed like a follow-the-dots drawing. If error is simply a property of points, then some counterintuitive results follow: most importantly, the uncertainty of position of a line is at a minimum in the middle, between the two end points. Mathematically, this result is correct, but it distorts the cartographic circumstance. The points selected as a digital approximation must forcibly simplify the detail of the earth's surface. The points measured must be more reliable than the straight line presumed between them.

Remote sensing, like surveying, looks for a testable, empirical basis for map error. Unlike surveying, the emphasis is on the success of classification. Surprisingly, there is little other literature on issues of attribute accuracy. Although remote sensing products are usually categorical coverages, routine statistical treatment typically treats them as traditional aspatial samples. Methods commonly used in remote sensing are discussed in the Section 2 below, since they relate to measures of map comparison.

The most theoretical work on spatial error processes has developed in mathematical statistics under the title of stochastic geometry (Harding and Kendall, 1974). A variety of applications for this work have emerged in the fields of quantitative ecology and geology, as well as geography. Point patterns have remained the easiest to model (Rodgers, 1974). But the models available for the treatment of areas are restricted to two cases (Getis and Boots, 1978): Thiessen polygons built from random points, and random infinite lines. The first model has been developed for a few cases of nodal regions, but it is clearly unsatisfactory for the "formal" regions of a categorical coverage. Since Thiessen polygons are always convex, the model forms a poor simulation of a categorical coverage. The infinite line model produces the wrong type of geometry because every node is of degree four, while degree three is much more common. The goal of the work in stochastic geometry has been to build a model of spatial distributions with a completely random origin. The goal of this research is somewhat different, because the random element may act to obscure otherwise non-random delineations.

Another mathematical concept, fractal dimensionality, has recently swept through academic geography, and other disciplines (Mandelbrot, 1977). A number of attempts have been made, starting from Mandelbrot but now expanded to include LucasFilms, to generate "random" surfaces and planetary topography from fractal forms. The results of these simulated landscapes are rather impressively realistic, showing that the fractal model holds some interesting components. However, fractals only provide a measure of texture, a relationship between local variation and the more global. The work of Mandelbrot and his followers requires that fractal dimensionality is constant for a given feature, but very little evidence supports this sweeping assertion (Goodchild, 1980).

A specifically cartographic theory of information content was presented by Poiker (formerly Peucker, 1976). This bandwidth decomposition of a line, based on Douglas's algorithm for line simplification, has been important to many software systems. Recent work in cartography (for example, McMaster, 1986) has validated Poiker's theory as a means to detect important detail. However, Poiker's theory of the cartographic line describes the geometric salience of a point, not the error in its location.

Chrisman (1982a; 1982b) has proposed a method to describe some forms of cartographic error under the term "epsilon distance" to recognize the work of Polish research of thirty years ago (Steinhaus, 1954; Perkal, 1966). This work provided some of the mathematical precursors to fractals, but it is applied not as a measure of texture, but as a measure of potential error. The epsilon error model works backward from the boundaries as delineated. The true position of each line will occur at some displacement from the measured position. Geometrically, the line spreads out to a zone shaped like a sausage, contouring a probability density function of the line's true location. When the width epsilon of the band is set to the standard deviation of the uncertainty of the line, the sausage represents some form of mean error in area. Goodchild and Dubuc (1987) [and others since] criticize the epsilon model for setting a sharp edge for spatial error, but the intention of the model is to provide the spatial expression and topological constraints that they also expect from a model. The epsilon band should not be used to define error in the sharp sense of "buffer" corridors used in GIS applications. The epsilon band was intended to describe a mean probable location of uncertainty for the line, to measure probable errors in area measurements. The model is more than geometric. The epsilon zones can be summarized in a square matrix of the categories mapped, because any displacement induces a gain and loss from a specific pair of categories. This matrix summarizes the spatial structure as the weight matrix does in spatial autocorrelation models. The epsilon model, as currently formulated, provides a plausible framework for certain components of positional accuracy, mostly the "process errors" introduced by cartographic processing, not fundamental errors in the phenomenon. Further development is required to encompass all sources of variation, but no form of error should be neglected.

Recently, Goodchild and Dubuc (1987) have presented a model for random categorical coverages (they call them natural resource maps, but the term should be more generic). They reject the epsilon model for some reasons which are plausible and for other reasons which are less so. Their model presumes a continuous phase space in which the classifications can be distinguished. The categorical map is created from a set of pixels approximating the continuous distributions. This model does provide a method to generate a form of random categorical coverage which exhibits the proper topological structure, unlike previous attempts mentioned above. As the authors admit, however, not all taxonomies derive from such continuous phase spaces. Some of the most common processes (streams, lakes, glaciers, cities, etc.) have edges. Their model would operate only inside one regime. Their goal is to produce random categorical coverages in order to understand the operations of a GIS, and for this purpose it may be useful.

A framework for measuring error in a categorical coverage

Before a complete stochastic model can be developed, the first step is to define the error to be modelled. The various disciplines discussed above have used widely varying concepts of error, but it may be possible to borrow concepts from all of the approaches mentioned. The fundamental issue in statistics is understanding deviations. The deviations in a categorical coverage involve distinct components. In particular, there are positional (geometric) issues and attribute issues. The concept of deviation used for these two are usually quite different, but, in a categorical coverage, the various error effects may combine. Goodchild and Dubuc (1987) reject the separation, but there are strong suggestions that parallel treatment is a useful fiction. This section describes a mechanism to deconvolve spatial error into identifiable processes, each modelled separately.

Instead of positing some "random" map, this presentation works from the simplest situation: the comparison of maps assumed the same. Two categorical coverages purporting to map the same phenomenon are overlaid comprehensively, and the results form a test of accuracy. If one source is assumed "correct", it is a test of the other, but it could also be a test of repeatability. As in many statistical applications, a test pairs measurements. Unlike the standard applications, a spatial test pairs each point in the map by location on the ground. Such an arrangement, with an infinite number of points, requires a different error model than a "case" oriented approach. This framework is described incrementally, starting from some simple cases, then providing more complexity.

The most common form of error in overlaid maps is called a "sliver". As demonstrated in Figure 3, a simple sliver occurs when a boundary between two categories is represented slightly differently in the two source maps for the overlay. A small, unintended zone is created. Goodchild (1978) reports that some systems become clogged with the spurious entities that provide evidence of autocorrelation at different levels. These reports are a part of the unwritten lore of GIS, because most agencies are unlikely to report on failures. Some algorithms for overlay include a filter to remove the smallest of these, up to the level a user is willing to tolerate (Dougenik, 1980). The availability of the filter makes it important to understand its relation to theory.

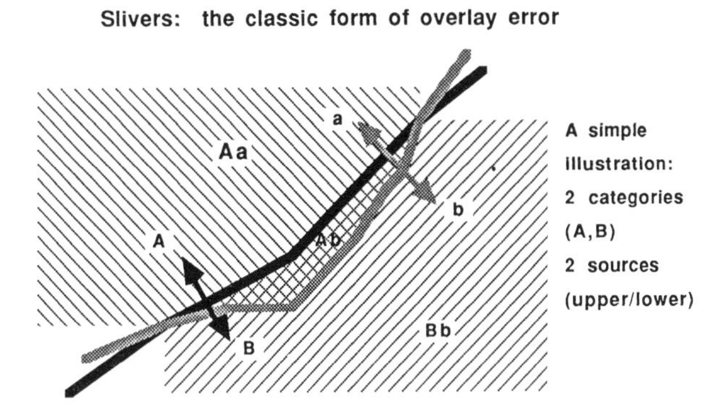

Figure 3 Slivers: the classic form of overlay error.

Although sliver error is the most frequently mentioned, overlay can produce other forms of problems. To follow the example described above (comparison of two maps assumed to be the same), it would be possible to have a feature on one map source which is completely missing on the other, as shown in Figure 4. While the sliver error seems to arise from positional error, such an error is caused by classification and depends on taxonomic similarities of the two categories. This taxonomic similarity could be modelled in some continuous phase space, or otherwise.

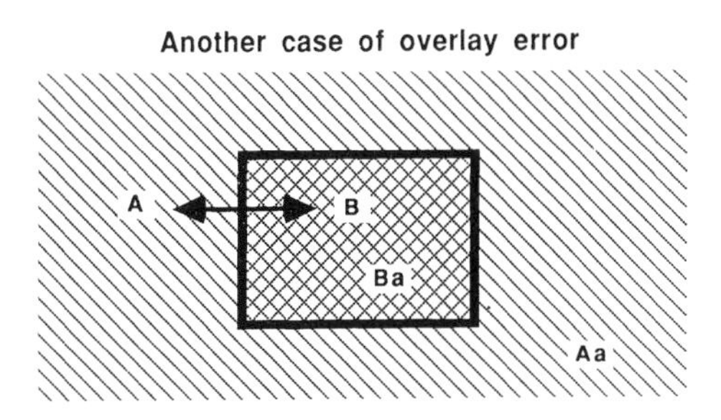

Figure 4 Another case of overlay error.

As extremes, the positional sliver and the attribute classification error seem perfectly distinct. But the two are quite difficult to disentangle in practice. For instance, a sliver error might arise from an interaction of positional error and difficulty in discriminating the classifications (more of an attribute problem than an error in positioning technology). Furthermore, not all error falls perfectly into the two cases presented in Figures 3 and 4. The sliver involves roughly the same contribution of linework from each source, while the classification error has all the linework from one source. As Figure 5 shows, there is a continuum possible between the two extremes which might be hard to classify. While it is easy to develop anecdotes about this kind of error, there is no workable theory.

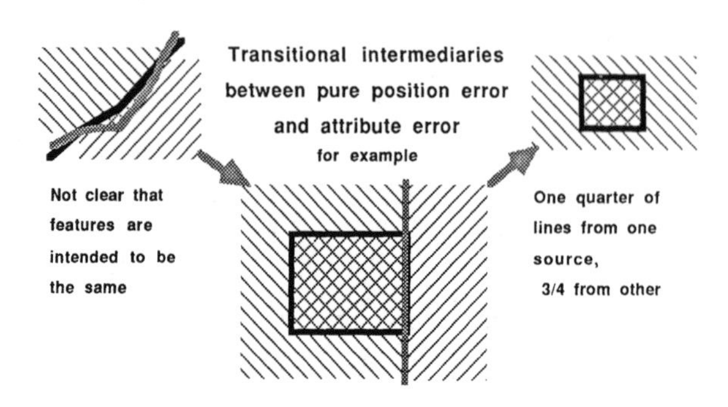

Figure 5 Transitional intermediaries between pure position error and attribute error.

The previous argument dealt with the existence of both positional and attribute error, but ignored the issue of scale. In spite of the power of modern GIS software, the basic information is still strongly dependent on scale. Positional accuracy of lines is expected to be linked to scale, but the amount is rarely specified or measured. Even more so, attribute error is linked to scale. At some scales, features like farmsteads are consciously removed from land use maps. Scale involves a distortion of the information, but a distortion that is tolerated and expected. To develop a framework, Figure 6 shows how positional and attribute error might interact schematically with scale issues.

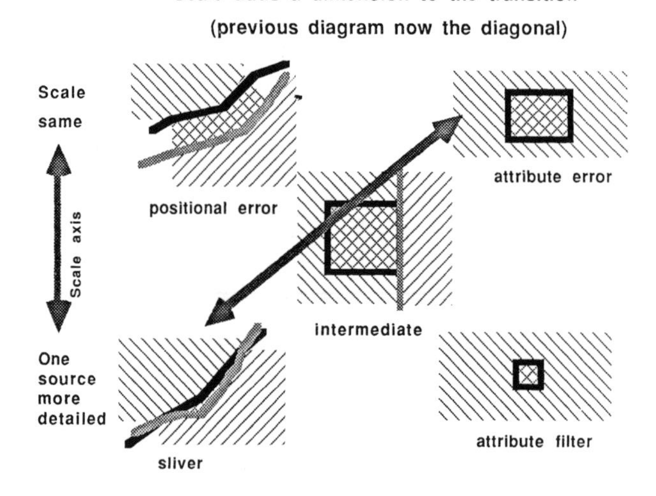

Figure 6 Scale dimensions to the transition (previous diagram now the diagonal).

The framework presented above is more than a diagram. It provides the basis to construct a mathematical model where the total error is decomposed into a set of stochastic processes operating simultaneously. The stochastic process for boundary error will have to reflect the geometric impact of cartographic representation and processing, while the attribute error will have to reflect taxonomic similarity of classes. For the technical "process" errors that simply degrade the positional accuracy, the epsilon model (Chrisman, 1982a; 1982b) may provide a useful start. For the taxonomic problems, some modification of Goodchild's phase spaces will be required. To further complicate affairs, these two processes will operate inside scale-dependent rules that can be modelled as filters and other constraints.

This framework may explain the results obtained from empirical accuracy experiments (eg. why Turk, 1979 could not fit a quasi-independence model with the expected residual structure). Empirical results measure the total error from all processes, and there is no guaranteed mechanism to deconvolve them. Each of the individual components above will be easier to model in isolation, then the error components can be combined.

Analytical measures for overlaid maps

While this discussion has centered on error, the error model is not simply an end in itself. The fundamental purpose of an error model is to support analysis. This section reviews previous work, particularly concentrating on measures used to describe overlaid map results.

Fundamentally, the overlay of two coverages produces a new coverage whose categories can be represented as a crosstabulation of the categories of the two input coverages. In statistical textbooks, it is common to use the geometric form of Venn diagrams to explain the operation of crosstabulations. While the metric of the Venn diagram is arbitrary, an overlaid map produces the same relationships in a metric space. Through iterative overlay, tables of arbitrary dimensionality can be constructed. In practical GIS use, the crosstabulation table is not as heavily used as the plotted graphic, but the information content is identical. There is a developing literature on the analysis of discrete multivariate data that is come into use for certain geographic problems, but not yet for map data. Before covering these promising new areas, it is important to review the simpler approaches which have been applied to maps.

In the description of attribute accuracy, and others aspects of map quality, there is a need to summarize the agreement of two maps. This problem can be characterized as "maps assumed the same", the case discussed above. Viewed in attribute terms, it creates a crosstabulation where the rows and columns have the same categories. During the early phase of quantitative methods in geography, this map overlay problem was raised (Zobler, 1957). As was common during this period, a statistical method (chi-square) was simply imported without careful examination of its error model. Chi-square is inappropriate for map overlay because it presumes that the crosstabulation contains counts of statistically independent cases. Mapped area (the most common measure of frequency for an overlaid map) has no clear, statistically independent unit "case". Other geographers recognized the flaws in the proposed procedure (MacKay, 1958; 1959), but no alternative procedure developed. Other attempts to provide an index for map overlay have been published (Court, 1970, Adejuwon, 1975), but these measures are rarely used. The statistical basis for them is weak, too (Chrisman, 1982a). The need for a map comparison index is much more common in the field of remote sensing. A few years ago, when the most common measure was "percentage correct", a number of authors took issue with this approach (Turk, 1979; Chrisman, 1982c; Congalton and Mead, 1981). The measure suggested by most (including the author) was Cohen's Kappa, a measure the deflates the percentage correct for "correct" answers that might occur through an independent process. This measure is simple and may gain some currency, however, the nature of independence may not operate in such a simple manner for overlaid categorical coverages. Even in the unlikely case that attribute error is independent of spatial structure, some interdependence must be expected due to the effects of boundary error discussed above.

Instead of a search for a magic index to encompass the results of map overlay, it is important to examine the methods developed by statisticians to treat categorical data. These models developed more slowly than methods for continuous data, but recently both forms of data have been unified in a general linear model. The general linear model (Bishop, Fienberg and Holland, 1975; Bibby, 1977) can be interpreted in a number of forms, although it is solved identically. For categorical data, the linear models can be "log-linear" (linear in the logarithms of the frequencies), logistic regression or logit/probit (linear in logarithms of odds ratios). This proposal will refer to the loglinear formulation, although the others might have potential utility. There are plentiful examples adapting these procedures in geographical research when the estimation models apply (Wrigley, 1985).

The loglinear model provides a method to decompose a contingency table into a heirarchical model of "effects". These effects follow the general approach of Analysis of Variance (ANOVA), but interpretation differs for a few reasons. First, the model is linear in the logarithms because the statistical definition of independent probabilities is multiplicative (as in Bayes' Law). Second, the terms of ANOVA include an error term for each "case" or observation, while the loglinear model of contingency tables only deals with the cells of the table. This second issue makes the procedure a good choice for categorical coverages, because there is no identifiable independent case to model. The cell of the table represents the Venn diagram combination produced by map overlay. A loglinear model estimates parameters that try to duplicate the cell frequencies with fewer terms - essentially a method of data reduction. The "saturated" model with the same number of terms as cells will always fit the data exactly. An analyst should select a more parsimonious model based on goodness of fit. The more parsimonious model can select a whole effect (interaction term) or select the role of a few key categories, while dropping out less significant ones (Aitken, 1979). This process offers a much more sensitive answer to the questions of association between categorical variables. A trial application for historical land use change demonstrated interpretable results (Chrisman, 1982a), although the estimation was known to be based on an inappropriate error model. The general framework of discrete multivariate analysis can be adopted, but many crucial details (including the measures of goodness-of-fit, and the error models) must be replaced. Major effort will be required to adapt the log-linear model to treat overlaid categorical coverages.

Questions for future research

Future research on this topic should develop analytical methods for overlaid maps that recognize the inherent uncertainties. With the restrictions proposed above, the goal reduces to methods to make honest statements about the areas measured after polygon overlay. Considering the variety of map information, many error models will be required to cover the whole problem. A few key research questions that can be addressed in the short term are described below.

Developing error models

Decompose map error

One task is to elaborate and validate the model presented above that integrates positional, attribute and scale effects on error. The first step is to demonstrate the existence of the various geometric forms of error in empirical data. This demonstration will begin to set some bounds on the interaction of positional and attribute accuracy. Then, this task must switch to simple spatial patterns (rectangles, circles) in order to demonstrate the workings of the different components through perturbation and simulation. Each error process will be simulated, and the results compared with the originals. For simple geometric forms, some of the error effects can be anticipated so that simulation will be unnecessary. The error distributions for the simulations can be developed from results of tests performed by surveyors, photogrammetrists and other mapping scientists.

The purpose of the simulation is to demonstrate that separate error processes can produce effects similar to empirical results. This work is required to settle the issue of separate treatment of positional and attribute effects (Goodchild and Dubuc, 1987). This task will result in rules to distinguish positional error from attribute error. At a minimum, it will apply to exhaustive tests performed according to the Proposed Standard for Digital Cartographic Data Quality. It may also permit some measures of texture without exhaustive testing.

Reporting probable bounds on area

The simplest use of geographic information systems reduces the spatial information to a tabulation of area. Area figures are used in tax assessment procedures, forest management and a number of other applications. Area calculations are routinely reported with many digits of spurious precision and no estimate of uncertainty. A clear taxonomy of error, combined with testing, will make it easier to report meaningful error rates.

The most complex component of positional accuracy is to confront some of the problems associated with ill-defined features - the fuzzy boundary problem. Some very rare field tests quantified the width of the transition zone and the ground location of the transition of environmental features to compare with their digital representation. Each point was chosen by specialists in the particular resource by examining a cross-section at right angles to the boundary. Such information has not been fully exploited.

Correlation of error

A correlated error structure can develop through the lineage of cartographic products which tie back to identical sources for original information such as topographic maps or geodetic control. Correlated error may make some issues less important. Notably, the use of area summary measures and correlated geodetic references will remove most effects of absolute error in positioning, although relative errors will remain. In other aspects, such as correlated errors of interpretation, the importance may be most critical.

In map overlay, one form of correlated error can arise from the process of combination. Registration, the process which establishes the identity of two locations, is critical. Registration error often has a strong systematic component, but other distortions in the sources may produce some other effects. The impact of registration error can be determined using simulation. The registration of layers can be experimentally altered to measure the distribution. Given an assumption about registration error, the impact on overlay results can be simulated from a sufficient number of separate displaced overlays. This kind of simulation work has not been reported in the GIS literature, probably due to the intense computing resources required to perform a credible simulation of the distributions. Other simulation-based research seeks to create random maps from a few parameters, which is an entirely different approach.

Apply error model

Develop general linear model

Analytical procedures should permit various statements about spatial data to be qualified by the amount of error expected. Such procedures are quite common in standard statistics, applied to random samples and other experimental designs. As discussed above, geographers and statisticians have developed some statistics that deal with particular forms of spatial error, but not those applicable to map overlay. Chrisman (1982a) has sketched a method to use general linear models (Bishop, Fienberg and Holland, 1975; Bibby, 1977), given the appropriate covariance matrix. Some effort, coupled with advances in statistics, should complete this work.

The covariance matrix required to estimate the general linear model has the same order as the misclassification matrix that results from empirical testing and from the epsilon model. Of course, there are major differences in interpretation and metric. The error distribution will be estimated from characteristics of the data itself, a relatively recent approach termed "bootstrapping" or "jackknifing" in the statistical literature. Statistical research has demonstrated that the general linear model can be applied to "double-sampling" problems similar to the requirements of this problem (Eckholm and Palmgren, 1987).

The measure of goodness of fit for discrete multivariate analysis is typically chi-square. However, such a measure is inappropriate for overlaid maps, since there is no "case" to count (MacKay, 1958; Chrisman, 1982a). One subtask will be choosing an appropriate measure of goodness of fit. It would be convenient if chi-square could be retained as a relative distance measure, but the numerical ramifications must be inspected carefully. Some new measure may develop from the work on developing an error model which incorporates both positional and attribute components. One simple starting point is Cohen's Kappa, a measure of correspondence suggested for accuracy comparisons by a number of remote sensing papers (Rosenfield and Melley, 1980; Chrisman, 1982c). This measure can be interpreted as "proportional reduction of error", which seems to be a useful approach. However, the standard used for Kappa is cross-product independence which is rather hard to conceive as a credible spatial distribution. If it is not useful, some other measure of correspondence is required, probably closely related to the measure of fit used for the linear model.

The error matrix recorded by the epsilon model (Chrisman, 1982a) currently tabulates area in a zone covered by the standard deviation of the line. This measurement is not equivalent to the covariances of the area measurements that are required in the general linear model. Considerable mathematical work will be required to ensure appropriate treatment. Any estimation procedure also depends on the tractability of the covariance matrix. Categorical coverages with many categories may produce sparse matrices that are difficult to invert, however, the nature of the spatial model avoids some of the classic pitfalls in estimation. For instance, a row or column of a spatial error matrix is very unlikely to be a multiple of another one, since the diagonal represents area correctly classified and should be the largest cell in any row or column. Another issue is the treatment of "structural zeros" - cells in a crosstabulation which are not valid combinations (e.g. steelmills in protected conservancy zoning districts). Such zero cells have a large impact on the underlying multiplicative models. This problem cannot be solved in the generic case, since it arises from the specific application. Still, some general formulation should be able to provide the tools to handle this common situation.

Impact on policy results

All of the preceding questions concern the technical description of error in spatial data. It is crucial to place the estimates of error into perspective. Seemingly large quantities of error might not have significant impact on the intended uses of geographic information systems. Many applications convert spatial information into a much reduced form. For instance, agricultural land value is calculated in many states from an area-weighted average of crop yields assigned to soils. Predicted productivity is then used to impute land value to agriculture by calibration with known farm sale figures. The final result may be much less sensitive to spatial error than the original figures for soils or parcels might suggest. The equity of land assessment is a major issue of public interest, but the taxpayer should not have to spend excessive sums for data collection for this process. Other policy applications convert information to ordinal classes (high priority, medium, low) or to nominal classes (eligible, ineligible). These transformations may contribute to reducing the impact of certain kinds of spatial error.

Conclusions

Considering the public and private investment in geographic information systems, additional research on the error of overlaid maps is required. Categorical coverages must be approached with a dual error model, separating the distinct forms of error commonly termed positional and attribute accuracy. The focus on categorical coverages will complement other strengths in spatial analysis of surfaces and other spatial distributions.

References

Adejuwon, O., 1975, A note on the comparison of chorochromatic surfaces. *Geographical Analysis* **7**, 435-440.
Aitken, M., 1979, A simultaneous test procedure for contingency table models. *J. R. Stat. Soc.* **C28**, 233-242.
Berry, B. J. L., 1964, Approaches to regional analysis: a synthesis. *Annals AAG* **54**, 2-11.
Bibby, J., 1977, The general linear model: a cautionary tale. In *The analysis of survey data*, vol. 2, O'Muircheartaigh, C., and Payne, C., (eds.), pp. 35-80. (New York: John Wiley).
Bishop, Y., Fienberg, S., and Holland, P., 1975, *Discrete multivariate analysis: theory and practice*. (Boston, MA: MIT Press)
Chrisman, N. R., 1982a, Methods of spatial analysis based on error in categorical maps. PhD thesis, University of Bristol.
Chrisman, N. R., 1982b, A theory of cartographic error and its measurement in digital data bases. *Proceedings AUTO-CARTO* 5, 159-168.
Chrisman, N. R., 1982c, Beyond accuracy assessment: correction of misclassification. *Proc. International Society of Photogrammetry and Remote Sensing Commission IV, 24-IV*, 123-132.
Chrisman, N. R., and Yandell, B., 1988, A model for the variance in area. *Surveying and Mapping* **48**, 241-246.

Congalton, R. G., and Mead, R. A., 1981, A quantitative method to test for similarity between photointerpreters. *Proceedings, American Society of Photogrammetry*, 263-266.
Court, A., 1970, Map comparisons. *Economic Geography*, **46** (supplement), 435-438.
Dougenik, J. A., 1980, WHIRLPOOL: a processor for polygon coverage data. *Proc. AUTO- CARTO IV*, 304-311.
Eckholm, A., and Palmgren, J., 1987, Correction of misclassification using doubly sampled data. *Journal of Official Statistics*, **3**, 419-429.
Getis, A., and Boots, B. N., 1978, *Models of spatial processes*. (Cambridge: Cambridge University Press)
Goodchild, M. F., 1978, Statistical aspects of the polygon overlay problem. In *Harvard Papers on Geographic Information Systems. 6* , Dutton, G., (ed.), pp. (New York: Addison Wesley)
Goodchild, M. F., 1980, Fractals and the accuracy of geographical measures. *Mathematical Geology* **12**, 85-98.
Goodchild, M. F., and Dubuc, O., 1987, A model of error for choropleth maps with applications to geographic information systems. *Proc. AUTO-CARTO 8*, 165-174.
Guevarra, J. A., 1983, A framework for the analysis of geographic information system procedures: the polygon overlay problem, computational complexity and polyline intersection. PhD thesis, SUNY at Buffalo.
Harding, E. F., and Kendall, D. G., 1974, *Stochastic Geometry*. (New York: John Wiley)
Hopkins, L. D., 1977, Methods for generating land suitability maps: a comparative evaluation. *American Institute of Planners Journal* **43**, 386-400.
MacDougall, E. B., 1975, The accuracy of map overlays. *Landscape Planning* **2**, 23-30.
MacKay, J. R., 1958, Chi square as a tool for regional studies. *Annals AAG* **48**, 164.
MacKay, J. R., 1959, Comments on the use of chi square. *Annals AAG* **49**, 89.
Mandelbrot, B., 1977, *Fractals: form, chance and dimension*. (San Francisco: Freeman)
McMaster, R., 1986, A statistical analysis of mathematical measures for linear simplification. *American Cartographer* **13**, 103-116.
Neumyvakin, Y., and Panfilovich, A., 1982, Specific features of using large-scale mapping data in planning construction and land farming. *Proc. AUTO-CARTO 5*, 733-738.
Perkal, J., 1966, On the length of empirical curves. *Discussion Paper 10*, Michigan InterUniversity Community of Mathematical Geographers.
Peucker, T. K., 1976, A theory of the cartographic line. *International Yearbook of Cartography* **16**, 134-143.
Rodgers, A., 1974, *Statistical analysis of spatial dispersion: the quadrat method*. (London: Pion)
Rosenfield, G., and Melley, M., 1980, Applications of statistics to thematic mapping. *Photogrammetric Engineering and Remote Sensing* **46**, 1287-1294.
Sauer, C. O., 1918, Mapping utilization of land. *Geographic Review* **8**, 47-54.
Steinhaus, H., 1954, Length, shape and area. *Colloquium Mathematicum* **3**, 1-13.
Steinitz, C., and others 1976, Hand-drawn overlays: their history and prospective uses: *Landscape Architecture* **66**, 444-455.
Tomlin, C. D., 1983, Digital cartographic modeling techniques in environmental planning. PhD thesis, Yale University.
Turk, G., 1979, GT index: a measure of the success of prediction. *Remote Sensing of Environment* **8**, 65-75.
Warntz, W., 1968, On the nature of maps. *Discussion Paper 12*, Michigan InterUniversity Community of Mathematical Geographers.
Wrigley, N., 1985, *Categorical data analysis for geographers and environmental scientists*. (London: Longman)
Zobler, L., 1957, Statistical testing of regional boundaries. *Annals AAG* **47**, 83-95.

Chapter 3

User considerations in landscape characterization

Stephen J. Walsh

Abstract

Total error within a GIS analysis can be minimized by recognizing inherent errors within the source products and the operational error created through various data manipulation functions. Design specifications followed in the generation of thematic files need to be thoroughly evaluated by users in order to assess the "fitness of use" of products for intended applications. Conspicuous and subtle spatial considerations impact on the quality of generated data sets for GIS analysis. Landsat digital classifications and enhancements, terrain characterization through Digital Elevation Models, and Soil Conservation Service soil type information are briefly discussed to demonstrate that the selection of even subtle alternative design specifications and construction methodologies can significantly affect whether a digital data set is judged suitable by the user for effective landscape characterization. Technical and practical problems affect questions of data quality. Decision-makers need to be aware of the need for establishing audit trails of error sources throughout the generation and manipulation of spatial information within a GIS environment.

Introduction

Organizations that have a planning, research, and/or regulatory responsibility at the county, state, or federal level are increasingly turning towards geographic information system (GIS) technology as an approach to data integration, synthesis, and modelling. The lack of spatially trained and sophisticated GIS specialists, however, is adversely impacting upon the level and effectiveness of GIS use by decision-makers. Too many users are permitting their GIS software to drive their spatial analyses without sufficiently appreciating and incorporating spatial considerations that represent a blending of spatial and natural sciences combined to mold and shape the analyses and interpret the results. In general, users are not sufficiently concerned with sources of error within their base products nor the appropriateness of generated and acquired data integrated within the GIS. GIS users, representing a wide spectrum of spatial sophistication, need to be cautious and suspicious of products utilized within a GIS and the specifications followed in their creation, since data quality can be affected by conspicuous and subtle, but nonetheless important, error considerations.

Conspicuous sources of inherent error confronted by the user include errors in attribute tagging and spatial position. Less obvious considerations include the appropriateness of source products whose data have been affected by interpolation procedures, classification alternatives, transformations, and format translations to name but a few issues in which the user needs to be conversant. User requirements of the data and the relationship between the suitability of the collected information to the application, and the acceptable level of inherent and operational error of the base products and of the overall analyses also are important considerations that impact on the quality of the decisions made by the GIS users. Users need to become more knowledgeable regarding the creation of inherent error in source products and the operational error propagated through the analysis as a consequence of poor quality and/or inappropriately designed source products.

This paper discusses some basic spatial considerations that affect the quality of digital data used to characterize the landscape through thematic overlays representing landcover, soil type, and terrain orientation. Obvious and less obvious spatial considerations will be addressed regarding satellite digital characterization of landcover, SCS county reports of soil type, and USGS digital elevation models of terrain orientation.

Basic spatial considerations

Inherent and operational errors contribute to a reduction in the accuracy of data contained within a geographic information system. Inherent error is the error present in source documents, and includes the design specifications utilized to generate the source material. Operational error is produced through the data capture and manipulation functions of a GIS (Walsh *et al.*, 1987).

Every map contains inherent error based upon the nature of the map projection, construction techniques, design specifications, and symbolization of the data. The amount of error is a function of the assumptions, methods, and procedures followed in the creation of the source map (Vitek *et al.*, 1984; and MacEachren, 1985). Data represented as discrete or continuous, spatially and/or temporally interpolated from point to area measures, classified through various organizational approaches, represented by spatial and/or temporal resolution considerations and resampling schemes, and characterizing phenomena that maintain a varying level of surface complexity further add to the level of inherent error within GIS base maps and documents (Willmott *et al.*, 1985). User evaluations of GIS products should include both an appraisal of the integrity of the data set and the design specifications followed to generate the data set (Mead, 1982).

In addition to the user evaluation of technical considerations that impact upon data quality, the user must deal with an array of practical problems that affect how the user will deal with both inherent and operational error. Some practical considerations include: what is the relationship between collected data and the intended application?; what is the acceptable level of error for each source product and of the final analysis?; what is the financial cost and time investment for improving the accuracy of the base products?; what are the trade-offs of utilizing acquired data or generating the data for a specific application?; what is the required spatial precision needed of the products generated as part of the final analysis?; and how can inherent and operational error be calculated, and how can error information be incorporated into the analysis to improve the quality of the final product? These types of questions bear on the realization that error cannot be eliminated but only minimized within a GIS analysis, and that the amount and spatial variability of error should be assessed by the user through technical approaches that maintain a practical perspective.

Operational error, categorized as positional and identification error, is introduced during the process of data entry and occurs throughout data manipulation and spatially modelling. Positional errors stem from inaccuracies in the horizontal placement of boundaries. Identification errors occur when there is mislabeling of areas on thematic maps. The upper and lower accuracy limits of a composite map indicate the range of accuracy probabilities when analyzing two or more thematic overlays, given the accuracies of the source map (Newcomer and Szajgin, 1984). Operational error reduces the accuracy of a GIS analysis and output product from its theoretical best. The highest accuracy possible in any GIS analysis can never be better than the accuracy of the least accurate individual map layer (Newcomer and Szajgin, 1984).

Landscape characterization and data quality considerations

The digital characterization of the physical landscape is an essential component of most GIS analyses. The quality of that representation relates to the concept of GIS error by requiring the user to determine the "Fitness of Use" of methodologies and products generated to characterize the physical landscape (Chrisman, 1987). Landcover, soil type, and terrain orientation, common to many GIS analyses, need to be evaluated regarding the complexity of the phenomena being represented in a digital format and the appropriateness of the specifications and tools utilized to generate the thematic information to meet user needs and to understand the earth science system being characterized.

Remotely-sensed landcover information

Effective landcover mapping through remote sensing techniques is related to the spatial, spectral, radiometric, and temporal resolution of the sensor systems being applied, and the spatial and compositional complexity of the target phenomena. Supervised and unsupervised classification techniques are the principal approaches to landcover categorization from spectral response values. Digital enhancements also are available to the user to accentuate subtle differences in spectral properties that may be critical in landcover differentiation. Both approaches will be discussed to demonstrate that user decisions as to processing specifications are important considerations in the production of landcover maps and in their evaluation from a data quality or accuracy perspective.

Digital classification

The success of a supervised classification of landcover is directly related to the quality of the training sets used to relate spectral responses to specific landcover types. The training areas should have precise spatial correspondence to their location in the field and their position on the digital data; compositional consistency should be as uniform and homogeneous as possible; size and spatial distribution of the control areas should be considered; a methodological approach to field verification of cover type location, identification, and categorization needs to be designed; and it should be realized that compositional categorization in the field and sensor sensitivity to that field categorization may be in some level of disagreement. The relationship of the spatial and compositional complexity of the landcover unit to the resolution specifications of the reconnaissance platform must be evaluated. A thorough understanding of the biophysical systems at work in the study areas, that affect spectral responses and landcover distributions, should be assessed, as well as the geometric fidelity of the satellite data (Ford and Zanelli, 1985; Welch *et al.*, 1985).

While the unsupervised classification approach theoretically requires less *a priori* knowledge about the study area, user decisions must be made regarding the appropriate classifier, statistical distance between clusters and composition of individual clusters, number of clusters to be formed, and the evaluation of statistical output needed to evaluate the relatedness of statistical clusters, and their relationship to landcover categories.

The selection of the temporal window utilized to map the landcover is another consideration. Landcover classification can be enhanced by selecting the optimum time period when biophysical conditions best serve for spectral differentiation of the target phenomena. Multi-temporal remote sensing has proved to enhance the possibility for landcover separation due to phenological differences between landcover (Badhwar, 1984). Digital classifications of multi-temporal scenes can improve the compositional separation of phenomena and the differentiation between phenomena.

Digital classification algorithms, such as maximum likelihood, nearest neighbor, nearest mean vector, are known to be sensitive to specific data clustering problems. Selection of the most appropriate classifier can be important to the success of a landcover mapping exercise. Once the classification has been executed, the approach followed to assess the classification accuracy can provide a range of error measures due to the methodologies followed and the assumptions accepted (Congalton *et al.*, 1983; Story and Congalton, 1986; Rosenfield, 1986; Rosenfield and Fitzpatrick-Lins, 1986). The number and distribution of control areas used to evaluate the classification are important considerations. As the size of the instantaneous field of view (IFOV) increases for a sensor system, the level of spatial aggregation of landscape components increases, as well as the difficulty in assigning point measures of landcover control information to specific pixel areas. Transects through the pixels, with multiple point samples, can be used to determine the compositional matrix of the landcover within the pixel for classification accuracy assessment (Woodcock and Strahler, 1987).

In short, the quality of landcover information, made part of a GIS analysis, depends upon the sophistication of the remote sensing analyst: both technical and environmental considerations relate to the appropriateness of the digital characterization of landcover. The type and quality of the software and hardware are certainly related to the success or failure of generating a landcover map that meets certain accuracy considerations. The desired level of landcover categorization, surface roughness of the landscape, spatial and compositional complexity of the landcover, and the selection of the sensor system further add to the level of uncertainty in achieving an acceptable digital classification.

Walsh *et al.* (1987) found that landcover classification accuracies, for a classified Landsat Multispectral Scanner data set aggregated to 100 sq.m and 200 sq.m cells, were 57% and 43%, respectively. The error matrices were generated through the distribution of 35 sample points that were compared to the results of the unsupervised classification for USGS level I landcover categories. Subjectivity of field verification, the integrating nature of the pixel spectral response, surface to area transformations involved in remote sensing, and the relationship between size and compositional complexity of the landscape and the spatial resolution of the sensor system all contribute to the observed classification error.

Digital enhancements

Ratio, filtering, and Principal Components Analysis are common digital enhancement techniques useful in highlighting and accentuating subtle differences in spectral responses that characterize landscape components. Enhanced remotely-sensed images can be directly interpreted through a manual process and entered into a GIS through digitization. The results of digital enhancements also can be merged into a classification or used as a stand-alone processed-image.

Ratio

Ratios are transformations employed to reduce the effect of shadows, seasonal changes in solar angle and intensity, and fluctuations in sensor-surface geometry caused by topographic orientation. The process of ratioing combinations of spectral channels reduces the impact of such environmental conditions on the brightness values of terrain units, and provides a mechanism for merging the information content of two spectral response channels into a single response channel that includes the information content of both channels. Ratioing spectral channels for feature enhancement or merging of ratioed spectral channels with raw spectral channels within the classification process can improve landcover differentiation in complex environments. The selection of remote sensing channel combinations that represent various wavelength regions, and hence physical attributes of the landscape, are important user considerations.

Spatial filtering

High frequency spatial filters can be constructed to emphasize directional and non-directional differences in brightness values summarized through a user-defined kernel that is passes over the satellite spectral responses and summarized through assigned filter weightings. The size of the kernel is related to the complexity of the terrain (Chavez and Bauer, 1982). High-pass filters, low-pass filters, Laplacian non-directional edge enhancements, and directional filters are but some of the filtering strategies that can be applied to the highlighting of specific landscape elements. High-pass filters accentuate overall detail within the image; Laplacian non-directional filters reduce to zero all pixel values except those associated with edges in the data; low-pass filters reduce noise in the data resulting in a moderately smoothed surface that can then be applied to high-pass filtering in order to draw out the remaining detail contained within the image; and directional filters are used to accentuate specific orientations. User considerations include those of kernel size, element weighting, and the design of filtering strategies that utilize a combination of techniques and directional and non-directional elements. An understanding of the surface roughness of the phenomena under study and the spatial resolution of the sensor system are important factors in achieving a successfully filtered image.

Principal Components Analysis

Principal Components Analysis (PCA) is a useful multivariate statistical technique for the analysis of multispectral data. PCA transforms the highly correlated remotely-sensed digital data into statistically independent orthogonal axes on which the original satellite data are reprojected. This capability to compress multiple spectral channels is a useful data reduction technique and has direct application in pattern recognition because differences between similar materials may be more apparent in PCA images than in the individual remote sensing channels. Fung and LeDrew (1987) indicate that PCA is a scene dependent technique that requires a visual inspection of the image to assign specific landscape elements to specific components. This assignment process necessitates an understanding of the area under investigation, its vegetation, hydrography, and geomorphic character. User considerations in using PCA relate to the approach for calculation of PCA (correlation or covariance based); knowledge of the landscape under study in order to determine the landscape significance of each component; and the ability of the user to correctly interpret the eigenvectors generated as part of the analysis.

Digital Elevation Models

Digital elevation models (DEM) have become an important source of information for analyses within a GIS framework. They are especially useful in earth science investigation where terrain orientation is a critical concern in biophysical modelling. Often, elevation, slope angle, and slope aspect information become stand-alone thematic overlays within a GIS or are combined to form environmental indices. DEM information also is merged with spectral response data from satellites in order to stratify the landscape by selected terrain units for further digital processing and/or for environmental interpretation. Users of DEM information, however, must be cognizant of the design specifications of the digital product: scale of source map, sampling interval, relationship between the precision of the base map and the complexity of the terrain, and the method of data representation within the DEM format. For instance, a map with a 10 foot contour interval and the assumption of linear gradient do not enable a map user to accurately recreate the original surface. On a base map having a contour interval of 40 feet, the assumption of linear gradient is less accurate in the recreation of the original surface. The portrayal of only selected data points with lines of equal value generalizes reality and increases the amount of error inherent to the product. While USGS topographic maps are accurate, they also possess inherent error because of the design and construction of the map. DEMs, based on such products, therefore, maintain a similar type of inherent error.

The sampling interval utilized in the creation of the DEMs is critical to its capability to portray the landscape as close to reality as possible, particularly in complex and rugged terrain. High sampling frequency of information represented on larger scale maps better represents the landscape terrain orientation for the user. Data resampling and rescaling techniques can produce apparently detailed information for the user from small scale terrain data, but their validity would certainly be of questionable value.

The user needs to evaluate the appropriateness of DEM products for specific applications. The relationship between the minimum mapping unit or terrain unit to be investigated, and the spatial sensitivity of the digital terrain data needs to assessed. Spatial integrity of the DEMs must be below the smallest terrain unit for which the user requires information.

The Altitude Matrix (AM) and the Triangulated Irregular Network (TIN) are two approaches of representing digital terrain data. The AM is a regular rectangular grid of point measures that forms the most common form of DEM. Its limitations primarily are in its inability to represent terrain complexity through a variable grid size. Therefore, areas of high terrain variability are represented by the same cell size used to characterize rather homogeneous landscapes. If the cell size is altered to become more spatially sensitive to areas of terrain variability, data redundancy causes problems in storing and processing the needlessly small grid size for the homogeneous portions of the study area.

The TIN approach reduces the data redundancy problem by offering to the user the capability of varying the spatial resolution where appropriate. The facets of the TIN model can be varied in size to accommodate the need for greater spatial resolution over highly complex environments. The vector topological structure of TIN regards the points of the facets as the primary entities within the data base.

Walsh *et al.* (1987) report that errors in the categorization of slope angle and slope aspect, summarized at 100 sq.m and 200 sq.m cells, were consistently over 50%. Error matrices were generated by comparison of surveyed terrain orientations at 35 field points to the digital characterization secured from the Altitude Matrix representation of topographic data obtained from the 1:250,000 scale USGS base maps. At the larger cell size, slope angle and slope aspect classes were eliminated from consideration due to the spatial aggregation problem.

Soil type

Soil information mapped as part of the USDA Soil Conservation Service county soil reports offers the user a wealth of data concerning soil type and many characteristics of soils and their ratings. Users, however, need to be acquainted with the boundary problem of mapping soils, soil inclusions, and reporting of soil attributes by horizons.

Soil boundaries shown on the maps are more realistically described as zones of transition rather than abrupt boundaries. Their cartographic representation and their environmental interpretation is not as straightforward as the maps would leave the user to believe. Lines of transition (probability, confidence, etc.)may be more appropriately used to denote soil boundaries than the current cartographic convention.

The method of map construction needs to be evaluated and related to the desired specificity of the final GIS products that utilize the soils information. Point samples collected along transects are a common technique of creating soils maps. The spatial frequency of the point samples is an obvious consideration. A less obvious consideration by users is that of soil inclusions. Although inclusions are acknowledged within the soil surveys, they are not mapped because of their small areal extent relative to the scale of the soil maps. The soils maps exclude soil inclusions as large as four acres.

The soil characteristics rated for a host of attributes are often misinterpreted by the casual user of soil information. Often times, one thematic overlay of soils is contained within the GIS. Depending upon the application, a thematic overlay for each of the soil horizons may be warranted, or a composite map might be generated that more appropriately integrates the behavior of the entire soil profile.

Finally, the geometric integrity of the photo mosaic, aerial photograph, etc., used as the base map of the county soil reports, should be evaluated by the user. If the base products are not controlled, placement of soil polygons on those map can further add to the lost of spatial integrity of the soils information.

Walsh *et al.* (1987) report that encoded digital representations of soil type, as compared to 35 field sites, showed an overall accuracy of 94%, when soil inclusions were not treated as distinct soil mapping units. When the soil inclusions were represented as distinct soil mapping units, the overall accuracy of the soils overlay decreased to 60%.

Conclusions

Questions regarding inherent and operational types of errors and the implementation of recommended data quality standards helpful in documenting and evaluating elements of the data base are the responsibility of both the producer and the user of the data. Literature has suggested conceptual standards such as lineage, positional accuracy, attribute accuracy, logical consistency, and completeness, and methods of documenting and evaluating information - reliability diagrams, header information, statistical overlays of probability and confidence, numerical thresholds that relate data quality levels to specific products and to possible applications, and logical examinations of spatial co-occurrence (Walsh *et al.*, 1987; Chrisman, 1984).

Users are involved in developing research-specific thematic files and acquiring maps and digital files presumed to be appropriate to their analyses. Information regarding the quality of the thematic overlay needs to be available to the user for self-evaluation of the appropriateness of the file given certain applications and accuracy demands. Mean accuracy statements for the digital thematic files are of little use in trying to understand the spatial variability of error throughout the product suitable to a variety of different applications. Data linked to the digital file that is capable of relating specific sampling information to a minimum acceptable accuracy and to producer and user risk would add to the assessment of overlay quality by indicating the degree of quality (Aronoff, 1985). Statistical surfaces (probability, covariance, standard deviation) need to become a standard thematic overlay in a GIS analysis, since these measures relate to the concept of "fitness of use" decided upon by the user. User sophistication in evaluating the "fitness of use" will impact on the appropriate use of prepared overlays in GIS analyses. Therefore, producers should include both reliability-type diagrams and header information on the design specifications of the digital file in order to accommodate users of different spatial sophistications.

Thematic overlays in a GIS can be produced in a manner that meets technical accuracy standards but still fails to transmit the needed information to the user. The choice of appropriate spatial interpolation methods, for example, must be decided upon by specific knowledge of the anticipated spatial analysis, level of accuracy required, sensitivity of the system being investigated, and the spatial sophistication of the tools being applied to the generation of the data set. Conspicuous and subtle factors can introduce error into the source documents created for integration through GIS technology. Decision-makers need to appreciate sources of inherent and operational error, and design and implement recommended approaches for inventorying such factors that impact on data quality considerations. Landcover, soil type, and terrain orientation thematic overlays can never "truly" represent the landscape surface because of the problems in surface representation. Design specifications decided upon by the producer and accepted or rejected by the data user, however, will directly affect the level to which the landscape surface can be characterized for GIS modelling and informed decision-making.

References

Aronoff, S., 1985, The minimum accuracy value as an index of classification accuracy. *Photogrammetric Engineering and Remote Sensing* **51**(1), 99-111.

Badhwar, G. D., 1984, Classification of corn and soybeans using multitemporal Thematic Mapper data. *Remote Sensing of Environment* **16**, 175-182.

Chavez, P., and Bauer, B., 1982, An automated optimum kernel-size selection technique for edge enhancement. *Remote Sensing of Environment* **12**, 23-38.

Chrisman, N. R., 1984, The role of quality information in the longterm functioning of a geographic information system. *Cartographica* **21**(2), 79-87.

Chrisman, N. R., 1987, Obtaining information on quality of digital data. *Proceedings, Auto-Carto 8* , 350-358.

Congalton, R. G., Oderwald, R. G., and Mead, R. A., 1983, Assessing Landsat classification accuracy using discrete multivariate analysis statistical techniques. *Photogrammetric Engineering and Remote Sensing* **49**(12), 1671-1678.

Ford, G. E., and Zanelli, C. I., 1985, Analysis and quantification of errors in the geometric correction of satellite images. *Photogrammetric Engineering and Remote Sensing* **51**(11), 1725- 1734.

Fung, T., and LeDrew, E., 1987, Application of principal components analysis to change detection. *Photogrammetric Engineering and Remote Sensing* **54**(2), 167-176.

MacEachren, A. M., 1985, Accuracy of thematic maps, implications of choropleth symbolization. *Cartographica* **21**(1), 38-58.

Mead, D. A., 1982, Assessing data quality in geographic information systems. In *Remote Sensing for Resource Management*, Johannsen, C. J., and Sanders, J. L., (eds.), pp 51-59. (Ankeny, Iowa: Soil Conservation Society of America)

Newcomer, J. A., and Szajgin, J., 1984, Accumulation of thematic map error in digital overlay analysis. *The American Cartographer* **11**(1), 58-62.

Rosenfield, G. H., 1986, Analysis of thematic map classification error matrices. *Photogrammetric Engineering and Remote Sensing* **52**(5), 681-686.

Rosenfield, G. H., and Fitzpatrick-Lins, K., 1986, A coefficient of agreement as a measure of thematic classification accuracy. *Photogrammetric Engineering and Remote Sensing* **52**(2), 223-227.

Story, M., and Congalton, R. G., 1986, Accuracy assessment: a user's perspective. *Photogrammetric Engineering and Remote Sensing* **52**(3), 397-399.

Vitek, J. D., Walsh, S. J., and Gregory, M. S., 1984, Accuracy in geographic information systems: an assessment of inherent and operational errors. *Proceedings, PECORA IX Symposium*, 296-302.

Walsh, S. J., Lightfoot, D. R., and Butler, D. R., 1987, Recognition and assessment of error in geographic information systems. *Photogrammetric Engineering and Remote Sensing* **53**(10), 1423-1430.

Welch, R., Jordan, T. R., and Ehlers, M., 1985, Comparative evaluations of the geodetic accuracy and cartographic potential of Landsat-4 and Landsat-5 Thematic

Mapper image data. *Photogrammetric Engineering and Remote Sensing* **51**(11), 1799-1812.

Willmott, C. J., Rowe, C. M., and Philpot, W. D., 1985, Small-scale climate maps: a sensitivity analysis of some common assumptions associated with grid-point interpolation and contouring. *The American Cartographer* **12**(1), 5-16.

Woodcock, C. E., and Strahler, A. H., 1987, The factor of scale in remote sensing. *Remote Sensing of Environment* **21**, 311-332.

Chapter 4

Knowledge-based approaches to determining and correcting areas of unreliability in geographic databases

Peter F. Fisher

Abstract

Errors of commission or omission are endemic to almost all maps that form the source of much of the spatial data input to GIS. This reflects a number of factors, and is well known among the surveyors and cartographers involved in production of maps, as well as among sophisticated users, including many of those using GIS. The surveyors and cartographers often attempt to compensate for the inaccuracy of the map by preparing informative reports and legends to accompany the maps. The information in these legends and reports, more often than not however, are ignored in the process of digitization of the map data prior to incorporation in a GIS. It is the contention of this paper that employing knowledge of experts in the domain of the phenomenon being mapped (soils, population, etc.) can improve the reliability of the resulting GIS database, and its products.

Introduction

It is axiomatic to say that all maps are an abstraction of reality (Robinson *et al.* 1978). They are not exhaustive in their coverage (not all phenomena are shown on all maps), nor are they precise in the locational information they portray. Thus, in both location and in attribute they are no more than acceptable representations of the mapped phenomenon, be it physiography, topography, soil, population or some other phenomenon (Chrisman 1987; Hsu and Robinson 1970; Jenks and Caspall 1971; MacDougall 1975; Morrison 1971; Monmonier 1977; Muller 1977). This being so there is inevitably a significant level of unreliability in the resulting map product. This unreliability is naturally incorporated in any geographic database for which the map may be an input.

This paper addresses some of the knowledge-based approaches that have been, and could be, used in the recognition and handling of this unavoidable and necessary, production- and design-created, unreliability in maps. The first section documents why it is, indeed, necessary to make a map unreliable, and makes the point that even as we move into a fully digital age, this form of unreliability will not disappear. The second section explores ways in which surveyors and cartographers have attempted to compensate for unreliability. There follows a discussion of how unreliability has been accommodated in statistical mapping, and how that approach may be extended into the digital context and to other data types. A description of the future research that is required to implement the approach advocated here, and other research that is subsequent to it, form the conclusion.

Map production and unreliability in spatial data

Any number of causes of error and unreliability in mapped data can be identified, including cartographic production, natural complexity, human error, and temporal change. In this section the influence on the cartographic process is briefly discussed, recognizing the importance of the map as the major source of input to geographic databases.

As representation scale drops below 1:1, it is inevitable that symbols and lines used to depict mapped phenomena start to merge and overlap (Tobler 1988). This is aesthetically and technically unacceptable, and so some form of generalization is adopted. A number of different approaches to generalization have been identified in the cartographic literature, including simplification, reclassification, selection, and displacement (Rhind, 1973; Robinson *et al.* 1978). In the design of any particular map some combination of these is used so that those phenomena that are to be shown in a map are distinguishable to the map reader. Such parameters as production medium, size and scale of the finished product, and required background information are used in deciding which combination of generalization strategies to adopt and how to implement them.

With processes such as displacement and reclassification being used, it is inevitable that the accuracy of the original data is degraded by some amount, such that either the phenomenon shown are not those present at a location, or else the location of a phenomenon is imprecisely represented. The "accuracy" may even be published in the specifications for a map series (e.g. Thompson 1983), although designers of any particular map in the series may not be required to adhere to them. When there are no published standards, the problems become particularly acute (see Rhind and Clark 1988, p. 93). Even the most detailed plans actually contain considerable generalization which is often undocumented (Croswell 1987).

Some suggest that the advent of digital collection and processing of spatial data effectively makes it possible to map at a scale of 1:1, and so eliminate generalization, and the concomitant degradation in accuracy. This may only be the case in the built environment, where it may indeed be possible to precisely survey a line in the landscape. Many natural and man-made phenomena are, however, not delimited by precise lines (Robinson and Frank 1985). Thus the boundaries between vegetation types and soil types are actually both zones of spatial intergrade (Burrough 1986; Campbell 1978a, 1978b; Conacher and Dalrymple 1977) and subject to change over time. Indeed, Pomerening and Cline (1953) show that repeated surveying of an area, to the level of detail and using methods common in the Soil Conservation Service, produces as many different maps as there are surveyors, with various levels of agreement between surveyors. Increasingly precise representation of the boundary does not alter the fact that the boundary is an artifact of human perception, not of the real world. Indeed, the positions of some humanly imposed phenomena such as property boundaries can also be vague (Dale and McLaughlin 1988).

In summary, many processes are at work to ensure that no matter the means or scale of mapping, neither the representation of spatial data, nor often the data itself, can be reliable or certain. Indeed, it is more usually the case that the data is inherently unreliable or uncertain.

Recognition of unreliability

Symbolism and legends

The major vehicles used by the cartographer to acknowledge the unreliability of their map are symbolism and the map legend. Various methods have been used to present the map user with a visual display of uncertainty with respect to either a map unit or boundary. In geological maps, for instance, it is common practice to show observed or precisely inferred boundaries by continuous lines, but to change to a dashed symbol where the boundary is imprecisely inferred (how often have such line symbols been recorded in a GIS?). Similarly it is possible to map composite soil units, such as the soil association or soil complex. Here the mapped unit represents two or more soil types intricately interwoven, such that attempting to distinguish them individually on the map is not possible since any particular area of each, or of one, would be visually indistinguishable (Soil Survey Staff 1975b). Similarly, it is possible to use interlaced colors to fill an area, such that each color represents one phenomenon present in the area (soil, rock, vegetation, etc.), and the proportion of the area is related to the extent of that unit. It is also possible to adopt map unit types which describe the nature of the spatial relationships of the several components of the mapped units (Campbell and Edmonds 1984).

The map legend is a second vehicle by which the cartographer or surveyor may convey their doubt as to the identity of a region. An area designated with a particular symbol may be shown as being principally Cover Type A, but including Cover B, and C. Not uncommonly such recognition is accompanied by quantification as to the extent of each; thus, Cover Type A may be 50%, Type B 30% , and Type C 20%. This approach to the caption is commonly used on small scale mapping, most notably on national maps (Soil Survey Staff 1975a, p. 412; FAO/UNESCO 1974), but also is common for identifying facies variation in the most detailed of geological maps.

Reports

The reports that commonly accompany a thematic map are another vehicle for conveying information as to the purity of the mapping units, and the reliability of the associated boundaries. Here, however, the information is reported by the surveyor who may or may not have been involved in the map production. Thus in many recent U.S. Soil Survey Reports, such textual comments as:

> Included in mapping are a few areas of the well drained Wooster soils, mainly where the slope ranges from 4 to 6 percent. Also included are small spots of the somewhat poorly drained, less sloping Ravenna soils and a few areas that are moderately eroded and have a plow layer that is a mixture of the original surface layer and some subsoil material. (Ritchie. *et al.* 1978, p. 68, comment on areas mapped as Canfield silt loam 2 to 6 percent slopes)

As pointed out by Campbell and Edmonds (1984) this information does not accompany all soil maps, and is often ignored by users. It was not commonly given by the Soil Conservation Service Reports when those authors were writing. Similar textual comments are, however, included in reports associated with many types of survey. For example geologists may be expected to report on all the facies encountered in mapping a particular formation.

Assessing the extent

When map reliability has been assessed, the information is more often than not buried in the specialized literature. Some results of investigations of the reliability of soil maps, for example, are presented in Table 1 (see also Fisher, in press).

Study		Error
Pomerening and Cline (1953), New York		
	Series:	10-21%
	Phase:	14-28%
Powell and Springer (1965), Georgia		
	Series:	26%
	Types (Phase):	17-40%, 36% overall
Wilding et al. (1965), Ohio		
	Series:	42%
	Type:	39%
Courtney(1973), U.K.		
	Series:	30%
Bascomb and Jarvis (1976), U.K.		
	Series:	40%
Campbell and Edmonds (1984), Virginia		
	Series:	82%

Table 1 Examples of measurements of error in soil maps.

The Soil Survey Manual (Soil Survey Staff, 1975b) of the Soil Conservation Service of the USDA acknowledges the problem of reliability by defining a *consociation* as the essential mapping unit. Here a consociation is the soil mapping unit, which is named after a soil series and where at least 85% or 75% of pedons in the mapping unit are of that series, depending on whether inclusions constitute a limiting factor for management.

No other single soil type may occupy more than 10% of the pedons. This definition fails to accommodate most cases quoted in Table 1, but it should be noted that the definition is more recent than most of the studies quoted, and not all studies are subject to USDA-SCS guidelines. Only in the study of Campbell and Edmonds (1984) is the soil variability attributed to any specific reason; there the high soil variability is caused by frequent changes in bedrock lithology (the Rome Formation of Virginia).

Using expert knowledge to handle unreliability

In statistical mapping

The dasymetric map is one way in which the cartographer has accommodated the fact that statistical information is usually collected by enumeration units defined independently of the distribution of that variable. The technique appears to have been pioneered by Wright (1936), in a study of population change on Cape Cod, Mass (Monmonier and Schnell, 1988; see also Robinson *et al.* 1978, p. 207, p. 246; Cuff and Mattson 1982, p. 40). The basis of the dasymetric map is the recognition that statistical information is presented for census areas, but the parameter measured is not evenly distributed across those areas. A choropleth map corresponding to the census divisions presents the information as if the variable mapped (e.g. population, crop type or income) were distributed evenly across the area. In very few instances is this the case. If the map units occupy a sufficiently large portion of the map that sub-areas may be depicted, then it may be possible to infer boundaries for the sub-areas, and how the variable may be distributed among those sub-areas, from knowledge of how that variable is actually distributed with respect to some related parameter or parameters (e.g. land use or urban areas).

Wright (1936), for example, first took a population map by township, and then identified those areas on USGS topographic maps that might be expected to be largely uninhabited, and re-calculated the population densities in the inhabited parts of the township. The resulting map shows a more likely distribution of actual population densities. He went on to break down the inhabited areas into areas of different population density, and produced a final map with considerable variability within township showing a very detailed map of probable population density.

A standard method of dot map production may be viewed as using the same general approach (Robinson *et al.* 1978, p. 204; Cuff and Mattson 1982, p. 31). Here the size of dot and the number of individuals it represents is identified. Dots are then placed in the map space in locations where the individuals are likely to occur. Thus farm animals are not represented as being in woodland or on steep slopes.

In short, both the dasymetric and the dot maps employ knowledge of how the variable being mapped is distributed with respect to some controlling variable(s), to produce a map that may be a more reliable representation of the distribution of that variable. Within a GIS, production of a dasymetric map is not a trivial problem, but equally is not impossible. Effectively all the tools required are available, so long as the GIS is vector based, and/or it is possible to address polygon attributes.

Application to environmental data

The knowledge-based approach to mapping is applicable to non-statistical (categorical) data as much as to statistical data. It is not, however, well established in the literature, perhaps because of a lack of understanding of the nature of unreliability of the categorical maps.

In a soil map of Portage County, Ohio, for example, when Canfield silt loam with 2 to 6 percent slopes is mapped, but the slope increases to 4 to 6 percent, it is likely that the soil is actually Wooster series. If the soil is less poorly drained, and the slope is less steep, it may be Ravenna series (see above, Ritchie. *et al.* 1978, p. 68). Thus using the textual, report information, and by accessing related spatial information, it is possible to delimit those areas that are likely to be imprecisely mapped in the original soil map; in this case the original mapping unit is unreliable where the slope is either steeper or flatter than usual. Furthermore, it is possible to suggest the soil that is more likely to occur in these areas.

In the case of some recent soil reports, there is well documented information relating to the reliability of the mapped units. Soils are, however, not only some of the most consistently mapped environmental parameters, but also the parameter mapped at the largest scale over much of the country (standard county soil maps are now produced at a scale of 1:15,840). Thus it may be even more appropriate to use the soil map to refine the reliability of other environmental maps at smaller scales. For example, in an area mapped as continuous woodland, which includes areas with soils mapped as, say, both Aquic Entisols (poorly developed waterlogged soils) and Aeric Alfisols (better drained soils), the two soil types may actually sustain completely separate assemblages of tree species.

Here then, it is suggested that knowledge of the interaction between environmental and other categorical parameters can be used to improve the reliability of maps. The number of related parameters may be large, and the nature of relationships cannot be specified here. In the examples quoted above, parameters are soil-relief and vegetation-soil, but detailed study would almost certainly reveal the relevance of many more parameters. This knowledge-based approach may also provide a methodology to enable the presentation of map data at a scale larger than that at which it was originally mapped and digitized, without reducing the apparent level of detail, and also, possibly, with a reasonable reliability (at least better than fractal methods).

In the case of categorical maps, as with dasymetric ones, the computer software tools are available to conduct the analyses required. Digital elevation modeling and overlay analysis would both be crucial, and this may therefore reduce the number of GIS that are usable. None is known, however, that can fully automate the functions suggested.

Deriving rules

Specific Rules

At the heart of the approach to handling map unreliability proposed here is the development of a rule base of how one parameter relates to others. Any rule is necessarily highly domain specific. Thus any rules relating to the occurrence of soils are unlikely to be useful in identifying the occurrence of vegetation types.

In the case of some soil maps the rule base is extant in the form of survey reports. Thus the comment on the Canfield series quoted above can be changed into the following rules:

1. IF the soil is mapped as Canfield silt loam 2 to 6 percent slopes,
   AND the slope ranges form 4 to 6 percent,
   THEN the soil may be Wooster series ;
2. IF the soil is mapped as Canfield silt loam 2 to 6 percent slopes,
   AND the slope is less than 2 percent,
   THEN the soil may be Ravenna series.

The veracity of these observations and the rules that can be derived from them have not been put to the test, however. They require rigorous testing to assess whether their implementation does indeed improve reliability. In reports associated with other types of maps, it may also be possible to derive rules directly.

In other cases, however, there may be no information to assist the delimiting of areas of alternate mappable units. This would typically be the case in a geological map and bulletin, where the alternate facies are reported, but little information is given as to how they are distributed. In these, and the majority of other maps, no information is given as to the distribution of impurities. In these cases there is no alternative to the time consuming process of knowledge engineering to elicit relationships such as the population densities on different land cover types in particular areas, and the effect of soil type on growth of tree species (both how well they grow and whether they grow at all). Knowledge engineering is now a well developed area of Artificial Intelligence and practiced widely in the development of expert systems (e.g. Waterman 1986; Hoffman 1987; Senjen 1988)

General Rules

The rules quoted above are specific to a particular soil type, in as much as they relate other environmental parameters to that soil type, and are irrelevant to other soil types, and even probably to the same soil type when it occurs in other areas. At its best, the vegetation-soil example would also be specific to the mapping unit. This example does, however, suggest that it is possible to identify rules at a higher, more general level that may be globally usable for at least a particular parameter.

The reliability of boundaries provides an example of such general rules. As stated above, mapped boundaries of natural phenomena are more often than not a function of human perception rather than of natural changes, although abrupt boundaries can certainly occur in nature. Furthermore, the detection of most soil or geological boundaries is a matter for inference and not observation, and so must always be a matter of locational uncertainty.

Environmental phenomena are, however, interrelated. Thus Jenny (1941) derived the equation of soil formation:

$$S = f(pm, r, om, cl, t) \qquad (1)$$

pm = parent material, r = relief, om = organic matter, cl = climate, and t = time.

Most of these can be related to other mappable environmental parameters. For example, relief is an inherent property of a DEM, and so if a soil boundary is mapped as being associated with an abrupt change in relief, it is likely to be more reliably identified, and so more correctly mapped than if there is no change in relief. Similarly if a change in vegetation is associated with a change in soil type then both boundaries may be relatively accurate. Conversely, if neither vegetation boundary nor relief change is identifiable in the vicinity of a soil boundary, that boundary should not be confirmed.

Geological materials are also related to relief, and are often mapped by observation of differential erosion. Therefore, if changes in elevation or relief can be associated with an inferred geological boundary, then it is possible to confirm the position within close limits, but if the terrain in the area is flat, the boundary is probably less certain.

Information on the classification of the mapped phenomena is widely coded in GIS, and may also be used to implement rules relating to boundary reliability. For example, Ultisols and Alfisols are distinguished primarily by the amount of carbonates in the lower soil horizons (Soil Survey Staff, 1975a), and so any boundary between soils of this order may be expected to be a broad intergrade, while an Alfisol and Histosol boundary is likely to be much more abrupt.

Conclusion and future research

In the spirit of this meeting, the concluding section of this paper is an agenda for fututre research. Some general observations can be made first, however, regarding the knowledge-based approach advocated here. Although not stated explicitly, two stages of analysis can be identified:

1. First is the need to locate those areas where a map is likely to be unreliable. The implication that logically follows from this is that the map units identified in other areas are probably reliable.
2. Subsequently we can revise the labels of map units in some areas .

Stage 1 may be achieved through the identification of areas as being of greater or lesser reliability. It may, however, be appropriate to investigate the use of a surface of some measure of reliability (probability, possibility, etc.).

To address all the issues discussed in this paper requires an extensive and complex research agenda.

1. It is necessary to establish whether the suggested knowledge-based approaches do indeed improve reliability, or just confound an already complex problem. This testing requires a thorough experimental design, and extensive field checking. Furthermore, it probably needs to be established as an appropriate approach for a number of different mapped phenomena.
2. Once proven, the extensive area of knowledge acquisition must be entered to develop the rule bases. This is, of course, specific to each map type and even each mapped area (e.g. county). It may, however, be possible to develop general rule bases pertinent to individual map types and broad areas.
3. Software back-up is required to implement the rule bases once established. The software will undoubtedly be a close relative of artificial intelligence programs, in some aspects, but should only need to implement existing GIS algorithms. Software needs to access and reason with the rule-base to develop reliability maps, and even labelled reliability maps.
4. The area of boundary unreliability introduces a whole new problem area, and although it has only been mentioned briefly here, some comment should be made. If boundaries in the natural environment are in fact intergrades, then the whole process of overlay analysis and spurious polygons must be questioned, and readdressed. The simple Boolean approach implemented in so much of our extant GIS software is actually inappropriate (Robinson and Miller 1988).

References

Bascomb, C. L., and Jarvis, M. G., 1976, Variability in three areas of the Denchworth soil map unit: I Purity of the map unit and property variability within it. *Journal of Soil Science* **27**, 420-437.

Burrough, P. A., 1986, Five reasons why geographical information systems are not being used efficiently for land resources assessment. Blakemore, M. (ed), *Proceedings Auto Carto Vol 2*, London, pp 139-148.

Campbell, J. B., 1978a, Spatial variation of sand content and pH within contiguous delineations of two soil mapping units. *Journal of Soil Science Society of America* **42**, 460-464.

Campbell, J. B., 1978b, Locating boundaries between mapping units. *Mathematical Geology* **10**, 289-299.

Campbell, J. B., and Edmonds, W. J., 1984, The missing geographic dimension to soil taxonomy. *Annals of the Association of American Geographers* **74**, 83-97.

Chrisman, N. R., 1987, The accuracy of map overlays: a reassessement. *Landscape and Urban Planning* **14**, 427-439.

Conacher, A. J., and Dalrymple, J. B., 1977, The nine unit landsurface model: an approach to pedogeomorphic research. *Geoderma* **8**, 1-154.

Courtney, F. M., 1973, A taxonometric study of the Sherborne soil mapping unit. *Transactions of the Insititute of British Geographers* **58**, 113-124.

Croswell, P. L., 1987, Map accuracy: what is it, who needs it, and how much is enough? *Papers from the Annual Conference of the Urban and Regional Information Systems Association Vol. II*, Fort Lauderdale, FL, pp 48-62.

Cuff, D. J., and Mattson, M. T., 1982, *Thematic Maps: Their Design and Production.* (New York : Methuen).

Dale, P. F., and McLaughlin, J. D., 1988, *Land Information Management.* (New York: Oxford University Press).

FAO/UNESCO, 1974, *Soil Map of the World 1:5,000,000 Volumes I to X.* (Paris: UNESCO).

Fisher, P. F., *in press*. The nature of soil data in GIS: error or uncertainty. *International Geographic Information System Symposium*, Washington, D.C.

Hoffman, R. R., 1987, The problem of extracting the knowledge of experts from the perspective of experimental psychology. *AI Application in Natural Resource Management* **1** (2), 35-48.

Hole, F. D., and Campbell, J. B., 1985, *Soil Landscape Analysis.* (Totowa, N.J.: Rowan and Allanheld).

Hsu, M. L., and Robinson, A. H., 1970, *The Fidelity of Isopleth Maps: An Experimental Study*. (Minneapolis: University of Minnesota Press).

Jenks, G. F., and Caspall, F. C., 1971, Error on choroplethic maps: definition, measurement, reduction. *Annals of the Association of American Geographers* **61**, 217-244.

Jenny, H., 1941, *Factors of Soil Formation* . (New York: McGraw-Hill)

MacDougall, E.B., 1975, The accuracy of map overlays. *Landscape Planning* **2**, 23-30.

Mader, D. L., 1963, Soil variability: a serious problem in soil site studies in the Northeast. *Soil Science Society of America Proceedings* **27**, 707-709.

Monmonier, M. S., 1977, Maps, distortion and meaning. *Resource Paper* **75-4**, Association of American Geographers, Washington, D.C.

Monmonier, M. S., and Schnell, G. A., 1988, *Map Appreciation.* (Englewood Cliffs, NJ: Prentice Hall).

Morrison, J. L., 1971, Method-produced error in isarithmic mapping. *Technical Monograph* **CA-5**, American Congress on Surveying and Mapping, Washington, D.C.

Muller, J. C., 1977, Map gridding and cartographic error: a recurrent argument. *Canadian Cartographer* **14**, 152-167.

Newcomer, J. A., and Szajgin, J., 1984, Accumulation of thematic map errors in digital overlay analysis. *The American Cartographer* **11**, 58-62.

Pomerening, J. A., and Cline, M. G., 1953, The accuracy of soil maps prepared by various methods that use aerial photograph interpretation. *Photogrammetric Engineering* **19**, 809-817.

Powell, J. C., and Springer, M. S., 1965, Composition and precision of classification of several mapping units of the Appling, Cecil and Lloyd Series in Walton County, Georgia. *Soil Science Society of America Proceedings* **29**, 454-458.

Rhind, D., 1973, Generalization and realism within automated cartographic systems. *Canadian Cartographer* **10**, 51-62.

Rhind, D., and Clark, P., 1988, Cartographic data inputs to global databases. In *Building Databases for Global Science*, Mounsey, H., and Tomlinson, R., (eds.), pp.79-104. (London: Taylor and Francis)

Ritchie, A., Bauder, J. R., Christman, R. L., and Reese, P. W., 1978, *Soil Survey of Portage County, Ohio*, U.S. Department of Agriculture, Washington, D.C.

Robinson, A., Sale, R., and Morrison, J., 1978, *Elements of Cartography*. 4th Edition. (New York: John Wiley).

Robinson, V. B., and Frank, A. U., 1985, About different kinds of uncertainty in collections of spatial data. *Proceedings of Auto-Carto 7,* pp 440-449.

Robinson, V. B., and Miller, R., 1988, Towards an expert polygon overlay processor for land-related information systems: an investigation of the spurious polygon Problem. *Working Paper No. 1/88,* Institute of Land Information Management, Toronto.

Senjen, R., 1988, Knowledgeacquisition by experiment: developing test cases for an expert system. *AI Application in Natural Resource Management* **2** (2), 52-55.

Soil Survey Staff, 1975a, Soil taxonomy: a basic system of soil classification for making and interpreting soil surveys. *Agricultural Handbook 436*, U.S. Department of Agriculture, Washington, D.C.

Soil Survey Staff, 1975b, *Revised Soil Survey Manual (draft).* U.S. Department of Agriculture, Washington, D.C.

Thompson, M. M., 1983. *Maps for America.* 2nd edition. (Reston, VA: USGS).

Tobler, W., 1988, Resolution, resampling and all that. In *Building Databases for Global Science*, Mounsey, H., and Tomlinson, R., (eds.), pp. 129-137. (London: Taylor and Francis)

Waterman, D. A., 1986, *A Guide to Expert Systems.* (Reading, MA: Addison-Wesley).

Wilding, L. P., Jones, R. B., and Schafer, G. M., 1965, Variation of soil morphological properties within Miami, Celina and Crosby Mapping Units in West-Central Ohio. *Soil Science Society of America Proceedings* **29**, 711-717.

Wright, J. K., 1936, A method of mapping densities of population with Cape Cod as an example. *Geographical Review* **26**, 103-110.

Chapter 5

Observations and comments on the generation and treatment of error in digital GIS data

Giulio Maffini, Michael Arno and Wolfgang Bitterlich

Background

The widespread availability of geographic information systems has enabled many users to integrate geographic information from a wide range of sources. An inevitable consequence of data integration has been the sober realization that geographic data is not of equal quality, that it can contain errors and uncertainty that need to be recognized and properly dealt with.

The body of literature devoted to describing the various aspects of spatial error is growing. Burrough (1986) (Chapter 6) provides an overview of important considerations in this field. An early conclusion that can be drawn from reviewing the published literature is that the error issue raises fundamental questions about the nature of observation in geography.

Purpose and scope

A complete overview of all the sources of error and uncertainty in geographic information is beyond the scope of this paper. The principal objectives here are to:

a) illustrate the size and range of some types of errors generated by GIS users in the digitizing process;
b) explore ways of dealing with these errors in derivative GIS products, and
c) make some observations and suggestions for research that may help to more comprehensively cope with the errors.

Sources of error

As developers of Geographic Information Systems, it is important to recognize the type, severity and implications of errors that are inherent in the use of a geographic information system. Figure 1 shows a schematic of the sources that contribute to likely error. The likely error can be thought of as occurring due to three major causes.

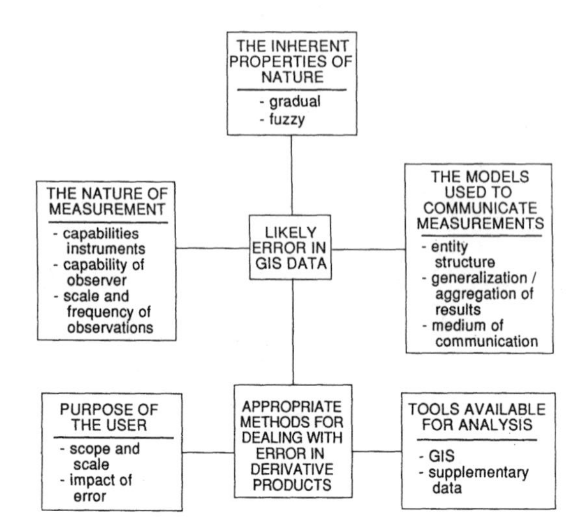

Figure 1 The error issue.

The first major cause of error is due to the inherent properties of nature. Unlike geographic information data structures, the real world is not always distinct and clear, but is frequently gradual and fuzzy. An aerial photograph or a satellite image taken from high altitude may appear to show a road as a distinct line feature on a plane, but as one approaches more closely the perception that the road is a single line becomes inappropriate. Similarly, for coverages, an area of grassland is not always distinct, it can shift into woods or desert in a very gradual manner. This inherent fuzziness will always be with us when we observe nature.

A second major source of likely error stems from the nature of measurement in geography. Any measurements that are acquired with instruments inevitably introduce error. The capability of the person using the measuring device can also clearly affect the amount of error introduced. Finally, the scale at which the measurements are made and the frequency of sampling will also introduce potential errors in geographic data.

A third source of error is due to the data models that we use to communicate our measurements. The very structure of the geographic model can be a source of error. In the case of vector, the representation of a line or an edge implies a level of certainty or precision that may not be discernable in the real world. In the case of raster, the aggregation or averaging of conditions by imaginary cells is also susceptible to error. Finally, the medium used to communicate measurements, itself may introduce errors. If a satellite image is converted into a photograph or transparency, the error properties of this product may be different from those of the same information retained in the original digital form.

Another important source of error in geographic data is, of course, the data processing and transformation. The more times a set of measurements are transformed through one process or another, the more likely new errors or uncertainty will be introduced into the derivative products. From a developer's point of view the error issue cannot be resolved just by understanding the causes or magnitude of geographic error. The important question is, what are the appropriate methods for dealing with error in derivative products? This leads to two other sets of questions which need to be considered when addressing this issue. These are:

a) what is the purpose that the user is trying to accomplish? What is the impact of varying levels of error on this purpose? For example, if we are trying to identify the location of an area suitable for growing a particular crop, should we be concerned with a potential error of 10 to 20 meters? Alternatively, if we are trying to locate a buried telephone cable, is a 10 to 20 meter level of error acceptable?
b) What tools do we have available to deal with likely error? Do we have a geographic information system to deal with the issue? What type is it? The answers to these questions may pose restrictions on how we can treat error in our spatial data. The availability of corroborative data from other sources to help cross reference and resolve uncertainty or error in spatial data may also be an important factor.

Background to digitizing trials

Simple trials were conducted in order to develop an appreciation of the range of digitizing errors that can be generated by typical GIS users. The trials were concerned with exploring positional errors. They excluded other types of errors, such as topological problems associated with data base creation.

In the trials an attempt was made to isolate the effects of two factors that one would expect to influence the propagation of error. The first was the scale of the source material used for digitizing, and the second factor was the speed with which the operator conducted the digitizing. Although efforts were made to be systematic the trials were not scientifically controlled experiments. It is also important to recognize that the trial results were based on the digitizing performance of one person.

Approach to the digitizing trials

The concept behind the digitizing trials is very simple: digitize entities whose location you can describe precisely and compare the results with the original entities.

This, at first, may seem somewhat simple. After careful consideration, however, the problem becomes a bit more complex. How can the real world entities be defined with absolute precision? Using an existing, already digitized product as the comparative base presents some difficulty since this is, itself, a representation of some other reality. To minimize this problem a set of mathematically definable geometric entities was selected as the digitizing material. The entities included a point, a straight line, a triangle, a square, a pentagon on through to a decagon. In addition to these discrete entities a set of continuous mathematical functions were also chosen. This included a circle and three sine functions of different amplitude. Figure 2 summarizes the approach to the digitizing trials. Figure 3 shows examples of the entities that were digitized.

These precise entities were generated in a CAD system and plotted in a large format. The large plots were then reduced in size to represent a range of scales (1:50,000, 1:100,000, 1:250,000). Each of the entities was then digitized by an operator at each of the three scales and at three different speeds (comfortable, hurried, very hurried). The speeds were governed by improving time limits on the operator.

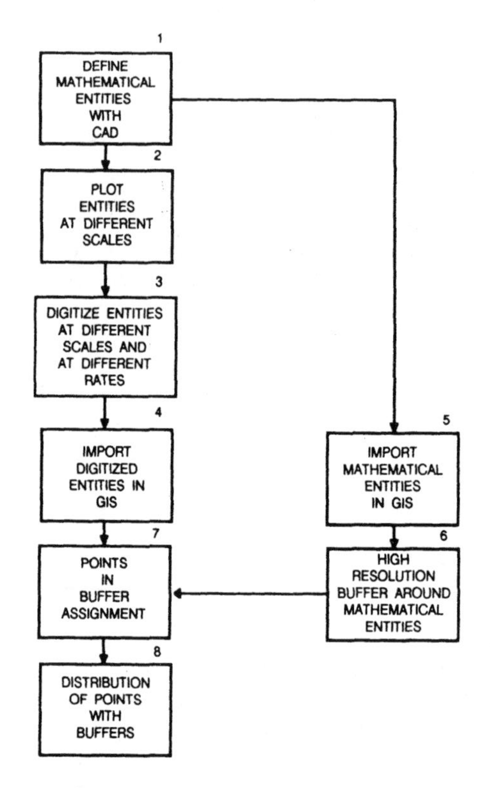

Figure 2 Digitizing trials approach.

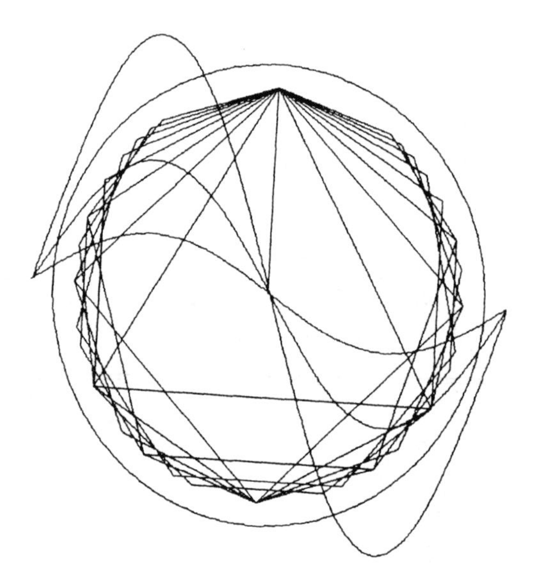

Figure 3 Examples of entities digitized.

In the case of the discrete entities, the purpose of the trial was simply to place a point at each of the vertices of the discrete entity. Identical registration points were also provided for each entity digitized. In the case of continuous functions, the operator attempted to approximate, as closely as possible, the continuous nature of the entity. Thus, the operator had the choice of determining how many points would be used to represent a particular segment of a continuous function. Digitizing in stream mode was not used.

After the entities were digitized they were imported into a common geographic information data base (universe) using a UTM projection. During this process obvious mistakes and errors associated with calibration or data entry were corrected so as not to improperly bias the results of the trials.

The mathematical entities as described in the CAD system were also imported into the GIS in the same UTM universe. Using the capabilities of the GIS a set of concentric buffers ranging from 10 meters through to 90 meters were propagated around each of the mathematically defined entities.

The points digitized in the trials were then mapped to the high resolution buffer around the mathematical entities and assigned the buffer interval class number in which they fell. This, essentially, assigned an error class interval to each point digitized in the trials where the class represented the extent to which the point was distant from the mathematical entity.

Results of digitizing trials

A number of summary statistics were calculated from the trials. These included:

a) the number of points generated to represent an entity, and
b) the frequency distribution of error classes for each entity.

A set of statistics was generated for each combination of speed and scale. Discrete and continuous entities were treated separately. A number of interesting observations can be made from the results. For example, it is apparent with the discrete entities that changes in the scale of the source document used for digitizing have a more significant impact on positional error than the time taken to digitize. For the continuous entities it is apparent that the number of points used to digitize an entity varied with both the time taken for digitizing and the scale of the source document.

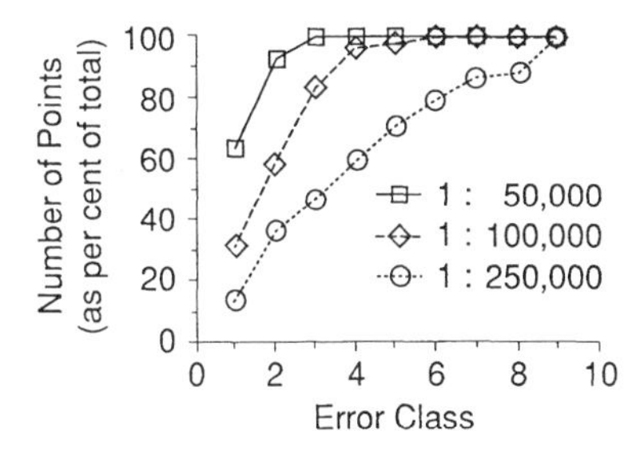

Figure 4 Results of digitizing trials.

Figure 4 shows a summary of the results of three trials for the discrete entities at the base digitizing speed. The graph clearly shows that the trials conducted using the 1:50,000 scale source documents have a higher level of accuracy. More than 90% of all the points for the discrete functions fell within the first two error classes (20 meters) of the mathematical entities. The same entities digitized at a comfortable digitizing rate but at the 1:250,000 scale produced a set of results that were more dispersed. Only about 33% of all points digitized fall within the two-error class range (20 meters).

Figure 5 presents a bar graph that illustrates the results of the digitizing trials for the continuous entities in terms of the number of points used to represent a continuous entity. As can be seen, as the scale of the source documents increased the number of points used for digitizing these entities decreased. Similarly, as the time available for digitizing diminished, the number of points also diminished.

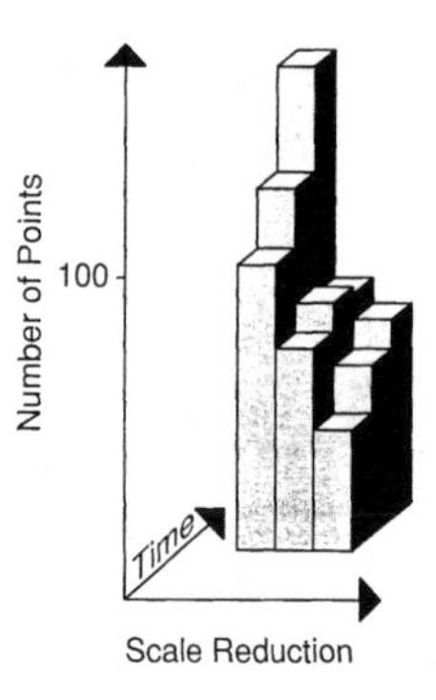

Figure 5 Continuous entities trials.

What does error mean?

The results of these digitizing trials confirms the obvious conclusion that an entity, discrete or continuous, cannot be captured absolutely correctly in the normal digitizing process. So what is this positional error? How can we try to understand it?

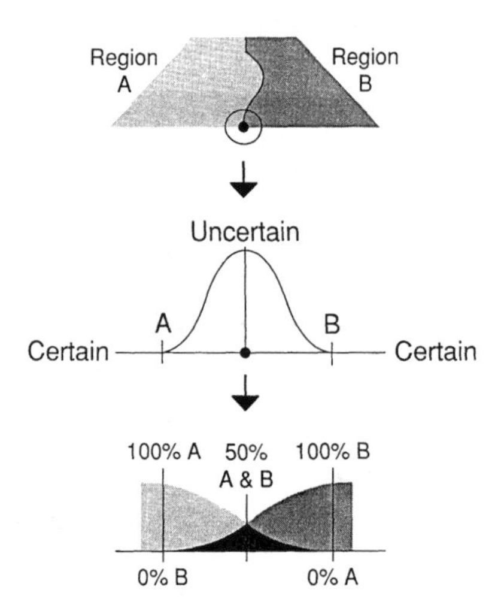

Figure 6 What does error mean?

Figure 6 illustrates how error or uncertainty as explored in these trials might be considered. Given a map made up of two discrete regions (A and B) which share a common boundary, repeated digitizing of the boundary line will generate a distribution of points around that line. All things being equal, there will be more points closer to the actual location of the line than further away from it. Conceptually, this distribution will probably take on the form of a normal distribution whose peak is centered on the line. This concept is closely related to the "epsilon" distance concept put forward by Blakemore (1984).The range and extent of the dispersion of points around the lines are likely to be associated with a number of variables including digitizing speed and, perhaps, the apparent clarity of the line. If the line is a thin, black line on white paper, it may be easier to determine its location than if the line is the edge of a field as seen in an aerial photograph.

The line, then, should represent the location of greatest probability that an edge occurs. Another way of thinking of this is to consider the line as the place of maximum uncertainty with regard to region containment. That is, on this line there is a 50/50 chance of being in either region. As one moves to the left of the line the probability of being within Region A increases rapidly. Similarly, starting from the line and moving to the right we increase the probability that we are in Region B. If the line is the breakpoint, integrating the normal distribution will produce two complimentary, cumulative probability S functions for Region A and Region B. These functions express the concept that within a certain range of the line there is a probability of being in either A or B.

How can the concentric error bound concept be used in GIS overlay operations?

Using the concept of error discussed previously and the results of the digitizing trials, some experiments were conducted to explore how positional error might be treated in typical GIS overlay operations.

A common problem often dealt with by GIS users is to select geographic areas that possess certain properties. For example, let us identify all of the locations within a target area that are within 100 meters of a road, have slopes greater than 15% and are contained within a certain drainage basin.

A simple example of this approach was applied to an area in Connecticut that had been digitized for a demonstration data base by the USGS and the State of Connecticut. The result of the search procedure outlined above, and the part of the input maps that was used to arrive at the selected areas, are presented in Figures 7 through 10.

Figure 7 Within 100 meters of Road.

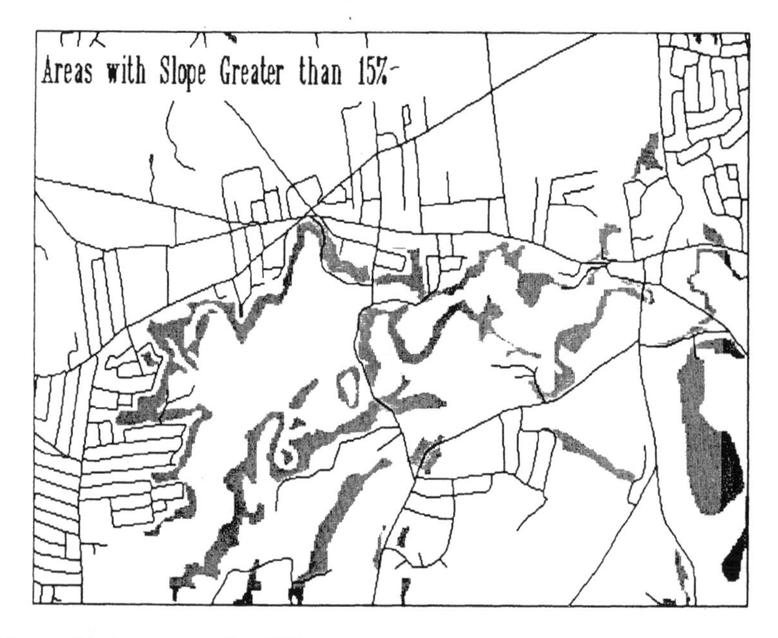

Figure 8 Areas with slope greater than 15%.

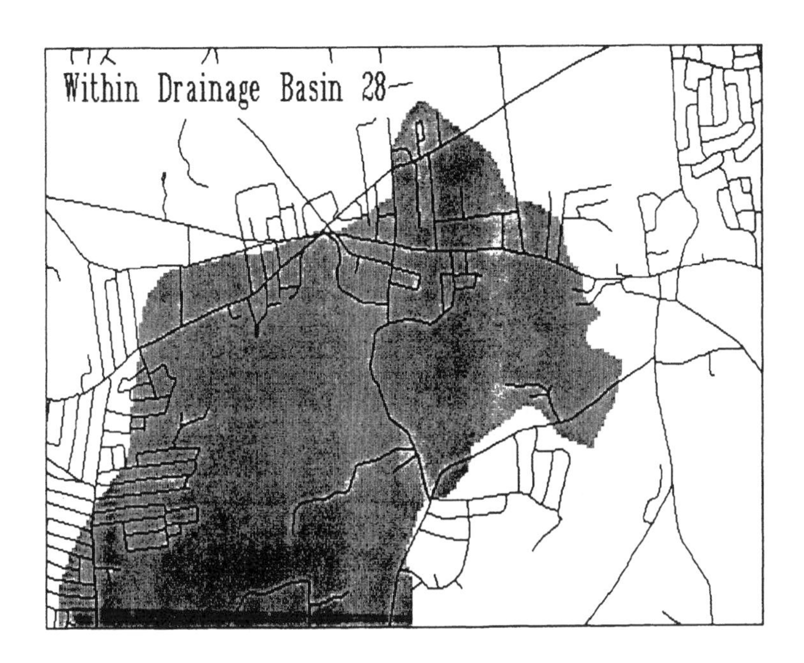

Figure 9 Within drainage basin 28.

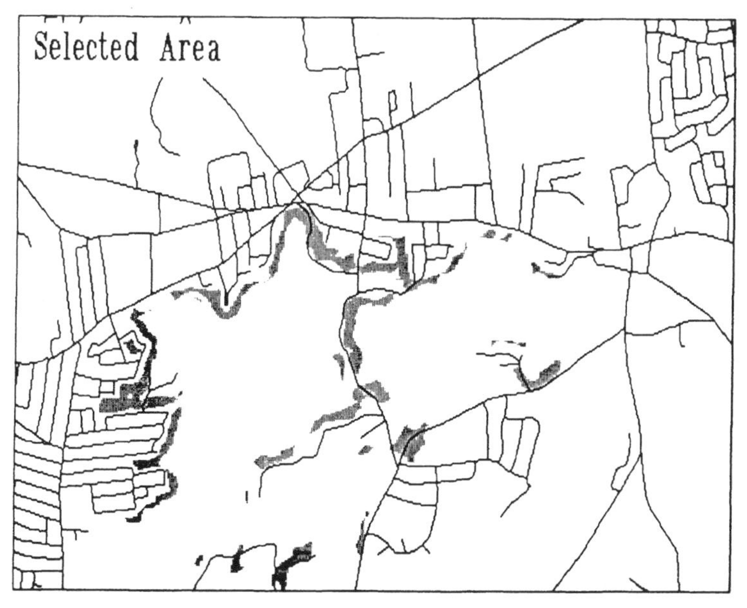

Figure 10 Selected area.

The approach used to generate a solution to this question can differ from GIS to GIS depending on its inherent approach and data structure. A vector-based GIS will handle this differently than will a raster-based, or quadtree-based GIS. Regardless of the method used in the GIS itself, however, it is apparent from the previous arguments in this paper that the geographic information possesses certain probable errors. For the sake of discussion, certain assumptions are made here regarding accuracy and error in the data. For instance the lines in the GIS that define the road network likely represent the most accurate data layer in the data set (depending upon the scale at which they were digitized). The location of the crest of the watershed is probably more uncertain than the location of the roads. The slope areas map is also probably not quite as accurate as the road, but it may be more accurate than the crests of watersheds.

Going back to the previously discussed model of error, these principles can be applied to attach different probabilistic distributions around each of the entities that are referred to in the analytical operation.

Figure 11 shows a set of probabilistic assumptions that were used regarding the location of each line defining the road, slopes and watersheds. It should be noted that these assumptions do not necessarily reflect the accuracy of the original source documents. They were selected to illustrate the concept in this paper.

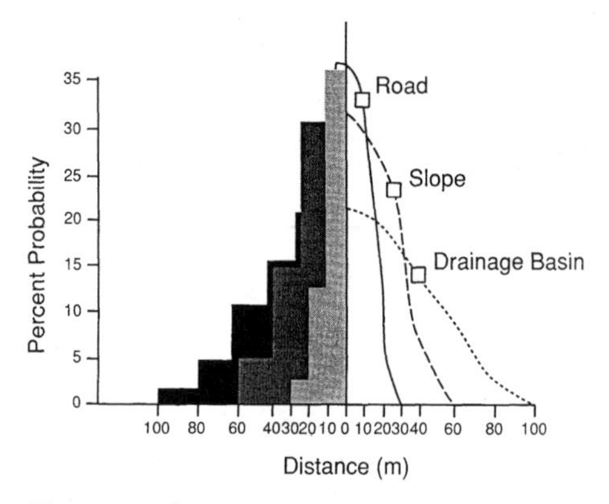

Figure 11 Error bounds used for case study.

Probability	Area found (km sq) Inside Selected area	Outside Selected area	Total
0	0.0000	18.4847	18.4847
0 - 0.1	0.0000	0.5759	0.5759
0.1 - 0.2	0.0000	0.4665	0.4665
0.2 - 0.4	0.0000	0.1525	0.1525
0.4 - 0.6	0.0041	0.5756	0.5797
0.6 - 0.8	0.5051	0.0000	0.5051
0.8 - 1.0	0.1744	0.0000	0.1744
1.0	0.0178	0.0000	0.0178
Total	0.7015	20.2553	20.9567

Table 1 Area cross tabulation.

Using the assumed distributions, concentric buffers were developed around each of the edges defining the classes that had to be used in the overlay process. Each buffer, or interval, was assigned a probability value of the class in question being found in that buffer interval. The procedure was completed for each of the three map layers and the overlay process was repeated. In this overlay a model was defined which calculated the combined probability of finding the desired combination of conditions as the product of the probabilities of each of the original input classes. A new map was produced to show a range of probabilities of finding the selected class (within 100 meters of a major road, above 15% slope and within drainage basin 28). The map in Figure 12 has some obvious similarity to the original selected area of Figure 10, but it also shows some important differences. The most notable difference is that the area of certainty for the desired combination of classes has been significantly reduced. Conversely the area which defines the limit of some minimal probability of the selected class being found is greater than the original selected area.

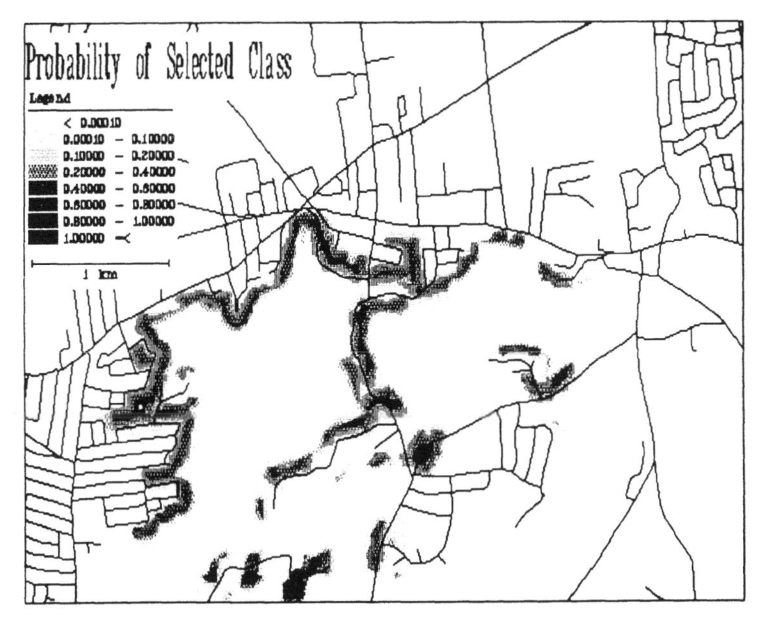

Figure 12. Probability of selected class.

A more quantitative comparison of the two maps is presented in Table 1 which shows a cross-tabulation of the map in Figure 10 with that of Figure 12. The table shows that the total area within the map is 20.9 square kilometers. Of that, approximately .7 square kilometers were found in the original search operation (Figure 10). Using the probabilistic approach the area of certainty for the selected class is reduced significantly to 0.0178 square kilometers. This is about 2.5% of the original search operation.

The table also shows, however, that despite this reduction of the area of certainty there is an additional significant area where there is some probability that the selected class could be found which was not visible in the original search. Depending upon the purpose and intent of the user in performing the search this information, both quantitative and spatial, could be of assistance in considering other desirable locations.

Observations and conclusions

The results of digitizing trials and our subsequent experiments with methods for dealing with error and uncertainty point to some interesting conclusions and observations.

a) First, it is clear that error in digital geographic data is inevitable. Errors cannot be eliminated by simply improving the performance of instruments and/or operators.
b) Many types of errors and uncertainties are a direct product of how we aggregate and represent natural phenomena in geographic information systems.
c) Attempting to eliminate error and uncertainty in our data products is a noble goal but one which may have practical limits.
d) Instead of improving the spatial representation and accuracy of our geographic information systems, we need to develop methods for users to explore and cope with error.
e) Our ability to cope with error means that we will need to be more explicit about the error or uncertainty contained in source geographic data and in derivative products.
f) The error or accuracy standards associated with geographic data should vary depending upon the nature of the entities being represented and the method used for describing them. Cartographic and digital products should therefore distinguish between the accuracy levels of different features in geographic data bases.
g) We need to explore alternative ways of packaging and distributing geographic data in products that are closer to the original source information and are less susceptible to new errors as a result of added processing or aggregation.
h) Geographic Information Systems should, and can be used to assess the consequences of error in geographic data. With GIS tools users will be able to make more informed judgements about the interpretation and integration of geographic data.

In terms of a research agenda for GIS investigators, we believe that there are some fertile and interesting areas which need further development. These include:

a) The design and execution of controlled scientific experiments to determine the error ranges propagated in commonly available geographic data products. A systematic investigation and documentation of these errors would provide valuable information to users. With the results of these experiments and investigations users could make more informed and appropriate use of the data.
b) The academic and GIS development communities need to develop practical guidelines for GIS users to help them assign error bounds to their data and to develop more comprehensive and versatile methods for assessing the impact of error bounds on derivative GIS products.

Finally, it is suggested that a fruitful area of enquiry is to consider the development of a different model for distributing geographic data which recognizes the existence and role of GIS. The unnecessary interpretation and generalization of original geographic data so that it can be packaged and distributed in a cartographic map model is a carry-over from a previous era and is no longer appropriate for many users. The custodians of our cartographic data bases should consider ways of broadening the range of data sets which are made available to the user community in a quality controlled, but essentially original, form.

Acknowledgements

The authors wish to acknowledge the assistance of Tim Webster who executed the digitizing trials and Glenn Holder who provided advice and suggestions on the interpretation of the digitizing trials.

References

Blakemore, M., 1984, Generalization and error in spatial data bases. *Cartographica* **21**, 131-139.

Burrough, P. A., 1986, *Principles of Geographical Information Systems for Land Resources Assessment.* (Oxford: Clarendon)

Chapter 6

Developing confidence limits on errors of suitability analyses in geographical information systems

Weldon A. Lodwick

Abstract

Confidence limits based on measures of sensitivity are developed for two of the most widely used types suitability analyses, weighted intersection overlay and weighted multidimensional scaling. Some underlying types of sensitivity analyses associated with computer map suitability analyses are delineated as a way to construct measures of these delineated sensitivities that in turn become the means by which confidence limits are obtained.

Introduction

This paper is concerned with developing approaches to limits of confidence in output attribute values given errors in the input attribute values. To do this we construct methods to measure the variation in the attribute values obtained from suitability analyses given variations in the attributes of the input source maps that comprise the suitability analysis of interest. Typically, it is very difficult to measure the actual errors associated with input maps. Therefore, to get an idea of how the resultant map attributes change with respect to variations of the attributes in the input map, perturbations of the attributes associated with the input maps can be imposed. We do not study variation due to perturbation of shapes of the polygons, only confidence limits in geographic suitability analysis due to input attribute error, both real and imposed. Errors in the shapes of the polygons themselves have been studied by Maffini *et al.* (1989), "Observations and comments on the generation and treatment of error in digital GIS data" (this volume) and by Honeycutt (1986).

The development of confidence limits due to input attribute errors whether real or imposed are handled in the same way. In order to develop confidence limit measures, it is necessary to perform a geographic sensitivity analysis for the suitability (geographic) analysis of interest. Geographic sensitivity analysis is implemented to discover to what degree the output map changes given perturbations of various types imposed on some or all the inputs involved in a geographic analysis. The full development of geographic sensitivity analysis can be found in Lodwick *et al.* (1988).

This paper is organized into five sections. The second section presents preliminary matters, definitions, mathematization of suitability and geographic sensitivity analysis. In the third section are found the sensitivity measures requisite for the development of confidence limits while the fourth section discusses the confidence limits associated with suitability analysis. The fifth section presents concluding remarks and points to directions for future research.

Preliminaries

The methods outlined herein can be applied to any geographic analysis for which a mathematical procedure (algorithm in the mathematical or computer science sense) can be defined. For definiteness, intersection overlay and multidimensional scaling suitability analysis are used as the prototypes for the development of confidence limits associated with errors in attributes. The suitability analysis we study will be couched as a mathematical function of the attributes of one or more primary maps to make the sensitivity measures and confidence limits computable and germane to computer resident GIS's.

The domain of inquiry pertinent to this study is depicted in Figure 1 below by phases 2, 3 and 4 where the suitability analysis process can be thought to move from phase 0 to phase 4.

	REALITY		PHASE 0
	CATEGORIES OF GEOGRAPHIC ENTITIES OF INTEREST		PHASE 1
	PRIMARY MAPS		PHASE 2
RANK MAPS	DERIVED MAPS	REFORMATTING	PHASE 3
	SUITABILITY MAPS		PHASE 4

Figure 1 The Suitability Process.

The basic concepts, relationships and associations of sensitivity analysis in the context of GIS are set forth next.

Definitions

The following concepts are used in the sequel.

Primary maps: maps of existing geographic entities of interest; e.g., a geologic or vegetation map.

Rank map: a map whose attributes are numeric (ordinal, interval, ratio) rankings and are mathematical relations (e.g. functions, transformations) of the attributes from one or more maps. A soils map could be ranked to indicate the potential it has to contaminate underground aquifers due to the dumping of hazardous wastes at the surface or the potential hazard that these soils may possess for landslides.

Geographic suitability analysis (or simply suitability analysis): as used here, means the resultants from an application of one or more mathematical relations (functions, transformations e.g.) to the attributes of one or more maps. Thus, a rank map is an example of a suitability analysis and an overlay is another example. To be specific and to set forth notation that will be used, suppose there are N (primary or rank) maps being overlaid. Let

P(N) denote the total number of polygons generated by the intersection overlay of N maps,
a_{pn} denote the attribute from rank map $1 \leq n \leq N$ contained in the pth polygon $1 \leq p \leq P(N)$ of the intersected overlay map, and
w_n denote the weight given to the nth rank map.

A suitability analysis generates for the pth polygon of the resultant (suitability) map a resultant attribute r_p which is a mathematical function of the weight vector $\mathbf{w} = (w_1,...,w_N)^T$ and the attribute vector $\mathbf{a}_p = (a_{p1},...,a_{pN})^T$ in the following way:

$$\begin{aligned} r_p &= f(w_1,\ldots,w_N, a_{p1},\ldots, a_{pN}) \\ &= f(w, a_p) \end{aligned} \qquad (1)$$

Thus, the attributes of the resultant suitability map are the components of the resultant vector,

$$\mathbf{r} = (r_1, \ldots, r_{P(N)})^T \qquad (2)$$

If $\mathbf{a}_p$ is replaced by point or line attributes instead of polygon attributes, then the formulation (1) and (2) attains the full generality including most commonly known and used geographic suitability analyses associated with commercially available GIS's. In particular the weighted intersection overlay has a function (1) represented by:

$$r_p = f(w, a_p) = \sum_{n=1}^{N} w_n a_{pn}, \text{ for } p = 1,2,\ldots, P(N) \qquad (3)$$

The multidimensional scaling approach obtains the resultant attribute, r_p, from the square root of the weighted sum of the squares of the difference of the rank map attribute from a given attribute value. That is, the attribute for the pth polygon of the resultant (suitability) map, is given by a function (1),

$$r_p = f(w, a_p) = \left(\sum_{n=1}^{N} w_n (a_{pn} - I_n)^2\right)^{1/2} \quad \text{for } 1 \leq p \leq P(N) \qquad (4)$$

where I_n is the most preferred (ideal, most suitable) attribute value or the least preferred (most hazardous, least suitable) associated with the nth rank map. Of course (4) is simply a weighted Euclidean distance of the vector of attributes (indices) contained in the intersection polygon p, a_p, to the vector of "ideal" attributes $\mathbf{I} = (I_1,...,I_N)^T$.

Suitability analysis (3) we call overlay suitability whereas (4) we call multidimensional scaling or target suitability. Clearly, we must assume that in forming the indices r_p, the functions that determine how these attribute values are formed are appropriate for the type of analysis being done with them. That is, for (3) and (4) the indices must be interval or ratio since multiplication is involved. These will be the two suitability analyses we use as prototypes in developing confidence limits. However, confidence limits can be obtained using any suitability analysis for which there exists an explicit mathematical algorithm for the right hand side of (1).

Overlay or target map: the map generated by (3) or (4) respectively.

Geographic sensitivity and error propagation analysis

Closely allied to sensitivity analysis is error propagation analysis, or as some authors call it, error analysis. Error propagation as used by some geographic information system researchers and practitioners (see Walsh [1989], "User considerations in landscape characterization," these proceedings, or Walsh, Lightfoot and Butler [1987]), measures the way known errors in a set of primary maps being used, affect the error in the final output map by actually verifying on the ground the veracity of what the output map indicates. The quality and correctness of the primary maps is determined beforehand by "ground truthing".

Sensitivity analysis on the other hand starts with the maps comprising the analysis, the primary maps, as given on the medium (mylar, paper, or computer file) with no presumption of error, and submits them to predetermined perturbations. Thus, changes or perturbations are imposed on the primary maps to see the effects on the final output map. The focus of geographic sensitivity analysis is on the resultant map generated from perturbing the inputs compared to the original unperturbed map of a suitability analysis and not on the actual errors that exist in reality for the given maps. That is, error propagation analyses as described by Walsh, Lightfoot, and Butler (1987), are measures of real errors that are propagated whereas the imposed perturbations on the suitability analysis (3) and (4) can be considered a theoretical error propagation analysis. In (1) error propagation analysis would investigate perturbations in the data, a_p. Sensitivity analysis is more interested in the function f itself which is given by (1) and on what the perturbations of a_p and w do to the value of r_p. Confidence limits for attribute errors indicate the range of validity in an output attribute value given the ranges of errors in the input attribute values that go into determining the corresponding resultant output attribute value.

The computational process and mathematical representation of suitability analysis

The process of obtaining a suitability map from primary maps for this study is as follows. Given M primary maps, N rank maps are created. If M is less than N, consider those maps that generate more than one rank map as distinct, separate maps. A geology map could be used to rank depth to ground water and also transmissivity of water flows. In this case the one geology map would be considered as two. On the other hand, if M is greater than N, two or more primary maps generate one rank map. Whatever method is used to combine the primary maps into one to generate a rank map, consider the combination of primary maps as the (one) primary map. In this way we will always have M equal to N. This simplifies our notation. With this understanding, the N primary maps generate the N rank maps. The polygons of these N rank maps are intersected yielding the P(N) polygons of the suitability map. The attribute values of each of these polygons are obtained by (3) or (4).

Two mathematical transformations are involved in obtaining a suitability map from primary maps. The first is from primary map attributes to attributes of the rank map that connote a ranking (interval or ratio). The second is transformation of the attributes of the rank map to attributes of the suitability map. The first transformation we represent as

$$T_n(b_{pn}) = a_{pn}, \text{ for } p = 1,\ldots, P(N), \text{ and } n = 1,\ldots,N \tag{5}$$

where b_{pn} is the attribute value for polygon p on primary map n. The second mathematical transformation is (3) or (4) (in general (1)).

Therefore, we can represent the p attributes p = 1,...,P(N) of the suitability map by the following two equations:

$$r_p = \sum_{n=1}^{N} w_n\, a_{pn} = \sum_{n=1}^{N} w_n\, T_n(b_{pn}) \tag{6}$$

for overlay suitability and

$$\begin{aligned} r_p &= \left(\sum_{n=1}^{N} w_n (a_{pn} - I_n)^2 \right)^{1/2} \\ &= \left(\sum_{n=1}^{N} w_n (T_n(b_{pn}) - I_n)^2 \right)^{1/2} \end{aligned} \tag{7}$$

for the target suitability. Equations (6) and (7) give the components of a P(N) dimensional vector so that we have the matrix/vector product,

$$\mathbf{r} = \mathbf{T}\mathbf{w} = \begin{pmatrix} T_1(b_{11}) & & T_N(b_{1N}) \\ \cdot & & \cdot \\ \cdot & & \cdot \\ \cdot & & \cdot \\ T_1(b_{P(N)1}) & & T_N(b_{P(N)N}) \end{pmatrix} \begin{pmatrix} w_1 \\ \cdot \\ \cdot \\ \cdot \\ w_N \end{pmatrix}$$

$$= \begin{pmatrix} a_{11} & \ldots & a_{1N} \\ \cdot & & \cdot \\ \cdot & & \cdot \\ \cdot & & \cdot \\ a_{P(N)1} & \ldots & a_{P(N)N} \end{pmatrix} \begin{pmatrix} w_1 \\ \cdot \\ \cdot \\ \cdot \\ w_N \end{pmatrix} \tag{8}$$

Likewise, if we let

$$\mathbf{t} = \left(r_1^2, \ldots, r_{p(N)}^2\right)^T$$

we have as in (8) a matrix/vector product,

$$
\mathbf{t} = \mathbf{Tw} = \begin{bmatrix} (T_1(b_{11}) - I_1)^2 & & (T_N(b_{1N}) - I_N)^2 \\ \cdot & & \cdot \\ \cdot & & \cdot \\ \cdot & & \cdot \\ (T_1(b_{P(N)1}) - I_1)^2 & & (T_N(b_{P(N)N}) - I_N)^2 \end{bmatrix} \begin{bmatrix} w_1 \\ \cdot \\ \cdot \\ \cdot \\ w_N \end{bmatrix}
$$

$$
= \begin{bmatrix} (a_{11} - I_1)^2 & & (a_{1N} - I_N)^2 \\ \cdot & & \cdot \\ \cdot & & \cdot \\ \cdot & & \cdot \\ (a_{P(N)1} - I_1)^2 & & (a_{P(N)N} - I_N)^2 \end{bmatrix} \begin{bmatrix} w_1 \\ \cdot \\ \cdot \\ \cdot \\ w_N \end{bmatrix} \quad (9)
$$

Measures for geographic sensitivity

The sensitivity of individual polygons can be obtained once an expression for the T_n is found. In this section we develop measures that pertain to the variations over the entire set of P(N) polygons of the suitability map. At least five geographic sensitivity measures for the entire suitability map as a whole can be defined. First there are measures of the overall magnitude of changes in the attribute values from their unperturbed values which we call **attribute sensitivity measures**. Second, there are measures of how many of the attributes change with respect to their rank order from their unperturbed position which we call **position sensitivity measures.**

For measuring the sensitivity associated with removing a set of maps from a suitability analysis, we develop a third category of sensitivity measures called **map removal sensitivity measures**. This measure is associated with perturbation of the weight vector **w**. In particular, a map removal is a geographic sensitivity analysis where the perturbation of the weight for the map being removed is equal in magnitude but opposite in sign to its value in the unperturbed suitability analysis. For the determination of the polygon(s) that is(are) most sensitive to perturbations, we develop a fourth category of sensitivity measures called **polygon sensitivity measures. Area sensitivity measures** compute the total area over which attribute changes have occurred. The full development of the theory and applications of geographic sensitivity analysis and sensitivity measures can be found in Lodwick *et. al.* (1988).

Attribute sensitivity measures

Let r_p denote the attribute resulting from no perturbation of (3) or (4), and $\underline{r}_p$ denote the attribute resulting from some perturbation of (3) or (4). For sensitivity analyses that do not involve the removing of sets of maps from the suitability analysis, the weighted attribute sensitivity measure is,

$$
m_1\,(\mathbf{s},\mathbf{r},\underline{\mathbf{r}}) = \sum_{p=1}^{P(N)} s_p \,|\, r_p - \underline{r}_p \,| \quad (10)
$$

where $\mathbf{s} = (s_1, \ldots, S_{p(N)})^T$. If $s_p = 1$, then the measure is unweighted. If $s_p = A_p$ where A_p is the area (this could be population for a demographic map or some other type of count geographic data) of the pth polygon, then we have an **areal weighted attribute sensitivity** measure. If

$$s_p = \frac{A_p}{\left(\sum_{p=1}^{P(N)} A_p\right)}$$

then we have a **normalized areal weighted attribute sensitivity** measure.

Position sensitivity measures

Let $R(r_p) = j$ (respectively $R(\underline{r}_p) = j$) be the function that assigns the rank order (from low to high) to all the P(N) attributes of the overlay (respectively target) suitability map. Then the weighted rank sensitivity measure is,

$$m_2(\mathbf{s},\mathbf{r},\underline{\mathbf{r}}) = \sum_{p=1}^{P(N)} s_p \,|\, R(r_p) - R(\underline{r}_p) \tag{11}$$

If $s_p = 1$, then the measure is unweighted. If $s_p = A_p$, then we have an **areal weighted rank sensitivity** measure. If s_p is the proportion of the total area (as described in the section on attribute sensitivity measures), then we have a **normalized areal weighted rank sensitivity** measure.

Map removal sensitivity measures

Removing one or more maps from a suitability analysis will in general lower the resultant perturbed attribute values $\underline{r}_p$ for all polygons compared to their value when all the maps are used in the suitability analysis. To remove this bias, we divide each of the suitability map attributes by the number of maps and use these values in the sensitivity measures. In particular, the weighted map removal sensitivity measure is,

$$m_3(\mathbf{s},\mathbf{r},\underline{\mathbf{r}}) = \sum_{p=1}^{P(N)} s_p \,|\, r_p/N - \underline{r}_p/K| \tag{12}$$

where as before N is the total number of primary maps used in the original suitability analysis and K is the number of maps remaining after the set of maps is removed from the suitability analysis. Other unbiasings could be used.

Polygon sensitivity measures

The attribute(s) undergoing the maximum change as a result of the perturbations, is obtained by finding the index or indices for which $|r_p - \underline{r}_p|$ is maximum. The corresponding magnitude(s) is simply the infinity norm of the difference vector $\mathbf{d} = \mathbf{r} - \underline{\mathbf{r}}$.

In this way, the polygon(s) which are the most susceptible to changing given perturbations or errors in the maps are the set of indices J such that

$$J = \{(j : |r_j - \underline{r}_j| = \max_{1 \le i \le P(N)} |r_i - \underline{r}_i|\}$$

where the magnitude of the maximum change is given by the value on the right hand side of the above. Thus the polygon sensitivity measure is given by,

$$m_4(\mathbf{s},\mathbf{r},\underline{\mathbf{r}}) = \max\{|r_i - \underline{r}_i|, i = 1, \ldots, P(N)\} \tag{13}$$

Area sensitivity measures

To get a measure of absolute total area that has undergone change in attribute values as a result of perturbations/errors, the **absolute area sensitivity measure** is defined as follows:

$$m_5(\mathbf{s},\mathbf{r},\underline{\mathbf{r}}) = \sum_{p=1}^{P(N)} s_p A_{\underline{r}} \tag{14}$$

$$s_p = \begin{cases} 1 \text{ if } r_p = \underline{r}_p \\ 0 \text{ otherwise} \end{cases}$$

If the weight s_p were set equal to the magnitude of the change and used in the above, we obtain a **relative area sensitivity measure**; that is, the weight that is used in (14) is,

$$s_p = |r_p - \underline{r}_p|$$

What is meant by a **geographic sensitivity analysis** is the computation of r_p and $\underline{r}_p$ followed by the application of one or more of the appropriate sensitivity measures, (10) - (14), to the suitability maps where the perturbed suitability map attributes $\underline{r}_p$ are obtained by subjecting the rank maps to perturbations of their attribute values and/or weights.

A procedure for implementing geographic sensitivity analysis

The implementation of a sensitivity analysis on a generic GIS requires that the GIS have the capability of performing an intersection overlay. Secondly, there must be a way to access the attribute values from each of the n = 1, ..., N maps being overlaid corresponding to the polygon of the resulting intersection. Assuming that our generic GIS has these two capabilities, then algorithms for the implementation of the sensitivity measures, (10) - (14), can be developed where their inputs would be the attributes associated with the P(N) polygons of the resultant suitability map. For the most widely used commercial GIS's, this is in principle a very simple procedure.

Thus, the way to perform a sensitivity analysis is to first obtain the perturbed resultant attribute $\underline{r}_p$ by using (8) or (9) with a perturbed value of the attributes and/or weights. To measure the sensitivity, one of the measures (10) - (14), appropriate to the type of suitability analysis, is applied. As long as access to the attributes is possible, the implementation is straightforward. At times the probability distribution function associated with the errors of the input map attributes is known. When this is the case, a

Monte Carlo simulation can be performed on the input attributes according to (8) for the overlay suitability analysis and (9) for the target suitability analysis. This produces resultant attributes whose statistics can be computed. It is emphasized that once the initial intersection overlay is performed (which is the most computationally intensive process), it need never be performed again for most generic GIS's since the information regarding which polygons went into forming the intersected overlay polygons is kept in a file. The sensitivity analyses and the variation attributes $\underline{r}_p$ values are obtained by a subroutine that reads the intersection overlay file and produces one which is like it except that the attribute values are the resultant of (8), (9) or the Monte Carlo simulated value using the perturbations.

Confidence limits associated with geographic suitability

Two approaches to confidence limits are developed next. Let $m_i(\mathbf{s},\mathbf{r},\underline{\mathbf{r}})$, $i = 1,\ldots,5$ be one of our sensitivity measures given by (10) - (14) respectively. The first approach to confidence limits is to use the sensitivity measures m_i to measure the magnitude of the variations associated with an imposed percent change in the variation of the input attribute values. The smaller the value of the resultant sensitivity measure, the greater its stability and hence the higher confidence it possesses. When the input attribute errors possess a known probability density function, a Monte Carlo simulation can be performed which results in a series of sensitivity measure values with associated computable statistics that can indicate the (statistical) confidence that can be placed on the attribute values.

A second approach to confidence limits supposes that the maximum variation that each resultant attribute value can sustain is known (or it is desired that the maximum variation on the resultant attribute value be within a given, imposed range). Let $\mathbf{v} = (v_1,\ldots, v_{P(N)})^T$ be the resultant attribute variation vector whose components are given by

$$v_p = r_p + u_p$$

Here u_p denotes the maximum variation (both positive and negative) that the unperturbed attribute value can sustain or the maximum variation desired. Obtain the sensitivity measure using the components of $\mathbf{v}$, v_p, in place of the perturbed resultant attribute value $\underline{r}_p$ so that we have the measure $m_i(\mathbf{s},\mathbf{r},\mathbf{v})$ for a particular sensitivity measure $i = 1,\ldots,5$ of interest. Next, impose a percentage perturbation on the input attribute value until a given percentage perturbation of the input attribute values generates a corresponding sensitivity measure $m_i(\mathbf{s},\mathbf{r},\underline{\mathbf{r}})$ which is greater than or equal to $m_i(\mathbf{s},\mathbf{r},\mathbf{v})$; that is, find a perturbation attribute $\underline{\mathbf{r}}$ which is such that,

$$m_i(\mathbf{s},\mathbf{r},\mathbf{v}) < m_i(\mathbf{s},\mathbf{r},\underline{\mathbf{r}}) \tag{15}$$

The percentage variation that was imposed on the input attributes can now serve as an upper bound to the confidence limit for the sensitivity measure of interest.

When the input attribute errors are known to be described by an associated probability distribution function, then a series of right hand sides of (15) are obtained as a result of the Monte Carlo simulation using the corresponding distribution where $\underline{\mathbf{r}}$ is the Monte Carlo generated (perturbed) attribute values. Next, $m_i(\mathbf{s},\mathbf{r},\mathbf{v})$ is used to separate the set of Monte Carlo generated $m_i(\mathbf{s},\mathbf{r},\underline{\mathbf{r}})$ into two classes, those whose values are lower than or equal to and those that are greater than $m_i(\mathbf{s},\mathbf{r},\mathbf{v})$. The usual statistical confidence limit can now be associated.

Conclusion

Two approaches to confidence limits in the attribute values generated by geographic suitability analyses were developed using suitability measures (10) - (14). Once an implementation of a reasonably efficient algorithm for GIS's is developed, the application of the two approaches to existing problems would be the next avenue for further research.

References

Blakemore, M., 1984, Generalisation and error in spatial data bases. *Cartographica* **21**, 131-139.

Brown, D., and Gersmehl, P., 1987, Maintaining relational accuracy of geocoded data in environmental modelling. *GIS'87,* 266-275.

Chrisman, N., 1982, A theory of cartographic error and its measurement in digital data bases. In *Proceedings of Auto-Carto 5, Environmental Assessment and Resource Management,* Foreman, J., (ed), American Society of Photogrammetry, 159-168.

Honeycutt, D., 1986,, Epsilon, generalization, and probability in spatial data bases, *Proceedings of the ESRI User's Group 1986*, Paper #18, ESRI, Redlands California.

Jenks, G., 1981,, Lines, computers and human frailties, *Annals of the Association of American Geographers*, March, 1-10.

Lodwick, W. A., Munson, W., and Svoboda, L., 1988, Sensitivity analysis in geographic information systems, Part 1: Suitability analysis. Research Paper RP8861, Mathematics Department, University of Colorado at Denver.

Smith, G., and Honeycutt , D., 1987, Geographic data uncertainty, decision making and the value of information. *GIS'87*, 300-312.

Walsh, S., Lightfoot, D., and Butler, D., 1987, Recognition and assessment of error in geographic information systems. *Photogrammetric Engineering and Remote Sensing* **52**, (10), 1423-1430.

Section III

Error in simple measurements, for example using a ruler or thermometer, is traditionally described using a form of the Gaussian distribution, and there are standard methods in science for combining the errors of different measured variables in equations. These techniques extend readily to the two-dimensional case of measurement of point location, and the theory of surveying errors is well developed in geodetic science.

However the "measurements" which populate a spatial database are far from analogous to length or temperature. If we limit our attention to the digitizing of a simple line, and assume that the map has been registered correctly, then the digitized version of the line will be subject to two error processes: the process of selection of points is not subject to error or variation between operators, so that uncertainty in the digitized line is entirely due to inaccuracy om cursor positioning.

With this preamble, some progress can be made in analyzing the effects of error, and this is the subject of Griffith's paper in this section. Essentially, Griffith assumes that a spatial database is a collection of pints, each of which has an uncertainty in both of its coordinates which can be described by a stochastic process. It is then possible to derive precise results of the effect of these errors on parameters and results derived from the database, such as the distance between points, or the location of the solution to a location optimization problem.

Guptill's paper takes a similar but non-analytic approach by asking how one would include information on accuracy within a spatial database populated by features, such as the digital cartographic database of the US geological Survey. Uncertainty may be described in terms much less precise than Griffith's, but the results may be equally informative to the user of digital data, particularly when that data has been generated and distributed by a distant agency.

While previous authors have concentrated on errors in points, line or area features, Theobald's chapter looks at error in surfaces, particularly topography. Developers of spatial databases have used many methods to represent topographic surfaces, including contours, grids of sample elevations, and triangulated irregular networks (TINs), each of which has different characteristics from the point of view of accuracy of representation.

The section ends with two papers which offer different views for the nature of spatial data error and appropriate responses to it. Goodchild argues that it is often useful to analyze error in cartographic objects not by looking at models of the distortion of the objects themselves, but by building models of errors in the fields of continuous variations from which many objects are actually generated. For example many maps which purport to show patches of homogeneous soil, vegetation, or land use were complied by interpreting aerial photographs or images, combined with spot ground checks. Error models of those images may be easier to formulate and analyze.

Tobler finds the roots of the error problem in the analyses which users are now expecting from their databases. Maps which were constructed as efficient stores of spatial information are now being used as models of spatial variation, for which they were never intended. Thus a patch of homogeneous soil on a map may be a very efficient way of storing and displaying an approximating to the true variation in soils, but problems of error arise when we try to use the patch as an object of analysis. Tobler argues that we must search for more appropriate methods of analysis which do not depend so closely on the objects being analyzed, since the objects are in most cases artificial constructs.

Chapter 7

Distance calculations and errors in geographic databases

Daniel A. Griffith

Introduction

Motivation for this preliminary research report comes from three sources, namely (1) joint work that is in progress with Amrhein (Amrhein and Griffith, 1987) concerning quality control issues associated with GISs, (2) recent discussions with Hodgson (1988) concerning the sensitivity of solutions for the p-median problem with respect to different sources of error, and (3) a personal concern with the expansion of analytical capabilities that is being undertaken by architects of current GISs.

The joint work with Amrhein reflects a number of points made by Maffini *et al.*, (1989), especially those error issues they raise pertaining to the overwhelming amount of manual digitization that is performed in today's numerous quests to build geographic databases. Specifically, without changing the scale of a geographic data set, we have been studying the error due to generalization of line complexity through the use of piecewise linear approximations and reduction in point density representations; sometimes this procedure is referred to as weeding. In addition, we have been concerned with the sampling error at any point along a theoretical line, or the elusive line from the real world that is being represented on a map. This avenue of research is reminiscent of the principal theme investigated earlier by Blakemore (1983).

The work Hodgson (1988) is engaged in relates to the data aggregation problem, where dispersed geographic facts within an areal unit are treated as though they are located at a single point, which is some sort of areal unit centroid. Within the context of the p-median problem, then, distance is measured from a p-median to these aggregate centroids rather than to the underlying dispersed individual geographic fact locations. In other words, measurement error, perhaps systematic in nature, is introduced into an analysis. Current and Schilling (1987) point out that this aggregation error is eliminated when distance from some areal unit centroid is replaced by a weighted average distance based upon the set of distances from individual geographic fact locations distributed within the areal unit; unfortunately, however, one often does not have the wealth of information necessary for making such calculations. Furthermore, digitization error also can be problematic here, even when the individual geographic fact locations are known.

Given the preceding background information, the purpose of this paper is to outline several useful future research topics, together with some rationale for their importance, regarding distance calculations and affiliated errors in geographic databases. Hopefully these themes will help guide those interested in improving quality control of geographic databases and/or expanding the analytical capabilities of existing and evolving GISs.

Spatial statistics and distances relating to the aggregation problem

First of all, consider the p-median problem. If the incorrect aggregate centroid is employed (spatial mean versus spatial median versus physical center, for instance), specification error can be introduced into an analysis. This notion of specification error is illuminated when one studies selected limiting cases. For example, as the number of medians, p, increases until it equals the number of areal units, n, then one median is assigned to each areal unit centroid. In this situation the appropriate areal unit centroid to use would be the spatial median; otherwise, centroids such as the spatial mean or the physical center will not minimize distance to all internally located geographic facts. But since most of the time internal areal unit geographic distribution information is unknown, the best assumption to invoke (a maximum likelihood type of result) is a uniform geographic distribution; such an assumption leads to choropleth maps. Assuming a uniform distribution is consistent with the calculation of physical centroids, whose use is widely popular. This class of specification error has been studied by Hillsman and Rhoda (1978), who employed the spatial mean, and who found that it is poorly behaved. Goodchild (1979) has found that this category of error produces much more dramatic effects on the resulting p-median locations than it does on the value of the corresponding objective function that is being optimized. Casillas (1987) disagrees with these earlier findings; he maintains that data aggregation may have little effect on the patterns or positions of the p medians.

Therefore, discussions and heavy utilization of three different areal unit centroids are found in the literature. The spatial median seems best suited for p-median problems. Specification error appears to be void of noticeable pattern when the spatial mean is substituted for the spatial median in these problems. But, particularly because of lack of detailed information, physical centroids are commonly used. Consequently,

Research topic 1

Which of these three centroids tends to minimize distance error impacts when aggregate geographic data are used? Can a typology of problems be developed for the suitability of these three centroids?

The seriousness of specification error questions is highly correlated with the spatial distribution of geographic facts within areal units. One spatial statistical measure of this dispersion is standard distance. Clearly, as the standard distance for an internal geographic distribution approaches zero, then aggregation error will decrease to zero, as long as either the spatial mean or the spatial median is used to calculate inter-areal unit distances, or as long as the cluster of geographic facts occurs at the physical center of the areal unit. Consequently,

Research topic 2

Can the standard distance be used effectively to establish thresholds for distinguishing between serious and trivial aggregation error situations. Can this dispersion measure be used productively to distinguish between areal unit centroid specification error situations?

These two themes argue for some imaginative and creative applications of spatial statistics.

These first two research topics describe specification error issues. Attention now will be turned to digitization sampling error issues.

Digitization error and the p-median problem solution

Mathematically speaking one simple version of the p-median problem may be written as follows:

$$\text{MIN:}\quad Z = \sum_{i=1}^{i=n}\sum_{j=1}^{j=p} w_i\lambda_{ij}\left[\left(U_j-x_i\right)^2+\left(V_j-y_i\right)^2\right]^{1/2}$$

$$\text{st:}\quad \sum_{j=1}^{j=p}\lambda_{ij} = 1, i=1, 2, \ldots, n, \text{ and}$$

λ_{ij} is binary (it only takes on the values of either zero or unity), (1)

where the centroid of areal unit i is denoted (x_i,y_i), the aggregate geographic facts for areal unit i are denoted by (w_i) and the location of median j is denoted by (U_j, V_j). The solution to this problem is well-known, based upon the pair of partial derivatives $\partial Z/\partial U\{\lambda_{ij}\}$ and $\partial Z/\partial V\{\lambda_{ij}\}$, and is obtained iteratively, for a given set $\{\lambda_{ij}\}$, from the pair of equations

$$U_j^{(\tau+1)} = \left[\sum_{i=1}^{i=n} w_i\lambda_{ij}x_i/d_{ij}^{(\tau)}\right] \Big/ \left[\sum_{i=1}^{i=n} w_i\lambda_{ij}/d_{ij}^{(\tau)}\right]$$

$$V_j^{(\tau+1)} = \left[\sum_{i=1}^{i=n} w_i\lambda_{ij}y_i/d_{ij}^{(\tau)}\right] \Big/ \left[\sum_{i=1}^{i=n} w_i\lambda_{ij}/d_{ij}^{(\tau)}\right] \qquad (2)$$

The optimal set of binary variables $\{\lambda_{ij}\}$ can be identified using numerical combinatorial methods, such as branch-and-bound. Here one can see that errors in distances have a multiplicative effect on the solution, since they appear in the denominator of fractions.

For simplicity as well as illustrative purposes, suppose that the error in centroid value i is uniformly distributed over the interval $[b_i,a_i]$, occurs only in the x-coordinate, and is additive. Then for all j the expected value of the statistical distribution of the p-median estimating equations is and is additive. Then for all j the expected value of the statistical distribution of the p-median estimating equations is

$$E(\partial Z/\partial V_j \mid \{\lambda_{ij}\}) \approx \sum_{i=1}^{i=n} w_i \lambda_{ij} (V_j - y_i) \left\{ \left[(U_j + b_i - x_i)^2 + (V_j - y_i)^2 \right]^{1/2} - \left[(U_j + a_i - x_i)^2 + (V_j - y_i)^2 \right]^{1/2} \right\} / (b_i - a_i)$$

$$\neq \sum_{i=1}^{i=n} w_i \lambda_{ij} (V_j - y_i) / \left[(U_j - x_i)^2 + (V_j - y_i)^2 \right]^{1/2} \quad (3)$$

Therefore,

Research topic 3

How does $E(\partial Z/\partial V_j \mid \{\lambda_{ij}\})$ behave with respect to different statistical error distributions? How is it affected by the presence of correlated errors in the x- and y-coordinate axes? How is it affected by the presence of autocorrelated errors?

Because of the nonlinear structure of the p-median problem solution, this research theme probably should be explored in terms of the estimating equations rather than the final estimates of spatial medians.

Several additional potential p-median outcomes attributable to the presence of error in geographic databases merit highlighting here. First, the set of allocations determined by the indicator variables $\{\lambda_{ij}\}$ could change; the optimal solution has a given set affiliated with it, and this set may change due to error. This possibility is consistent with the comments made in Goodchild (1979). Second, intuitively speaking the impact of error should be smaller when the number of medians is close to 1 or n, and should increase as this number approaches n/2.

Digitization error and area calculations

Consider a set of n digitized points $\{(x_i, y_i), i=1, 2, \ldots, n\}$, and construct a closed polygon with these points by setting $(x_{n+1}, y_{n+1}) \equiv (x_1, y_1)$,. Suppose the original n coordinates are placed into two n-by-1 vectors **X** and **Y**. Let the n-by-n matrix **C+** define the spatial structure of the sequence of these points such that $c_{ij} = 1$ if either $j = i + 1$ or both $i = n$ and $j = 1$, and otherwise $c_{ij} = 0$; moreover, the upper off-diagonal cells contain ones, the very extreme lower left-hand cell contains a one (this entry closes the polygon), and all other cells contain zeroes. Let the matrix transpose $(\mathbf{C}^+)^t = \mathbf{C}^-$, where $c_{ij} = 1$ if either $j = i - 1$ or both $i = 1$ and $j = n$, and otherwise $c_{ij} = 0$; moreover, the lower off-diagonal cells contain ones, the very extreme upper right-hand cell contains a one (this entry closes the polygon), and all other cells contain zeroes. Consequently, the matrix product $\mathbf{C}^+\mathbf{C}^- = \mathbf{I}$, the identity matrix.

From calculus the trapezoidal rule can be used to approximate the area of the polygon in question. Modified versions of the formulae together with this area calculation then can be used to compute an areal unit's physical centroid. Using the preceding matrix notation, these formulae can be written as

$$\text{Area} = \left(X^t C^+ Y - Y^t C^+ X\right) / 2 \tag{4}$$

which is the formula reported by Olson (1976). Since $\mathbf{Y^tC^+X}$ is a scalar,

$$Y^t C^+ X = \left(Y^t C^+ X\right)^t = X^t C^- Y \tag{5}$$

and hence

$$\text{Area} = \left(X^t C^+ Y - X^t C^- Y\right) / 2 = X^t \left(C^+ - C^-\right) Y / 2 \tag{6}$$

which is the formula reported by Chrisman and Yandell (1989). In fact, for the sake of completeness, this formula also could be written as

$$\text{Area} = Y^t \left(C^+ - C^-\right) X / 2 \tag{7}$$

Obviously these three different versions of the area formula can be used interchangeably, and are used so in the ensuing analyses by adhering to a criterion of convenience.

Assume (a) that the set of digitized points $\{(x_i^*, y_i^*)\}$ contains error, such that $x_i^* = x_i + \xi_i$ and $y_i^* = y_i + \eta_i$, (b) $\xi_i \sim F_\xi(0, \sigma_\xi^2)$, and (c) $\eta_i \sim F_\eta(0, \sigma_\eta^2)$. Three interesting cases of error structure will be inspected now for this collections of descriptors.

CASE I: $COV(\xi, \eta) = 0$, ξ and η not autocorrelated.

$$E(\text{Area}) = E\left\{\left[(X + \xi)^t C^+ (Y + \eta) - (Y + \eta)^t C^+ (X + \xi)\right] / 2\right\}$$

$$= \left(X^t C^+ Y - Y^t C^+ X\right) / 2 \tag{8}$$

meaning that the estimate of area is unbiased in this classical statistics setting.

$$VAR(\text{Area}) = VAR\left\{\left[(X + \xi)^t C^+ (Y + \eta) - (Y + \eta)^t C^+ (X + \xi)\right] / 2\right\}$$

$$= \left(\frac{1}{2}\right)^2 \left[\sigma_\eta^2 X^t \left(I - C^+ C^+\right) X - \sigma_\xi^2 Y^t \left(I - C^+ C^+\right) Y + n \sigma_\eta^2 \sigma_\xi^2\right] \tag{9}$$

meaning that the variance of area is not as simple as a standard distance measure.

CASE II: $COV(\xi, \eta) = \rho \sigma_\eta \sigma_\xi$, ξ and η not autocorrelated.

$$E(\text{Area}) = E\left\{\left[(X + \xi)^t C^+ (Y + \eta) - (Y + \eta)^t C^+ (X + \xi)\right] / 2\right\}$$

$$= \left(X^t C^+ Y - Y^t C^+ X\right) / 2 \tag{10}$$

meaning that the estimate of area remains unbiased in this traditional multivariate statistics setting.

$$\text{VAR(Area)} = \text{VAR}\left\{\left[(X+\xi)^t C^+(Y+\eta) - (Y+\eta)^t C^+(X+\xi)\right]/2\right\}$$

$$= \left(\frac{1}{2}\right)[\sigma_\eta^2 X^t(I - C^+C^+)X - \sigma_\xi^2 Y^t(I - C^+C^+)Y +$$

$$2\rho\sigma_\eta\sigma_\xi X^t(I - C^+C^+)Y + n(1-\rho^2)\sigma_\eta^2\sigma_\xi^2] \qquad (11)$$

which is equivalent to the result reported in Chrisman and Yandell (1989).

CASE III: $\text{COV}(\xi, \eta) = 0$, ξ and η autocorrelated.

Suppose that the autocorrelation (which should be serial in nature) is characterized by the spatial linear operators $(\mathbf{I} - \mathbf{A}_x)$ and $(\mathbf{I} - \mathbf{A}_y)$, which represent moving average correlational structures.

$$\text{E (Area)} = \text{E}\left(\left\{\left[X + (I - A_x)\xi\right]^t C^+\left[Y + (I - A_y)\eta\right] - \right.\right.$$

$$\left.\left.\left[Y + (I - A_y)\eta\right]^t C^+\left[X + (I - A_x)\xi\right]\right\}/2\right)$$

$$= \left(X^t C^+ Y - Y^t C^+ X\right)/2 \qquad (12)$$

meaning that the estimate of area still is unbiased in this conventional time-series statistics setting.

$$\text{VAR (Area)} = \text{VAR}\left(\left\{\left[X + (I - A_x)\xi\right]^t C^+\left[Y + (I - A_y)\eta\right] - \right.\right.$$

$$\left.\left.\left[Y + (I - A_y)\eta\right]^t C^+\left[X + (I - A_x)\xi\right]\right\}/2\right) \qquad (13)$$

$$= \left(\frac{1}{4}\right)\{\sigma_\eta^2 X^t(C^+ - C^-)(I - A_y)(I - A_y)^t(C^- - C^+)X - \sigma_\xi^2 Y^t(C^- - C^+)(I - A_x)$$

$$(I - A_x)^t(C^+ - C^-)Y + \text{tr}\left[(I - A_y)^t(C^- - C^+)(I - A_x)(I - A_x)^t(C^+ - C^-)(I - A_y)\right]\sigma_\eta^2\sigma_\xi^2\}$$

where tr(.) denotes the matrix trace operation. If $\mathbf{A}_x = \mathbf{0}$ and $\mathbf{A}_y = \mathbf{0}$, the case where errors are not autocorrelated, then this solution reduces to that for Case I; $\text{tr}(\mathbf{I}) = n$, and $\text{tr}(\mathbf{C}^+\mathbf{C}^-) = 0$. Furthermore, if the two spatial linear operators are replaced by their respective inverses, namely $(\mathbf{I} - \mathbf{A}_x)^{-1}$ and $(\mathbf{I} - \mathbf{A}_y)^{-1}$, then an autoregressive rather than a moving average correlational structure could be explored. These are the sorts of models Keefer, Smith, and Gregoire (1988) are alluding to; their work recommends that the entire body of time series analysis theory can shed light on the specification, estimation, and diagnostic evaluation of manual digitization error models.

These three cases suggest that complications introduced into the calculation of areas impact upon the variance of these calculations from situation to situation, but on average the area calculations are correct. These findings encourage one to propose the following research theme:

Research topic 4

Can areas obtained from secondary sources be used effectively to evaluate areal unit boundary digitization by comparing area calculations based upon these digitized points with those found in the secondary sources?

This comparative approach has been used by Griffith (1983); a scatter diagram was constructed for the predicted municipio areas from the digitized boundaries versus the municipio areas reported by the Department of Agriculture, Commonwealth of Puerto Rico, and the pair of regression null hypotheses, stating $\alpha = 0$ and $\beta = 1$, was tested. A second research theme stemming from the above findings is

Research topic 5

What are the expected value and variance results for this area calculation when the more realistic mixture situation of $COV(\xi, \eta) = \rho\sigma_\eta\sigma_\xi$, ξ and η autocorrelated holds?

Digitization error and physical centroid calculations

The physical centroids of areal units use the area computation from the preceding section in their calculations. Let the x-coordinate of this centroid be denoted by x_c, and the y-coordinate be denoted by y_c. Since these two values are calculated in a parallel manner, only x_c will be treated in this section; results for it extend directly to those for y_c.
The value of x_c may be calculated using the following formula, recalling that there are n distinct points, with the first point also being attached to the end of the sequence as the (n+1)th point:

$$\sum_{i=1}^{i=n} \left(y_i - y_{i+1}\right)\left(x_i^2 + x_i x_{i+1} + x_{i+1}^2\right) / \left(6*\text{Area}\right) \qquad (14)$$

Because the area calculation can be factored out and hence separated from this summation (its summation is independent of the one being performed here), only error in the numerator of this formula will be investigated now. One should understand, however, that while these two estimates are being treated separately, when they are put together in the above fraction, impacts of error become multiplicative. This promotes the research theme of

Research topic 6

Are the area and centroid calculations statistically independent?

Following the error definitions appearing in the preceding section, the numerator of this foregoing fraction may be rewritten as

$$\mathrm{NUM} = \sum_{i=1}^{i=n} \left[(y_i - \eta_i) - (y_{i+1} + \eta_{i+1}) \right] \left[(x_i - \xi_i)^2 + (x_i - \xi_i)(x_{i+1} + \xi_{i+1}) + (x_{i+1} + \xi_{i+1})^2 \right]$$

(15)

Again three cases can be explored.

CASE I: $\mathrm{COV}(\xi, \eta) = 0$, ξ and η not autocorrelated.

$$\mathrm{E(NUM)} = \sum_{i=1}^{i=n} (y_i - y_{i+1})(x_i^2 + x_i x_{i+1} + x_{i+1}^2) \tag{16}$$

meaning that the estimate of the numerator is unbiased. This result, when combined with that for area, implies that estimates of (x_c, y_c) should be unbiased.

$$\mathrm{VAR(NUM)} = 2\sigma_\eta^2 \sum_{i=1}^{i=n} \mathrm{VAR}\left[(\xi_i + \xi_{i+1})^2 - (x_i + \xi_i)(x_{i+1} + \xi_{i+1}) \right] \tag{17}$$

which involves third and fourth moments, and so requires some knowledge of the statistical distribution function in order to obtain a more simplified solution (this will be true for the variance expressions of Cases II and III, too).

CASE II: $\mathrm{COV}(\xi, \eta) = \rho\sigma_\eta\sigma_\xi$, ξ and η not autocorrelated.

$$\mathrm{E(NUM)} = \sum_{i=1}^{i=n} (y_i - y_{i+1})(x_i^2 + x_i x_{i+1} + x_{i+1}^2) + 2\sum_{i=1}^{i=n} \mathrm{E}\left[\eta_i (x_i - \xi_i)^2 \right] \tag{18}$$

meaning that the estimate of the numerator is biased, implying that estimates of (x_c, y_c) will be biased. This result also involves higher moments of the statistical distribution, suggesting that supplemental knowledge of the statistical distribution function is needed in order to obtained a more simplified solution.

CASE II: $\mathrm{COV}(\xi, \eta) = 0$, both ξ and η autocorrelated.

As before, assume a simple moving average serial correlational structure for the error terms, such that each digitized coordinate is rewritten as $[x_i + (\xi_i - \rho\xi_{i-1}), y_i + (\eta_i - \rho\eta_{i-1})]$.

$$\mathrm{E(NUM)} = \sum_{i=1}^{i=n} (y_i - y_{i+1})(x_i^2 + x_i x_{i+1} + x_{i+1}^2) \tag{19}$$

meaning that the estimate of the numerator is unbiased. This result, when combined with that for area, implies that estimates of (x_c, y_c) should be unbiased in the presence of serial correlation. One should note that both $\xi_1 = \xi_n$ and $\eta_1 = \eta_n$, introducing a boundary complication into the autocorrelation cases.

These three cases suggest that complications introduced into the calculation of centroids not only impact upon the variance of these calculations from situation to situation, but on average may fail to yield correct physical centroid calculations.

These findings give rise to the following research themes:

Research topic 7

What are the expected value and variance results for physical centroid calculations when the more realistic mixture situation of $COV(\xi, \eta) = \rho\sigma_\eta\sigma_\xi$, ξ and η autocorrelated holds? What are meaningful statistical distribution functions to use in order to obtain closed-form solutions of the variance statements for physical centroids? How serious is the expected value bias when errors are correlated?

Concluding comments

Six research topics are advocated in this paper. Pursuit of these topics should prove fruitful, and render valuable knowledge for handling the impact of selected error types upon distance calculations in GISs. Surely these preliminary findings coupled with their accompanying implications, even though based on a somewhat cursory treatment of the topic, are both disturbing and foster some anxiety about the accuracy of results obtained with current geographic information and analysis (GIA) analytics, such as shortest network paths. Likewise, sundry convolutions of specification and sampling error deserve to be explored, too. We should be compelled to understand the impact of these different sources of error on analytical model solutions before incorporating them as standard options in GIS software.

References

Amrhein, C., and Griffith, D., 1987, GISs, spatial statistics, and statistical quality control. Paper presented to the *International Geographic Information Systems Symposium*, (The Research Agenda), Crystal City, Virginia, November, 1987.

Blakemore, M.,1983, Generalization and error in spatial data bases. Wellar, B. (ed), *Proceedings, AUTO-CARTO Six*, Hull, Quebec: Imprimerie Carriere, pp. 313-322.

Casillas, P. ,1987, Data aggregation and the p-median problem in continuous space. In *Spatial Analysis and Location-Allocation Models*, Ghosh, A. and Rushton, G.(eds), pp 327-344. (New York: Van Nostrand)

Chrisman, N., and Yandell, B., 1989, Effects of point error on area calculations: a statistical model. *Surveying and Mapping*, forthcoming.

Current, J., and Schilling, D., 1987, Elimination of source A and B errors in p-median problems. *Geographical Analysis* **19**, 95-110.

Goodchild, M.,1979, The aggregation problem in location-allocation. *Geographical Analysis* **11**, 240-255.

Griffith, D., 1983, Phasing-out of the sugar industry in Puerto Rico. In *Evolving Geographical Structures*, Griffith, D., and Lea, A., (eds), pp 196-228. (The Hague: Martinus Nijhoff)

Hillsman, E., and Rhoda, R., 1978, Errors in measuring distances from populations to service centers. *Annals of Regional Science* **12**, 74-88.

Hodgson, J., 1988, Stability of the p-median model under induced data error, Paper presented to the *IGU Mathematical Models Working Group*. Canberra, Australia, meeting, August 16-19.

Keefer, B. J., Smith, J. L., and Gregoire, T. G., 1988, Simulating manual digitizing error with statistical models. *Proceedings, GIS/LIS '88*. ASPRS/ACSM, Falls Church, VA, 475-83.

Maffini, G., Arno, M., and Bitterlich, W., 1989, Observations and comments on the generation and treatment of error in digital GIS data, (this volume).

Olson, J.,1976, Noncontiguous area cartograms. *The Professional Geographer* **28**, 371-380.

Chapter 8

Inclusion of accuracy data in a feature based, object-oriented data model

Stephen C. Guptill

Abstract

A digital spatial data base can be considered as a multifaceted model of geographic reality. In such a model, entities are individual phenomena in the real world; features define classes of entities; and a feature instance, one occurrence of a feature, is represented by a set of objects. One type of object, a spatial object, contains locational information. A second type, a feature object, contains nonlocational information. The objects have attributes and relationships. Attribute and relationship values can also have attributes. This recursive attribution scheme allows for the incorporation of accuracy information at any level of the geographic model.

Locational precision can be associated with each spatial object. Uncertainty measures can be assigned to any attribute value. Temporal attributes, along with uncertainty information, can also be included in the model. The mechanisms exist in data models and data structures to handle accuracy data. However, several underlying questions remain: what are the uncertainty measures, how are they collected, and how are they used?

Introduction

In his opening remarks at the Specialist Meeting for Research Initiative #1 (Accuracy of Spatial Databases) of the National Center for Geographic Information and Analysis, Michael Goodchild, initiative leader, noted the following major premises of the conference:

- all spatial data are of limited accuracy;
- precision in geographic information system (GIS) processing exceeds the accuracy of the data; and
- the means to characterize the accuracy of spatial data, track uncertainty through GIS processes, and compute and report uncertainty are inadequate.

Given these premises, the major goals of the research initiative are to examine:

- improvement of models of uncertainty,
- methods of encoding uncertainty in data bases,
- methods of tracking uncertainty,
- methods of computing and communicating error in products, and policies to encourage implementation of results.

This paper deals with the second topic on the list, and it describes a method of encoding uncertainty in the digital cartographic data bases being developed by the U.S. Geological Survey and discusses some remaining difficulties.

Deliberations on this topic have been conducted for a number of years. Working Group II (Data Set Quality) of the National Committee for Digital Cartographic Data Standards (NCDCDS) has identified five types of information to be provided to users to evaluate a data set. The five sections of a quality report are: lineage, positional accuracy, attribute accuracy, logical consistency, and completeness (NCDCDS, 1988, p. 131-135). Furthermore, the group notes this information on data quality can be carried at different levels of aggregation, referring to an individual cartographic object, a map, or a set of maps (NCDCDS, 1988, p. 79-81). A methodology is proposed, in accordance with these general guidelines, that encodes positional and attribute accuracy information with every feature contained in a digital spatial data base.

Digital line graph-enhanced

The U.S. Geological Survey has been producing digital cartographic data in its digital line graph (DLG) format for almost a decade. During this time, the tasks for which the data are being used have become increasingly diverse, placing information demands on the data that were not planned for in its initial design.

In response to these demands, the Geological Survey (as both a data supplier and data user) has designed an enhanced version of the digital line graph, termed Digital Line Graph - Enhanced (DLG-E). In simple terms, the DLG-E begins with the topological model now used in the Survey's present DLG data structure (U.S. Geological Survey, 1987) and builds a cartographic feature layer upon the topology. Features are the sum of our interpretations of phenomena on or near the Earth's surface. Buildings, bridges, roads, streams, grassland, and counties are examples of features. The feature definition is open-ended, allowing the definition of additional features of interest. The features are described using objects, attributes, and relationships.

Several other major agencies collecting digital cartographic data are also considering (or have implemented) new feature-based spatial data model designs. These agencies include the U.S. Bureau of the Census (Marx, 1986; Kinnear, 1987), the U.S. Defense Mapping Agency (1988, p. 99-100), the Institut Geographique National (France) (Bernard and Piquet-Pellorce, 1986; Salg and Piquet-Pellorce, 1986), the Landesvermessungamt Nordrhein-Westfalen (West Germany) (Barwinski and Bruggemann, 1986), and the military geographic services of the NATO allies (J. Garrison, oral commun., 1987). This trend toward feature-based data models signals an evolution in the design of future spatial data bases and GIS's.

In the DLG-E model, the phenomena represented by geographic and cartographic data are considered, in totality, as entities. Entities represent individual phenomena in the real world. A feature is an abstraction of a class of entities, with the feature description encompassing only selected characteristics of the entities (typically the characteristics that have been portrayed cartographically on a map). A feature instance, that is, one occurrence of a feature, is described in the digital environment by feature objects and spatial objects. Nonlocational characteristics of the feature instance are associated with the feature object. The locational aspects of the feature instance are represented by spatial objects. To link the locational and nonlocational aspects of the feature instance, a given feature object is associated with (or is composed of) a set of spatial objects.

Digital cartographic data consist of objects, attributes, and relationships. Objects are the basic units of representation of the encoded phenomena. Attributes are the locational and nonlocational characteristics of the entities represented by the objects.

Relationships are the topological and nontopological links between the objects. Nonlocational attributes and nontopological relationships are associated with feature objects. Locational attributes (for example, x,y coordinates) and topological relationships are associated with spatial objects. Five spatial objects are defined: points, nodes, singular chains, bounding chains, and areas. Taken collectively, these objects, attributes, and relationships comprise the DLG-E data model. For more information on the DLG-E design see Guptill *et al.* (1988) and Guptill and Fegeas (1988).

Nonlocational attributes

Attributes are the locational and nonlocational characteristics of the entities represented by the objects, or of an attribute value. The nonlocational attributes of an entity include such concepts as its operational status, name, and function. An attribute value is a measurement assigned to an attribute for a feature instance, or for another attribute value. These measurements are made from one of the four levels of measurement (nominal, ordinal, interval, ratio). The term "attribute accuracy" as used by the NCDCDS refers to the set of nonlocational attributes of a feature.

In the DLG-E design, a domain of nonlocational attributes is predefined for each feature instance. In addition, a domain of attribute values is specified for each attribute. An example of an attribute and its domain of attribute values is shown in Table 1.

Attribute: Operational status	
	Abandoned
	Dismantled
	Operational
	Proposed
	Under construction
	Unknown

Table 1 Domain of attribute values for the attribute "Operational Status"

For a particular feature instance, appropriate attribute values describing the instance are chosen for each attribute. An example of the attributes, with associated attribute values, for the feature instance of bridge are given in Table 2.

Feature: Bridge	Attributes:	Attribute values
	Covering:	Not covered
	Name:	Roosevelt Bridge
	Number of decks:	1
	Operational status:	Operational

Table 2 The encoding of attribute values for a feature instance.

Every attribute in the domain of attributes describing a feature must be assigned at least one attribute value. More than one value may be required for an attribute to describe a feature instance. For example, for an instance of the feature "road," the attribute "route designator" may have multiple values (route numbers).

Attributes may also refine other attribute values. For example, the feature "well" has the attribute "product." An instance of the feature might have the value "water" for this attribute. The attribute value "water" is further modified by the attribute "water characteristics," that could contain a number of values as shown in Table 3.

Feature: Well			
	Product:		
		Gas	
		Heat	
		Oil	
		Water: Water Characteristics:	
			Alkaline
			Hot
			Mineral
			Salt

Table 3 Example of attributes of attribute values.

Encoding uncertainty about nonlocational attributes

Attributes and attributes of attribute values provide a tool to encode uncertainty measures about spatial entities. Consider the example shown in table 4. The feature "lake/pond" has associated with it an attribute "categorical accuracy." This attribute makes a statement about the likelihood of a given entity in the real world being correctly classified as a "lake/pond." A variety of types of attribute values could be associated with this attribute, such as a simple probability estimate or a misclassification matrix.

Lake/Pond		
	Categorical accuracy	
		90% (possible confusion with reservoirs)
	Elevation	
		400 feet
	Vertical accuracy	
		3-feet rms

Table 4 Describing accuracy of nonlocational attributes.

Using attributes of attribute values it is possible to further describe other aspects of data quality. As shown in table 4, the elevation value (400 feet) has an attribute "vertical accuracy," that in turn has a value (3-feet rms). Alternate descriptions of vertical accuracy could be included.

Encoding uncertainty about locational attributes

Just as with the feature objects, the spatial objects are characterized by the same type of attribute/attribute value description scheme (table 5). The attributes and attribute values describe only locational aspects of the spatial object such as its geographic position (for example, x,y coordinates) or a geometric characteristic (for example, accuracy).

Spatial object type

Attribute
Attribute value

Attribute
Attribute value

{ Attribute
Attribute value }

Table 5 Locational attributes of spatial objects.

To characterize the locational uncertainty of spatial objects in the data base, a "horizontal accuracy" attribute could be attached to every point, node, and chain. This attribute would have values such as "xxx meters rms error" or whatever accuracy measure was appropriate. An example is given in Table 6.

Chain 101		
	Horizontal accuracy:	12.6 m rms
	Coordinates:	x,y,...x,y
Chain 102		
	Horizontal Accuracy:	17.8 m rms
	Coordinates:	x,y,...x,y

Table 6 Locational accuracy attributes and attribute values.

This scheme allows for multiple locational accuracy measures to be assigned to a given feature. For example, chains 101 and 102 in Table 6 could represent the shoreline of a pond. Chain 101 might be the segment of the shoreline along a steep, rocky embankment, where the shoreline is not subject to much lateral displacement. Chain 102 might be the segment of shoreline along a broad, shallow beach and be subject to greater locational variation. Thus one feature, the shoreline of the pond, is given several locational uncertainty estimates.

Remaining challenges

A data model has been presented that allows for the inclusion of uncertainty information with each object within a spatial data base. While this represents a first step in developing data base structures that can represent accuracy explicitly, a number of issues remain unresolved.

The first issue is what are the proper measures of accuracy? In the pond example in Table 4, how can the information be encoded for a high degree of certainty that the area in question is covered with water, and a lesser degree of certainty whether the body of water is a pond or a reservoir? To fire fighters looking for water sources to combat a canyon brushfire, misclassification between pond and reservoir is of much less concern than the fact that a body of water exists at that location (as opposed to a dry lake bed or gravel pit). This example seems to argue for multiple levels of misclassification matrices (Tables 7 and 8) to fully characterize the uncertainty of nonlocational attributes of certain features.

	Land	Water
Land	0.95	0.05
Water	0.05	0.95

Table 7 Misclassification matrix of land versus water for the lake/pond feature.

	Lake	Reservoir
Lake	0.85	0.15
Reservoir	0.15	0.85

Table 8 Misclassification matrix of lakes versus reservoirs for the lake/pond feature.

A second issue deals with assumptions about the homogeneous nature of various features. Consider a forest. Typically the boundaries of the forest are outlined, and perhaps an accuracy figure is calculated for the forest land use type. An accuracy measure, such as a misclassification matrix does not fully describe the heterogeneous nature of the forest. A method needs to be developed to describe characteristics such as the variation of tree density within the stand, the gradual transition on the north edge of the stand to brush vegetation, or the sharp, definite transition on the south edge where a road sharply defines the forest boundary. Similar conditions apply to many other nonhomogeneous features. Additional methods of description need to be developed to characterize such conditions.

A major practical challenge is to devise methods for collecting accuracy measurements for locational and nonlocational attributes in an economical fashion. While collection of accuracy data has a cost, so do incorrect decisions based on inaccurate data or data of unknown quality. Studies on the use and value of geographic information in decision making may provide insights into how data quality affects the quality of decision making. These studies could lead to quantification of the additional value of doing quality assessments of data sets, as well as the value of increasing the accuracy of the data.

GIS software needs to be designed to utilize the types of accuracy measures mentioned earlier in this paper. As GIS processes improve or degrade data quality, the appropriate attribute values should be modified. Accuracy figures for the end product of a GIS analysis need to be calculated. The GIS users need to know if they are victims of the "additional data paradox." Conventional wisdom would say that as you add more data to the solution of a problem, the likelihood of getting an accurate solution increases. However, if each additional data layer degrades the quality of the combined data set, and hence the accuracy of the solution, then adding data sets may be counterproductive. It may be the case that there is an optimal collection of data sets for a given problem; adding or subtracting data lowers the probability of a correct solution. Much more research needs to be performed on the interactions between data quality, GIS processes, and the quality of solution to answer this question.

**Publication authorized by the Director, U.S. Geological Survey.*

REFERENCES

Barwinski, Klaus, and Bruggemann, Heinz, 1986, Development of digital cadastral and topographic maps - requirements, goals, and basic concept. *Proceedings Auto Carto 1986* , 2, London, Imperial College, South Kensington, pp. 76-85.

Bernard, Antoine and Piquet-Pellorce, Daniel, 1986, A workstation for handling located data: PISTIL. *Proceedings Auto Carto 1986* **1**, London, Imperial College, South Kensington, pp. 166-174.

Guptill, S. C., and Fegeas, R. G., 1988, Feature based spatial data models--the choice for global databases in the 1990s? In *Building Databases for Global Science*, Mounsey, H., and Tomlinson, R., (eds.), pp. 279-295. (London: Taylor & Francis)

Guptill, S. C., Fegeas, R. G., and Domaratz, M. A., 1988, Designing an enhanced digital line graph. *1988 ACSM-ASPRS Annual Convention*, Technical Papers, **2**, pp 252-261.

Kinnear, Christine, 1987, The TIGER structure. *Proceedings, Auto Carto 8*, Eighth International Symposium on Computer Assisted Cartography, Baltimore, MD, pp 249-257.

Marx, R. W., 1986, The TIGER system automating the geographic structure of the United States Census. *Government Publications Review*, **13**, pp. 181-201.

Morrison, J. L., Callahan, G. M., and Olsen, R. W., 1987, Digital Systems Development at the U.S. Geological Survey. *Proceedings International Cartographic Association Conference*, pp 201-214, Morelia, Michoacan, Mexico, October 12-21, 1987.

National Committee for Digital Cartographic Data Standards (NCDCDS), 1988, The proposed standard for digital cartographic data. *The American Cartographer,* **15**.

Salg, Francois, and Piquet-Pellorce, Daniel, 1986, The I.G.N. small scale geographical data base (1:100,000 to 1:500,000). *Proceedings Auto Carto 1986* **1**, London, Imperial College, South Kensington, pp. 433-446.

U.S. Defense Mapping Agency, 1988, *Digitizing the future*, DMA Stock No. DDIPDIGITALPAC, pp. 104.

U.S. Geological Survey, 1987, Digital line graphs from 1:24,000-scale maps. *U.S. Geological Survey Data Users Guide 1*, pp. 109.

Chapter 9

Accuracy and bias issues in surface representation

David M. Theobald

Abstract

Surface representation is increasingly used in environmental modeling because of topography's dominant role in such processes. The accuracy of these elevation models is seldom addressed, and if it is, is usually constrained to an estimate of root-mean-square-error in the vertical measure or a description of noise or striping errors. Seldom are the errors described in terms of their spatial domain or how the resolution of the model interacts with the relief variability. Additionally, when using an elevation model the research objective must define the land features that need to be captured and the precision with which they need to be represented. These features are dependent upon the spatial frequency distribution of the terrain and how the resolution interacts with these frequencies. Thus, in defining the accuracy of a digital elevation model (DEM), one needs to ultimately know the spatial frequency distribution of the terrain and the bias in the resolution in addition to the type of data structure used and accuracy of the source document. A brief review of sources of DEMs, common data structures used in terrain representation, techniques used in the derivation of parameters (i.e. slope), and methods of interpolation will be given. The concept of accuracy will then be discussed in terms of the research objective, precision, resolution, and terrain variability.

Introduction

Information on the surface of an area under study is basic to nearly any type of environmental research such as hydrology, geomorphology, and ecology. With the advent of computerized mapping and analysis techniques, methods of digitally representing the surface, both the elevations and derivatives such as slope and aspect, have been developed. Advantages of digitally encoding elevation data include fast manipulation, the ability to handle large projects, and of course ease in integrating with automated techniques of computer mapping, geographic information systems (GIS) and remote sensing techniques. However, along with many other automated processes, there is a tendency to lose sight of the quality and accuracy of the original data and subsequent manipulation in making decisions based on this technique. Any abstract of reality will contain discrepancies from its source, and attention to these discrepancies and methods of minimizing their impacts must be considered.

Therefore, this paper will examine sources of elevation data, the methods or structures used to store this information, and the biases that occur in extending the elevation data to derived parameters. Because interpolation methods are required to extend any point-sampled data to a surface, applicable techniques of interpolation will also be discussed, along with how the research objective affects the resolution, precision, and accuracy of a DEM.

Data sources

There are a variety of sources to produce digital elevation models (DEMs). Within the United States, the US Geological Survey (USGS) is probably the largest producer of DEMs and as such, most of the following discussion pertains to their production techniques (other producers will most likely use the same methodology).

Perhaps the most common method is to derive the elevations from measurements of aerial photographs made with an analytical stereo-plotter (e.g. Gestalt Photo Mapper II or GPM 2). The GPM 2 measures the parallax at 2,444 points for each 8 x 9 mm patch on the stereo-photo model (using an aerial photo scale of 1:80,000), with a sub-set the size of which is selected to accommodate the terrain conditions. Within each patch, the points are spaced at about 47 feet on the ground. This process tends to have problems especially in forested areas, where tree-tops are taken as the ground surface, producing the effect of low-amplitude random noise.

A second approach is to use a semi-automatic profiling device. Here, stereoplotters, interfaced with digital profile recorders, are used to scan in the photo y direction, usually at intervals of 90 m with elevations recorded every 30 m along the profile (for a 30 m DEM). When these points are resampled to a 30m grid, the points along the profile lines have a much stronger autocorrelation than in the y direction, frequently resulting in stripes in the photo y direction (Mark, 1983). This method is not sensitive to the anisotropic nature of the data, and significant errors have been shown to occur in this type of DEM, mostly in the form of striping (Brabb, 1987). In fact, Elassal and Caruso (1983) estimate that about one third of the 30 m DEM series have this systematic error of striping.

Another method has been introduced which derives the altitude matrix from digitized contour lines, with hydrographic data augmenting the topographic information producing a much cleaner DEM.

Typically, the USGS's DEM products have been classified as having a vertical error either less than 7 meters or between 7 and 15 meters, being related to the amount of relief in the mapped area (i.e. generally consistent with the accuracy of the contours). The vertical accuracy achieved from these processes is measured as a weighted root mean square error (RMSE). Profiles taken in the La Honda, California Quadrangle, an early test site for the digitized contour/hydrographic approach, indicate the RMSE is about 4 m, well within the stated one half of the 40 foot contour interval (Brabb, 1987).

Data structures

Many data structures have been developed to try to capture the continuous nature of elevation data, each with its own advantages and disadvantages. Probably the most familiar method is a map of contour lines. Their popularity in topographic representation stems from their wide-spread use in manual cartography, though the limitations of that technology (i.e. the use of pen or lines drawn on a paper product) are presently being surpassed by digital techniques. The digitized contour method has generally been discounted because it requires much more storage space (as compared to a grid-DEM) and more time for retrieval of both a stored point, an interpolated point, and in deriving parameters such as slope and aspect.

The grid-based DEM, also known as an altitude matrix, is based on a regular sample of elevation points. Elevation values are stored in a matrix, with the horizontal location being implicitly stored by the ordering of the elevation values. It is probably the most common type of DEM in use today. Its advantages include a relatively compact storage structure and a form which lends itself to fast algorithms to manipulate the data, capitalizing on the power and wide usage of image processing techniques. This method also is easily integrated with raster-based GISs and remote sensing techniques.

Disadvantages of the grid DEM include redundancy in an area of uniform terrain and the inability to adjust the sampling scheme to different relief complexity (Burrough, 1986). Progressive sampling (Makarovic, 1973) has been proposed as a method to automate the sampling in response to a varied terrain, though there are still problems of redundancy associated with raster encoding. Other problems include ambiguous interpolation in the case where two diagonal points (of a four-some) have the same value. Also, a point which is surrounded by higher neighbors, referred to as a sink or pit, occasionally occurs because of signal to noise ratio effects, most occurring in flat areas (O'Callaghan and Mark, 1984). These are particularly noticeable, for example, when deriving the drainage network in hydrological modeling, as artificial drainage points create unnatural drainage patterns.

Another popular data structure is the triangular irregular network (TIN), which has been defined as a network of contiguous, planar triangles that vary in size, shape, elevation, slope, and aspect in response to local surface geometry (Peucker *et al*, 1978). These triangles are generally formed from a set of Delaunay Triangles, formed from a random distribution of points. Here, the coordinates for x and y and the z value need to be explicitly stored, along with a set of pointers to a node's neighbors. Until recently, a TIN was built up using manual methods (hindering repeatability and production) by selecting information rich points such as peaks, ridges, and streams directly from a contour map (Heil and Brych, 1974). However, commercial packages are now available which automate the derivation of a TIN model from digitized contour lines, irregular distributions of points, and matrices of elevation points. This can be done using user-selected very important points (VIPs) or on the basis of a defined error tolerance of the facets from the grid elevation data.

Perhaps the most important advantage of TIN is that because the sampling intensity can be varied to reflect the variation of the terrain thereby avoiding redundancy, a considerable savings in storage space, as well as representing the terrain at a consistent resolution, is realized. The points stored are surface specific and reflect actual points along stream channels, ridges and breaks in slope and therefore keep the geometry of the terrain intact as well as accurately representing a real hill rather than one of arbitrarily sized segments (Peucker *et al.*, 1978). There is no absolute limit of precision (i.e. both the rise and run can be defined arbitrarily), therefore areas of low slope may theoretically be depicted more accurately, though still controlled by the quality of the DEM. Also, this method can be easily used by a vector-based GIS.

Disadvantages of TIN models include the requirement of a more complex data structure and more involved algorithms to manipulate the explicitly defined topology. Because of the required topological linking of triangles, in a terrain with an abrupt transition zone, long, thin facets are often produced, limiting the reliance of these structures in representing these types of surfaces.

Additionally, a data source which complements the advantages of the TIN model is needed. Currently, the smallest resolution elevation data commonly available (and only in limited areas) is the 30 m grid DEM. Developing a TIN from this data will surely reduce the storage size, but the model can only be as good as its source, whose points were not chosen on the basis of their information richness, but arbitrarily in relation to the surface. This is short-changing the TIN methodology. Of course the TIN model could be developed from digitizing contour lines, though this data is three generations removed from reality, but also there are difficulties in representing the contour lines themselves. Perhaps a new product the USGS could develop would be a set of random elevation points specifically chosen for a TIN-based DEM, either gleaned from the original points found by the GPM process, or re-selecting points from the aerial-photo source document on the basis of their surface location. This would require the determination of the optimal set of points necessary to provide a desired goodness-of-fit with the original surface.

Besides the above models, there exist a few other less common methods of surface representation. In an attempt to find natural unit areas of terrain, methods based on digitized contours have been developed. These generally subdivide a catchment to represent the phenomenon of waterflow formed by equipotential lines (contours) and their orthogonals (streams), forming irregularly shaped polygons. Storing the elevation data in this format often requires an order of magnitude more storage space. Examples of this approach include the *RICHLINE* method developed by Douglas (1986) and a contour-based method developed by Moore, O'Loughlin and Burch (1988). Another approach is to use a quad-tree, where the model is recursively broken into quadrants in response to terrain variability. A mathematical surface (commonly a hyperbolic paraballoid) can be fitted to the "black node" of a sub-quadrant, the level of which is in response to a given error range and the variability of the terrain (Chen, 1984).

Interpolation

In order for a DEM to represent a continuous surface, it must not only have elevation values, but also a method to interpolate between these stored points and thus, these methods become very important. Schut (1972) provides an in-depth review of various interpolation techniques used in DEM.

Interpolation is necessary on two separate occasions. The first occurs in deriving the nodes of a DEM from its source map control points while the second occurs after the DEM is completed and subsequently used in an application.

The elevations which will represent a DEM need to be interpolated first from its source product. In using the GPM/stereo plotter method, over 1 million correlated coordinates (for a 7.5' Quad) are reduced to roughly 200,000 elevation points. For a TIN-based DEM, interpolation should be viewed more in terms of selecting representative or information-rich points.

Once the DEM has been produced, interpolation from these elevation values must occur to fully represent a surface. One common method of grid-based interpolation is to use an inverse weighted distance on 9 neighbors technique. For a TIN-based DEM, a surface can be described either at the nodes or the triangle facet itself. One method currently in use (System/9) uses surface normals at each node, defined by Zienkiewicz functions (Steidler, *et al.*). All the neighboring nodes are used and the normals are computed from tangent planes derived from least-squares fit. This has the characteristic of a smooth surface within a triangle and is continuous from triangle to triangle. A second approach is to define a mathematical function inside the facet. There are two options here, either using a linear constant (i.e. defining the triangle as a plane) or finding an equation to define a more complicated surface. One approach (ARC/INFO TIN) uses a bivariate quintic surface. The bivariate palettes are splined smoothly into neighbors, honoring the node points, resulting in a continuously smooth faceted surface (ESRI, 1987).

Derived parameters

So far, the term DEM has referred strictly to models of elevation points, but in addition to representing the altitude points of a surface, derivations of elevation are very useful. This broader category can be termed Digital Terrain Model (DTM)--slope, aspect, and curvature being familiar examples. For a grid-based DEM, these are usually found by a differencing operation, taking the partial derivatives with respect to x and y which are found by:

$$\frac{\partial z}{\partial x} = \frac{\left[z_{i,j+1} - z_{i,j-1}\right]}{2\Delta h}$$

and

$$\frac{\partial z}{\partial y} = \frac{\left[z_{i,j+1} - z_{i,j-1}\right]}{2\Delta h} \qquad (1)$$

where Z is the elevation of the grid and Δh is the grid spacing. Thus slope or gradient (the maximum rate of change of altitude) is given as (from Goetz, 1985):

$$s = \left[\left[\frac{\partial z}{\partial x}\right]^2 + \left[\frac{\partial z}{\partial y}\right]^2\right]^{1/2} \qquad (2)$$

and exposure (the compass direction of the maximum rate of change) is calculated as

$$\tan E = -\left[\frac{(\partial z / \partial x)}{(\partial z / \partial y)}\right] \qquad (3)$$

(where E is $\pm$ 0 to 180°, positive to the east from the south). This technique uses the four immediate neighbors and thus an individual slope value is based on a horizontal distance of twice the cell size, causing smoothing (Evans, 1972).

Horn (1982) shows two other approaches for finding slope. The first is by using the weighted average of three central differences, using the surrounding cells in a 3 x 3 neighborhood, smoothing out peaks and pits by virtue of its exception of the central point. The second method is to use a 2 x 2 window, finding the slope of a quadrant, rather than a central point. This would seem to reduce the bias of slope, using only 1 grid space distance between points, but would shift the grid by 1/2 a pixel. A similar method is to fit a quadratic surface to a 3 x 3 pixel window and measure slope and aspect in the center. Since there is a limit in both the rise and run in a grid-based DEM (based on the size of a cell), the slope angle has a precision of 1.91 ° in a 30 m spacing. Often times in areas that are nearly level, the signal to noise ratio becomes too low, resulting in aspects that frequently cross. Generally, these derived maps are noisier than the original surface and consequently, smoothing of the data (e.g. with a low-pass filter) is a common pre-processing step.

Another derivative of elevation, most useful in radiation modeling, is the slope orientation or southness, a combination of slope and aspect. This can be found by:

$$\text{southness} = \sin(s) * \cos(E) \qquad (4)$$

where aspect E is in degrees from south. Thus, a vertical south-facing plane would have an index of 1, whereas a 45° slope facing south-east would be 0.5.

Other measures that can be derived from DEMs and DTMs (after Pike, 1988) are mean, maximum, and minimum altitude; relative relief (range in altitude); slope frequency distribution (giving information on the processes involved); hypsometry; and coarse and fine texture. Evans (1972) found that the standard deviation of elevations provides one of the best statistics to describe the relief or dispersion of altitude and is also very stable because it is based on all values, not just the extremes. Steps are frequent artifacts in statistical distributions of grid-based DEMs and DTMs.

The derivation of slope and aspect from a TIN-based DEM is a more straightforward technique. Here, the trigonometric computation using three nodes will find the slope and aspect of the contained triangular plane.

Accuracy

From the above discussion, we can now categorize into two separate groups, the processes where errors in DEM occur. The first is a function of the source document's ability to capture the real surface. This occurs during the production of surface points from the source document where errors are caused by inadequacies of the documents themselves and the instruments required to manipulate them. For example, in some types of terrain, such as a heavily wooded ravine, there is great difficulty in determining exactly where the surface lies, and commonly the surface is found at the canopy top, displacing the surface a few meters above the actual ground. Other artifacts cause systematic errors in the DEM, such as striping in the profiling-technique. Note that to compare this model with reality is quite difficult, indeed ground surveying (one of the better ways of verification) is a very expensive and time intensive process. Therefore, at this point, the accuracy of the DEM is basically limited by the source documents and devices used in their production.

The second group occurs when the data derived from the source document is to be resampled to compress it into a manageable size, while maintaining a reasonable approximation of the surface. This is a distinct process because the information extracted can be largely controlled by the sampling interval and interpolation method used. There is, therefore, a convenient benchmark in which subsequent resolutions of DEMs can be compared (i.e. the original source document points), and now the question becomes one of how well the sampling interval and interpolation methodology extract elevation information. This introduces the issues of sampling and resolution and their interaction with terrain variability.

Accuracy of a DEM entails not only the average departure of points on the DEM from the real ground surface, but also involves the distribution and the non-random spatial component of errors. What largely dictates whether a DEM is acceptable for a given application is dependent upon the objective of the research and the precision required as well as the resolution of the sampling method and its sensitivity to the variation in terrain. Accuracy requirements vary tremendously depending upon the purpose and type of study and in turn affect the required resolution and precision. The precision with which the data is collected will also reflect the accuracy of the model. For example, elevation of a 30 m USGS product is measured in 1 meter increments and thus, for any modeling which requires elevation data at a finer precision than this (e.g. wetlands mapping), this DEM would be inappropriate. Similarly, a global topographic database would be ill-designed to use 1 m or even 10 m vertical precision.

The resolution of a DEM should also be viewed in relation to the variability of the terrain. The interaction between sampling frequency and terrain variation is important because the type of land feature captured at a certain resolution is site specific, being dependent upon the scale of the terrain. Weibel and DeLotto (1988) and Pike (1988) discuss this interaction of resolution and sampling in terms of the geometric signature of terrain. In the case of a grid-DEM, depending upon the distribution of spatial frequencies of the terrain and the sampling interval, certain landforms will be captured in some areas, while landforms of a differing frequency may be missed in others. Sampling theory dictates that a grid DEM can represent terrain features no smaller than twice the spacing (the Nyquist frequency) and therefore for a 30m DEM, the minimal feature resolution is 60m. Conversely, the TIN-based DEM theoretically responds to the variation change in a terrain, with the sampling intensity increasing in areas where terrain variability increases. Thus, this model tends to maximize the goodness-of-fit of the DEM surface with the real surface. The measure of the accuracy must be defined in terms of the objective or purpose of the study and the required precision, and the resolution or sampling interval and the variation of the terrain.

Conclusion

Surface representation is an important variable in nearly all aspects of environmental processes and as such, the use of DEMs in environmental modeling, especially with computerized methods, has become very popular. Frequently, however, these DEMs are misused as the limits and biases of the model are either not clearly understood or are ignored. This paper has attempted to illustrate the sources of DEMs, the techniques in capturing and producing a DEM, and also the various methods of structuring the data, all in terms of how bias and errors are produced. The objective of the research must be clear as it defines the landforms being studied. This in turn dictates the precision required and with knowledge of the terrain variability, the resolution of a DEM. Finally, the sensitivity of resolution of a DEM with the terrain variation has a large effect on the resulting accuracy and fidelity of the model.

References

Brabb, E. E., 1987, Analyzing and portraying geologic and cartographic information for land use planning, emergency response, and decisionmaking in San Mateo County, California. *Proceedings GIS '87*, San Franscisco, 1987, pp.362-374.

Burrough, P. A., 1986, *Principles of Geographical Information Systems for Land Resources Assessment.* (Oxford: Clarendon Press)

Chen, Z. T., 1984. Quad-tree spatial spectrum: its generation and application, *Proceedings of the International Symposium on Spatial Data Handling*, Zurich, August 1984, pp. 218-237.

Douglas, D. H., 1986, Experiments to locate ridges and channels to create a new type of digital elevation model. *Cartographica* **23** (4), 29-61.

Elassal, A. A., and Caruso, V. M., 1985, Digital elevation models. *Geological Survey Circular* 895-B. USGS Digital Cartographic Data Standards.

ESRI, 1987, *TIN User's Manual.* (Redlands, CA: ESRI)

Evans, I. S., 1972, General geomorphometry, derivatives of altitude, and descriptive statistics. In *Spatial Analysis in Geomorphology*, Chorley, R. J., (ed.), pp 17-90. (London: Methuen and Co.)

Goetz, S., 1984, Digital Terrain Representation and Applications, Unpublished Masters thesis, Department of Geography, University of California, Santa Barbara.

Heil, R. J., and Brych, S. M., 1978, An approach for consistent topographic representation of varying terrain. *Proceedings of the DTM Symposium*, St. Louis, May 1978, pp. 9-11, 397-411.
Horn, B. K., 1982, Hill shading and the reflectance map, *Geo-Processing* **2**, 65-146.
MacEachren, A., and Davidson, J. V., 1987, Sampling and isometric mapping of continuous geographic surfaces. *The American Cartographer* **14** (4), 299-320.
Makarovic, B., 1973, Progressive sampling for digital terrain models. *ITC Journal* **3**, 397-416.
Mark, D. M., 1983, Automated detection of drainage networks from digital elevation models. *Proceedings Auto-Carto 6*, 168-178.
Moore, I. D., O'Loughlin, E. M., and Burch, G. J., 1988, A contour-based topographic model for hydrological and ecological applications. *Earth Surface Processes and Landforms* **13**, 305-320.
O'Callaghan, J. F., and Mark, D. M., 1984, The extraction of drainage networks from digital elevation models, *Computer Vision, Graphics, and Image Processing* **28**, 323-344.
Peuker, T. K., Fowler, R. J., Little, J. J., and Mark, D. M., 1978, The Triangulated Irregular Network. *Proceedings of the DTM Symposium*, St. Louis, May 9-11, pp. 516-540.
Pike, R. J., 1988, The geometric structure: quantifying landslide terrain types from digital elevation models. *Mathematical Geology* **20** (5), 491-511.
Pike, R. J., Thelin, G.P., and Acevedo, W., 1987, A topographic base for GIS from automated TINs and image-processed DEMs. *Proceedings GIS '87*, San Franscisco.
Schut, G., 1976, Review of interpolation methods for digital terrain models. *Canadian Surveyor* **30**, 389-412.
Steidler, F., Dupont, C., Funche, G., Vuatoux, C. and Wyatt, A., Digital terrain models and their applications in a database system. *Personal correspondence*.
Weibel, R. and DeLotto, J. S., 1988, Automated terrain classification for GIS modeling, *Proceedings GIS/LIS '88*, San Antonio, pp. 618-627.

Chapter 10

Modeling error in objects and fields

Michael F. Goodchild

Introduction

The current interest in spatial databases stems largely from their role in supporting geographic information systems, and the rapidly growing GIS industry. GISs are powerful systems for handling spatial data, and in recent years they have found application in fields as different as transportation, forestry and archaeology. Yet the power of a GIS to input, store, analyze and output geographic information of all kinds is at the same time a major liability. To the database, the structure used to store a polygon or pixel has almost no connection to the real meaning of the polygon, as a parcel of land, object on a topographic map or stand of timber. The analyst making use of a polygon overlay operation has similarly little pressure to be sensitive to the interpretation of the data layers being overlaid. In reality a GIS may encourage poor analysis by separating the data collection, compilation and analysis functions, and failing to make the user aware of the possible dangers of indiscriminate use of such functions as scale change, reclassification and overlay.

One of the more obvious issues from this perspective is the existence of two traditions of GIS analysis. The distinction between raster and vector is often seen as a problem of system design, but actually presents a major issue of data interpretation. To emphasize this difference, and to stress the context of data interpretation rather than system design, we will use the terms field and object in this paper, although they are to some extent synonymous with raster and vector respectively. Some spatial databases represent the world as if it were populated by objects - points, lines and areas - with associated attributes, continuing a tradition developed in cartography. Others represent the world as fields, or arrays of pixels, again with associated attributes. The choice between the two representations has variously been seen as depending on the method of data collection (satellites generate fields,cartographers generate objects), the degree of spatial resolution required (objects appear to imply higher levels of spatial resolution, whereas pixels imply a level which is fixed by the pixel size), and the efficiency of algorithms (for example, the widely held perception that overlay is faster in raster). However we will argue in this paper that fields and objects represent fundamentally different forms of abstraction of geographical reality.

This paper examines the relationship between fields and objects from the perspective of database error. It is clearly possible to represent a given set of data in either form, and to derive one from the other by a simple GIS operation such as raster/vector conversion. But conversion must be sensitive to the nature of the data and its uncertainty if subsequent analysis is to be successful.

Errors in objects

Consider the common process of creating a spatial database of objects by digitizing a topographic map. Attributes will likely be entered from a keyboard, and provided they are keyed correctly, we can reasonably expect them to be perfectly accurate. The locations of objects will be obtained by digitizing or scanning, and will be subject to assorted errors. A number of factors may contribute to distortion of the source document, including folding and stretching, changes in humidity, copying processes etc., and the process of map registration will introduce additional error. If digitizing is used, positional accuracy will be affected by the operator's precision in positioning the cursor, and by the rules used to select points to be digitized from line or polygon objects, whether in point or stream mode. Finally the positional accuracy of a scanned line will be affected by the resolution of the scanner.

Of all of these errors, only cursor positioning has been subject to successful analysis. Keefer, Smith and Gregoire (1988) have described a model in which each digitized point on a line or area object is distorted from its true position by a bivariate distribution. If each point were distorted independently we would reach the unreasonable conclusion that the expected error at digitized points is greater than between digitized points, but it is likely that errors are positively correlated between adjacent points along the digitized line. Chrisman and Yandell (1988) and Griffith (this volume) have obtained useful estimates of the expected error in polygon area measures based on this type of model of digitizing error. However it is much more difficult to devise a reasonable model of the process of point selection,which varies substantially between digitizer operators and types of lines, and likely contributes at least as much to error in area estimates. Unfortunately any such model would have to be sensitive to the type of line being digitized, as meandering rivers clearly present very different problems from topographic contours or highways, whereas this seems less important for modeling cursor positioning error.

In the case of a county boundary or the outline of a building, the database object corresponds directly to a clearly defined object in the real, geographical world. But in many cases the object in the database is an abstract model of real, continuous and complex geographic variation. Although counties and buildings are frequently found in spatial databases, many GIS applications have been developed for abstracted objects. In the forest industry, an inventory of forest resources is commonly maintained in digital form by dividing the forest into area objects or "stands" with descriptions which are attributed homogeneously to the entire stand. In reality the boundaries of stands are transition zones, and the attributes are heterogeneous. Despite the popularity of vector databases, the object model is often a poor representation of geographic variation.

Greater difficulty may arise if the attributes assigned to an object are themselves abstractions. For example the term ""old growth"" may be attached to a forest stand, but ambiguities in the definition of old growth may make it impossible to resolve whether a particular point within the stand is or is not covered by old growth. Similarly it may be necessary to observe a significant area in order to assign a classification such as"aspen parkland", which is not strictly an attribute of a point,but of an extended area.

Various terms have been used to distinguish between the two types of error implied by this argument. In this paper we use the term processing error to describe the uncertainty introduced by digitizing and any subsequent form of digital processing, such as vector/raster conversion. The term source error is used to describe the differences which may exist between the object model and the geographical truth which it represents. While source errors may be absent in representations of counties or buildings,we conjecture that they will exceed processing errors in representations of such geographical variates as vegetation, soil or land use. Since the results of a GIS analysis will be interpreted by reference to geographic truth rather than to the source documents used to create the database, the existence of large source errors is a problem of major significance for the field. Several early papers (see for example McAlpine and Cook, 1971; MacDougall, 1975) drew attention to the importance of source errors.

The distinction between processing and source error is critical in estimating the accuracy of measures derived from a database. The error in the estimated area of a forest stand can be obtained by analyzing the errors introduced in processing its area object, and is independent of source errors implicit in the object. The area of a given species can be estimated by summing the areas of the objects classified as containing the species, but will be subject to source errors if the species attribute does not apply homogeneously to the area within each stand. Since an object database commonly contains no information about heterogeneity, it may be difficult or impossible to estimate source errors.

Maps of forest stands or soil types are compiled by a complex process which combines two forms of data. Information is first obtained on the ground from a series of point samples or transects, and then extended spatially using an aerial photograph or remotely sensed scene, or similar image. The area objects are compiled by interpreting a field, augmented by point attributes, so in these examples the object representation is clearly a derivative of the field representation. In compiling the map from the image, the cartographer imposes his or her own expectations on the data. The boundaries between area objects will be smooth generalizations of complex transition zones whose width will likely vary depending on the classes on either side. The holes and islands which one would expect in and around the transition zone will be commonly deleted. Finally, in the case of forest stands the cartographer may impose some concept of minimum size, since it is difficult to administer stands of less than, say, 10 hectares. In essence the cartographer acts as a low-pass filter, removing much of the detailed geographical variation in converting from field to objects. Unfortunately this removes much of the information on which an error model might be based, since error is selectively deleted by a low-pass filter, and suggests that we might focus on modeling errors in fields rather than derivative objects.

While some progress has been made in developing error models for digitized lines and areas, it has proven much more difficult to construct comprehensive models of processing and source errors for complex geographical objects. Chrisman (1982), Blakemore(1984) and others have described the uncertainty in the position of a line using a band of width ε (Perkal, 1956), which is a useful model of certain kinds of processing and source errors but does not deal with the problem of heterogeneity of attributes. Moreover while the epsilon band can be used to give a probabilistic interpretation to the point in polygon operation (Blakemore, 1984) it is not adequate as the basis for estimating error in area measures, or for simulating error in object databases to benchmark GIS error propagation. At this point we have no fully satisfactory means of modeling error in complex spatial objects.

Contours provide another example of the difficulty of modeling error in objects derived from fields. A model of object distortion might take each contour line and displace it using a random, autocorrelated error process, but it would be easy to produce topological inconsistencies such as crossing contours or loops. However contours are derived from an elevation field: no matter how elevations are distorted, the contours derived from them must always be topologically consistent. This suggests a general proposition - that the solution to modeling error in complex spatial objects may lie in modeling error in the fields from which many of them are obtained. In the next section we explore this possibility in more detail.

Field models of error in area objects

The previous section concluded with the proposition that satisfactory models of error in complex spatial objects could be formulated as models of error in fields. In this section we consider two such models.

Goodchild and Dubuc (1987) proposed a model based on an analogy to the effect of mean annual temperature and precipitation on life zones. We first generate two random fields, using one of a number of available methods, such as the fractional Brownian process, turning bands or Fourier transforms (Mandelbrot, 1982). The autocovariance structure of the fields can be varied to create a range of surfaces from locally smooth to locally rugged. Imagine that one field represents mean annual temperature and the other, annual precipitation.

Holdridge *et al.* (1971) have proposed a simple two-dimensional classifier which, given temperature and precipitation, yields the corresponding ecological zone. By applying such a classifier to the two fields, we obtain a map in which each pixel has been classified into one of the available zones. If the pixels are now vectorized into homogeneous areas, the map satisfies the requirements of many types of area class maps: the space is exhausted by non-overlapping, irregularly shaped area objects, and edges meet in predominantly three-valent vertices. On the other hand if a single field had been used with a one-dimensional classifier, the result would have the unmistakable features of a contour or isopleth map. To simulate the influence of the cartographer, which we have previously compared to the action of a low-pass filter, Goodchild and Dubuc (1987) applied a spline function to the vectorized edges, and selectively removed small islands.

The model has interesting properties which it shares with many real datasets of this class. Because the underlying fields are smooth, classes can be adjacent in the simulated map only if they are adjacent in the classifier, so certain adjacencies are much more common than others. The response of an edge to a change in one of the underlying fields depends on the geometry of the classifier: it is maximum if the corresponding edge in the classifier space is perpendicular to the appropriate axis, and minimum (zero) if the edge is parallel to the axis. So distortion or error can be simulated by adding distortion to the underlying fields.

The model successfully simulates the appearance of area class maps, and is useful in creating datasets under controlled conditions for use in benchmarking spatial databases and GIS processes, and for tracking the propagation of error. On the other hand the large number of parameters in the model make it difficult or impossible to calibrate against real datasets. Parameter values would have to be established for the underlying fields, the distorting field(s), the classifier, and the spline function used to smooth vectorized edges.

Goodchild and Wang (1988) describe an alternative model based on an analogy to remote sensing. Suppose that the process of image classification has produced an array of pixels, and that associated with each pixel is a vector of probabilities of membership in each of the known classes. For example the vector{0.3,0.3,0.4} would indicate probabilities of 0.3, 0.3 and 0.4 of membership in classes A, B and C respectively. In practice we would expect the proportion of non-zero probabilities to be small, allowing the vectors to be stored efficiently.

We now require a process of realizing pixel classes such that (1)the probabilities of each class across realizations are as specified, and (2) spatial dependence between pixels in any one realization. The degree of spatial dependence will determine the size of homogeneous patches which develop. With high spatial dependence, a given pixel will belong to large patches in each realization: in 30% of realizations the example pixel will be part of a patch of class A, 30% B and 40% C.

Goodchild and Wang (1988) described a simple process which satisfies only one of the requirements. Initial classes were assigned to each pixel by independent trials, and a 3x3 modal filter was then passed over the array, replacing the value of the central pixel by the modal value of the 3x3 window. While this ensures spatial dependence, the posterior probabilities are not equal to the priors except in special cases.

More recently, we have experimented with two methods which satisfy both requirements. In the first, we first generate a large number of realizations using independent trials. On each simulated map, we count the number of 4-adjacencies between unlike classes. We next execute a number of cycles to induce spatial dependence without at the same time changing the posterior probabilities. The two realizations with the lowest levels of spatial dependence (highest number of 4-adjacencies between unlike classes) are selected in each cycle. A random pixel is selected, and the contents of the pixel are swapped between the two selected realizations if the result would yield a higher level of spatial dependence. After examining a large number of pixels, another pair of realizations is selected. Because the method conserves the numbers of each class of pixel across realizations, while increasing spatial dependence, it clearly satisfies both of our requirements. Finally the simulated maps are vectorized and smoothed to create area objects.

In the second method, we make use of a simple spatially autoregressive process (Haining, Griffith and Bennett, 1983), to generate a random field of known distribution, and classify the result by comparing the value in each pixel to the prescribed probabilities. Since only two classes can be simulated, the method must be repeated n-1 times to develop a map of n classes.

Let $\mathbf{x}$ denote the random field. The spatially autoregressive process is defined as:

$$\mathbf{x} = \rho \mathbf{W} \mathbf{x} + \varepsilon \quad (1)$$

where $\rho<0.25$ is a parameter of spatial dependence, $W_{ij}=1$ if i and j are 4-adjacent, else 0, and ε_i is a normal deviate of zero mean. We find $\mathbf{x}$ by inverting the matrix $(\mathbf{I} - \rho \mathbf{W})$. Unfortunately an array of n by n pixels requires the inversion of an n^2 by n^2 matrix, but it is possible to do this for arrays as large as 64 by 64 by taking advantage of the block structure of the W matrix. Given the known distribution of x_i across realizations, we can compute $P(X<x)$ and compare it to the pixel's specified probability p_i.

Certain types of prior information can be introduced into both of these models in order to broaden their applications. For example, suppose there exists a predefined "parcel" of known boundaries, and it is suspected that the class or classes within the parcel are independent of those outside. The parcel boundary is adjusted to pixel edges: $W_{ij}=0$ if i is inside the parcel and j is outside, even though i and j may be 4-adjacent. This has the effect of removing spatial dependence between the parcel and its surroundings. The homogeneity of the parcel depends on the magnitude of ρ; if ρ is sufficiently large, the parcel will act as an object whose attribute is determined by a single trial. Geman and Geman (1984) have described an edge process with similar objectives.

The vectors of probabilities required by this process are readily obtainable from many remote sensing classifiers. Conventionally, pixels are classified by maximum likelihood even though the classifier yields a complete vector. This results in the loss of valuable information on uncertainty, and leads to severe bias in derived estimates of area, particularly for large patches. The model has only one parameter, ρ. Its value might be established by calibration against ground truth, or might be set to reflect expectations about patch size and map complexity. Inverses of $(\mathbf{I}- \rho \mathbf{W})$ might be precomputed for various values of ρ, and multiplied by various ε to obtain multiple realizations. In this way it would be possible to simulate error rapidly for any classified image. Although the first method above based on swapping between realizations is conceptually simpler, it does not lend itself as readily to precomputing, and so would be more difficult to implement in practice.

Discussion

The process by which a spatial database is created from a source map is complex, and error of various types is introduced at each step. Some of the error components can be modeled, and progress has been made in analyzing the errors due to cursor positioning, but others such as point selection are more difficult, and the goal of a comprehensive model of spatial data processing errors is still elusive. Yet despite this, for most types of spatial data the errors inherent in the source document are clearly more significant than those introduced by processing. This is particularly true when the source document contains objects which are approximate abstractions of complex and continuous spatial variation.

We have argued in this paper that many of the more abstract types of spatial objects have been obtained or compiled from raw data in the form of fields, and that the process of compilation often removes much of the information on which a useful model of uncertainty or error might be based. The role of the cartographer in compiling vegetation or soil maps was compared to the action of a low-pass filter in selectively removing the high spatial frequencies which contain diagnostic information on uncertainty, such as wiggly lines and small islands.

From a spatial statistical point of view, the central proposition of this paper is that uncertainty in the objects on a map results from different outcomes of a stochastic process defined for a field. This is substantially different from the approach which has underlain much work on stochastic processes for images. The problem of image restoration has the objective of finding the"true" value for each pixel given some distorted value. However in the geographical case there is usually no comparable notion of true value, because spatial variation extends to all scales, and because class definitions are frequently ambiguous. Similarly, although spatial statistics contains extensive literature on image segmentation, this problem is seen as one of a range of possible models for the cartographic process of forming objects from fields.

From the perspective of spatial databases and GIS, there are several possible roles for a model of error in spatial data. On the one hand it would be useful to have a calibrated model which could be used to describe error in a particular dataset, to track the error through GIS processes, and to report uncertainty in the results of processing. However such models may be unobtainable in many cases, due to the lack of adequate information for calibration, and to the complexity of the error process itself. There seems to be no equivalent with the generality of the Gaussian distribution for complex spatial objects. On the other hand models of error are useful for generating simulated datasets under known conditions, for benchmarking GIS processes and storage methods.

Abstract point, line and area models developed in cartography because of the need to represent complex spatial variation on paper using images which could be created with a simple pen (Goodchild, 1988). DEMs, TINs and quadtrees are a few of the new data models which have been developed for spatial databases to take advantage of the removal of cartographic constraints. The contour was devised as an efficient way to display spatial variation of elevation on a topographic map with a pen capable of drawing lines of fixed width, but DEMs and TINs have distinct advantages over digitized contours in terms of sampling efficiency and the execution of various analytic operations. We have seen in this paper that the selection of the cartographic model also has the effect of removing information on uncertainty, since it is easier to model error in fields than in derivative objects. A database populated by fields is more useful for modeling and tracking uncertainty than one populated by abstracted objects. From an accuracy perspective, then, it would be preferable if databases representing spatial variation of parameters such as soil type, vegetation or land use contained the raw point samples and images from which such maps are usually compiled. If the cartographic view of the database were required, it could be generated by interpreting areal objects, either interactively or automatically. But this strategy would avoid the common practice of imposing the cartographic view as a filter between the database and the reality which it represents.

References

Blakemore, M., 1984, Generalization and error in spatial databases. *Cartographica* **21**, 131-9.

Chrisman, N. R., 1982, Methods of spatial analysis based on errors in categorical maps. Unpublished PhD thesis, University of Bristol.

Chrisman, N. R., and Yandell, B., 1988, A model for the variance in area. *Surveying and Mapping* **48**, 241-6.

Geman, S., and Geman, D.,1984, Stochastic relaxation, Gibbs distributions and the Bayesian restoration of images. *IEEE Transactions on Pattern Analysis and Machine Intelligence* PAMI-**6**, 721-41.

Goodchild, M. F., 1988, Stepping over the line: technological constraints and the new cartography. *American Cartographer* **15**, 311-20.

Goodchild, M. F., and Dubuc, O., 1987, A model of error for choropleth maps, with applications to geographic information systems. Proceedings, *AutoCarto 8*. ASPRS/ACSM, Falls Church, VA, 165-74.

Goodchild, M. F., and Wang, M.-H., 1988, Modeling error in raster-based spatial data. *Proceedings, Third International Symposium on Spatial Data Handling*. IGU Commission on Geographical Data Sensing and Processing, Columbus, Ohio, 97-106.

Haining, R. P., Griffith, D.A., and Bennett, R.J., 1983. Simulating two-dimensional autocorrelated surfaces. *Geographical Analysis* **15**, 247-55.

Holdridge, L. R., Grenke, W. C., Hathaway, W. H., Liang, T., and Tosi, J. A., Jr., *Forest Environments in Tropical Life Zones: A Pilot Study*. (Oxford: Pergamon)

Keefer, B. J., Smith, J. L., and Gregoire, T. G., 1988, Simulating manual digitizing error with statistical models. Proceedings, *GIS/LIS '88*. ASPRS/ACSM, Falls Church, VA, 475-83.

MacDougall, E. B., 1975, The accuracy of map overlays. *Landscape Planning* **2**, 23-30.

McAlpine, J. R. and Cook, B.G., 1971, Data reliability from map overlay. In *Proceedings, Australian and New Zealand Association for the Advancement of Science*, 43rd Congress, Brisbane, Section 21, Geographical Sciences.

Mandelbrot, B. B., 1982, *The Fractal Geometry of Nature*. (San Francisco: Freeman)

Perkal, J., 1956, On epsilon length. *Bulletin de l'Academie Polonaise des Sciences* **4**, 399-403.

Chapter 11

Frame independent spatial analysis

Waldo R. Tobler

Abstract

The results of an analysis of geographical data should not depend on the spatial coordinates used - the results should be frame independent. This should also apply when areal units are used as the spatial data collection entity. Previous work has shown that some analysis procedures do not yield the same results under alternate areal aggregations, but some of these studies have used measures known to be inappropriate for spatial data, e.g., Pearsonian correlation instead of cross-spectral analysis. And there are some methods of analysis which do seem to yield frame invariant results, especially under alternate partitionings of the geographic space. In other cases it is appropriate to consider aggregations as spatial filters, with response functions which can be estimated *a priori*. There also exist linear spatial models which allow exact calculation of the effects of a spatial aggregation, so that consistent empirical and theoretical results can be obtained at all levels of spatial resolution. It is proposed that all methods of spatial analysis be examined for the invariance of their conclusions under alternative spatial partitionings, and that only those methods be allowed which show such invariance.

From a philosophical point of view it is important that spatial analyses not depend on the units used to identify the geographical location of the objects being studied. In its simplest form this is an assertion that it should not matter whether one is identifying places by rectangular coordinates or by polar coordinates. In this overly simplified example everyone would agree that the names used to identify the locations are irrelevant to the substantive analysis. This same point of view should prevail when areal units are used. But it apparently does not. Openshaw (this volume) quotes Kendall and Yule (1950, p. 313) who warn that one must not lose "... sight of the fact that our results depend on our units". The units in this case are areal units for which agricultural statistics (wheat and potato yields) are assembled. This, I think, is very poor science, and represents a misconception. It is not the areal units which are to blame. The difficulty is that the method of analysis used was inappropriate. This tautology is immediate. If the procedure used gives results which depend on the areal units used, then, *ipso facto*, the procedure must be incorrect, and it should be rejected *a priori*. As an aside, one of the reasons why tensors are used for many calculations in physics is that they give results which are independent of the particular place names chosen. For example, the components of a gradient vector depend on the system of coordinates used, but the gradient itself is a concept independent of these units. We aim for the same type of frame free analysis in geography.

In the particular instance Kendall and Yule were computing correlation coefficients between areal units. That correlations between data sets assembled by areal units are subject to fluctuation has long been known. Openshaw (1984) cites a fifty year old paper by Gehlke and Biehl (1934), in which these authors observed that a correlation coefficient increased when they aggregated the data to larger areal units. Openshaw (1984) goes into some detail here, citing further comments by Kendall and Yule (1950), by Robinson (1950) in his well known study of ecological correlations, and by Blalock (1964).

The fallacy in all of these studies is the assumption that the correlation coefficient is an appropriate measure of association amongst spatial units. Clearly it is not - the appropriate measure is the spatial cross - coherence function (see Rayner, 1971) and the association between the two variables may be different in different locations - but all of these authors put the blame on the spatial units. This fallacy is compounded when these authors do not recognize that the spatial frequency point of view quickly and easily predicts the types of results which they obtained. For example, Curry (1966) points out that "Administrative units having area dimensions represent a filtering out of wavelengths less than their size". And Casetti (1966) notes that "Aggregating smaller areal units into regions filters out the harmonics whose wavelengths are smaller than the size of the regions", and, if two (or more) "space series have harmonics which are filtered out by a given aggregation, the correlation and regression coefficients of the series before the aggregation will differ from the coefficient obtained after the aggregation." In spite of this clear theoretical understanding Openshaw (1984, p. 13 *et seq.*) feels compelled to perform extensive numerical and computer experiments with empirical data, and does manage to demonstrate that correlation coefficients do indeed perform in the expected unsatisfactory manner. Again the blame is put in the wrong place, on the areal units. Somewhat later Openshaw (1984) demonstrates that similar results hold for regression coefficients, and for a particular spatial interaction model as well as for the simpler measures of association. As noted above this was already anticipated theoretically. Other theoretical insights have not been pursued adequately either. For example Tobler (1969) suggested computation of the spatial response function of an areal partitioning as a method of adjusting for the filtering effects of the partitioning, and Moellering and Tobler (1972) demonstrated how to isolate the most important level(s) of an administrative spatial aggregation.

One of the difficulties, implicit in Kendall and Yule (1950) but explicit in Openshaw (1984), is that many kinds of geographical data inevitably seem to require reporting in some areal units, and that these units are always to some extent arbitrary. The inference is thus that the "modifiable areal unit problem" is unavoidable. This appears to be another fallacy, at least when stated in this naive way. Not all geographical problems are well posed. For example, did Kendall and Yule (1950) really need to use those areal units, or could the problem have been reformulated to be independent of the units? Could they, for example, have gotten the raw data on agricultural fields and used a form of near neighbor analysis (Getis and Boots, 1978), or could they have given the yields in the form of spatially continuous geographical probability density functions (see, e.g., Silverman, 1986)?

This last approach is itself not without problems. If one asks for the cancer rate (cancers per 100,000 persons) at a particular latitude and longitude one can get a different answer if the data are computed from national observations, or from state data, or from county data, or from city data, or from census tract data, or from data by city block, or by house. Does this process have a limit? Is there an actual cancer rate at this place? Note the similarity to Richardson's (1926) question "Does the Wind have a Velocity?". We are told that the air, that water, are made up of discrete particles but we also observe that aero- and hydrodynamicists use partial differential equations for the study of these systems, and not quantum mechanics. How can one use calculus in such a situation? The answer, the textbook answer, is in the continuum hypothesis. The books (e.g., Batchelor, 1967, pp.4-6) often have a diagram such as that in Figure 1, where the density of a gas (for example) is plotted as a function of the resolution. At some point the density oscillates

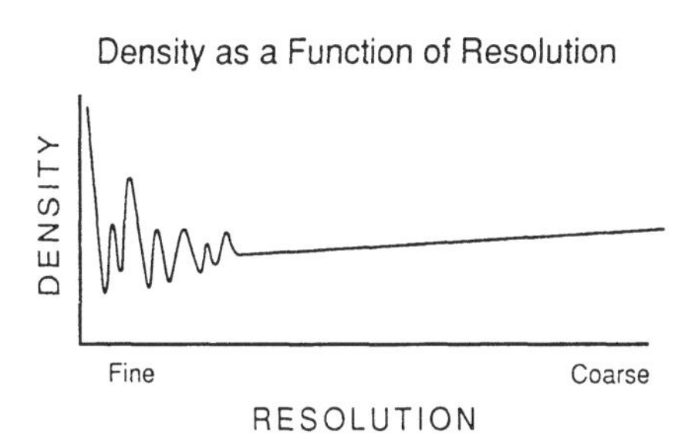

Figure 1 Density as a function of resolution.

erratically, and is not a well defined, useful quantity. The student is warned that the analysis procedures, theorems, and techniques which follow in the book do not hold in the vicinity of this region or below. Interestingly I have yet to find a book which is explicit and precise about where this region occurs. The conclusion which I draw from this is that there may well be problems of areal units, but they are not the ones which have been studied, and have almost nothing to do with correlation and regression.

From this discussion it is clear that the "modifiable areal unit problem" really consists of at least two distinct problems. The first I label the partitioning problem. It can be imagined in this fashion. In some piece of territory there exist discrete (immobile) individuals with attributes. Think of these as dots on a geographical map. This continuous piece of territory is then partitioned into a set of N areal units (put boundaries on the map inside the territory), and some procedure is used to summarize the attributes within each subunit, and to compute a measure of association between the summary attributes. Then a *de novo* different partitioning of the territory, again into N areal units, is undertaken, with a comparable summarization of the attributes. To what extent do the associations between the attributes differ for these two partitionings?

There are obviously arbitrarily many ways in which these partitionings can be performed, and the areal subunits can differ in size, shape, and orientation. Perhaps pentominoes (Buttenfield, 1984; Gardner, 1959; Golomb, 1960), which fix the size, can help to study some of the questions via simulations. Does the value of N make a difference? Suppose N is not the same for the two partitionings? The usual situation in practice is that one is given two different sets of data, assembled by two different partitionings and must work with these data, which are all that one can get. Much of the literature suggests that use of data from two such incompatible areal partitionings is only possible by aggregation to some larger unit sizes where the partitionings happen to coincide. Such coincidence often occurs in bureaucratic/political hierarchical spatial partitionings. Here again the conventional notion may not be correct. Pycnophylactic interpolation has recently been suggested (Tobler, 1979) and studied (Rylander, 1986) as a method for converting data from one set of areal units to another, and appears to work quite well. Conversion of data from latitude and longitude to transverse Mercator coordinates does not appear to cause any difficulties. Why should conversion from census tract to school district cause problems? It is helpful to organize geographic conversion problems into a square table, with point coordinates, line coordinates, and areal coordinates along both the side and top of the table. Now fill in the complete table by considering the conversions (and their inverses) between each method of data recording. Areal data are frequently converted to centroids (area -> point), lat/lon to UTM (point -> point), street addresses to State Plane Coordinates (point -> point) or to census tracts (point -> area), and so on. Most geographical information systems contain a number of such conversion possibilities. Further experimentation with the types of invariances which can be obtained under such transformations does not appear difficult (see Arbia, this volume, for example).

The second type of "modifiable areal unit problem" I refer to as a true aggregation. Here one starts from data in areal units and, for some reason, groups some of these units together into larger, and consequently fewer, units. This has the effect of coarsening the resolution of the data, where the average resolution is defined as the square root of the size of the territory divided by the number of areal units. The size of the smallest detectable pattern is of course twice that of the resolution. Most analyses are degraded by such a procedure, particularly if it increases the variance in resolution. Here we can point to some positive results, even though correlation and regression may be useless. In the migration model of Dorigo and Tobler (1983) it can be shown that it is possible to calculate exactly all of the model parameters when one combines data from areal units (Figure 2).

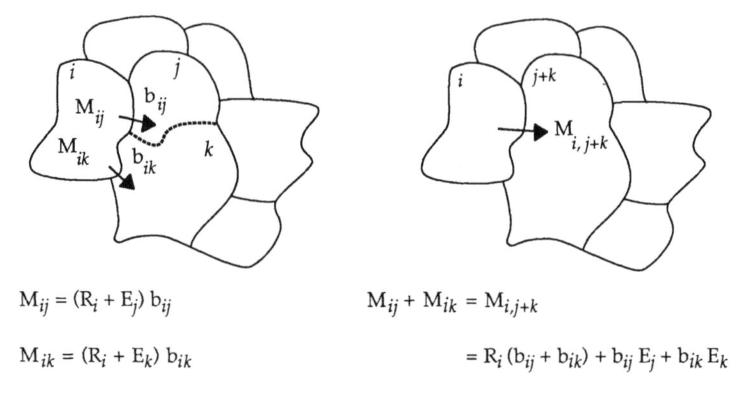

The model asserts that migration from i to j is equal to the push from i plus the pull from j times the length of the boundary between i and j. The pull value for the combined region $i + j$ is easily computed from the parameters estimated for the disaggregate model, as explained in Dorigo and Tobler, 1983.

Figure 2 Computing migration before and after aggregation.

Aggregation is thus not at all a problem in this model. The only question is why one would want to do it. Although not aggregation invariant the model changes in an exactly predictable manner. The popular entropy migration model suggested by Wilson (1967) does not have this property and must be recalibrated, with apparently unpredictable results, for every alternate aggregation of the data.

There is another sense in which aggregation is simpler than the partitioning problem. The spatial frequency response point of view allows one to consider the effects of an aggregation to be similar to that of a spatial filter, generally a low-pass filter (see Holloway, 1958; Burr, 1955). The accompanying figures illustrate this dramatically. In each instance the same analysis was performed on data given at different levels of resolution, and in each case the results are as if one had passed a low- pass filter over the results of the higher resolution analysis.

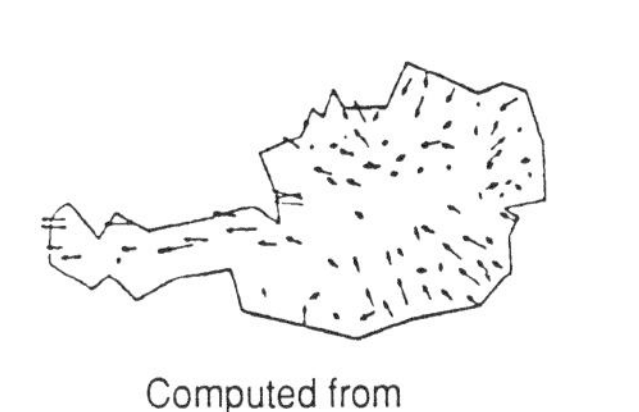

Figure 3 Migration in Austria.

In this analysis there is no "modifiable areal unit problem". The problem has gone away when we use the correct analysis procedure.

References

Batchelor, G., 1967, *An Introduction to Fluid Dynamics.* (Cambridge University Press)

Blalock, H., 1964, *Causal Inferences in Nonexperimental Research.* (Chapel Hill, NC: U. North Carolina Press)

Burr, E., 1955, The sharpening of observational data in two dimensions. *Australian Journal of Physics* **8**, 30-53.

Buttenfield, J., 1984, Pentominos: a puzzle solving program. *Compute!* May: 106-122.

Casetti, E., 1966, Analysis of spatial association by trigonometric polynomials, *Canadian Geographer* **10**, 199-204.

Curry, L., 1966, A note on spatial association. *Professional Geographer* **18**, 97-99.

Dorigo, G., and Tobler, W., 1983, Push-pull migration laws. *Annals, AAG* **73**, 1-17.

Gardner, M., 1959, *Mathematical Puzzles & Diversions,* pp. 124-141. (New York: Simon & Schuster)

Gehlke, C., and Biehl, H., 1934, Certain effects of grouping upon the size of the correlation coefficient in census tract material. *J. Am. Stat. Assn., Supplement* **29**, 169-170.

Getis, A., and Boots, B., 1978, *Models of Spatial Processes.* (Cambridge University Press)

Golomb, S., 1960, *Polyominoes.* (New York: Scribner)

Holloway, J., 1958, Smoothing and filtering of time series and space fields. *Advances in Geophysics* **4**, 351-389.

Kendall, M., and Yule, G., 1950, *An Introduction to the Theory of Statistics.* (London: Griffin)

Moellering, H., and Tobler, W., 1972, Geographical variances' *Geographical Analysis* **4**, 34-50.

Openshaw, S., 1984, *The Modifiable Areal Unit Problem,* CATMOG # 38. (Norwich, England: Geo Books)

Rayner, J., 1971, *An Introduction to Spectral Analysis.* (London: Pion Press)

Richardson, L., 1926, Does the wind have a velocity? *Proc. Roy. Soc, A* **110**, 709 .

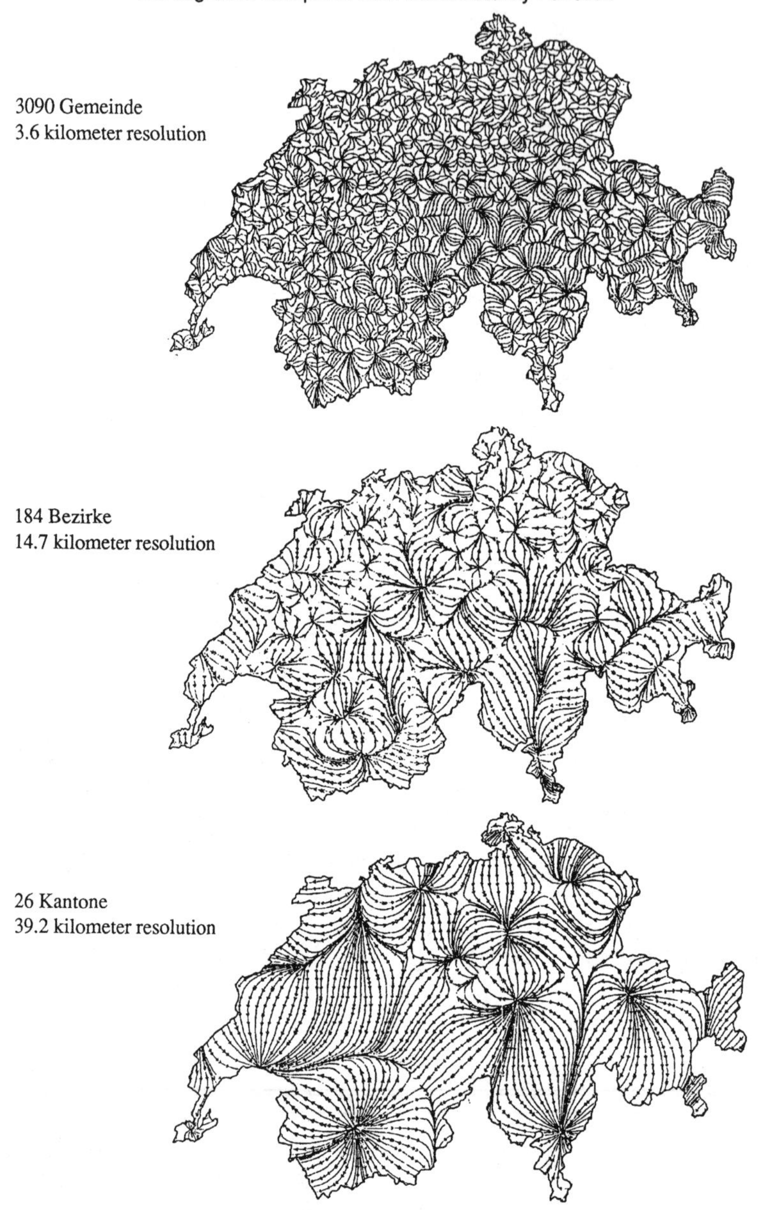

Figure 4 Migration patterns.

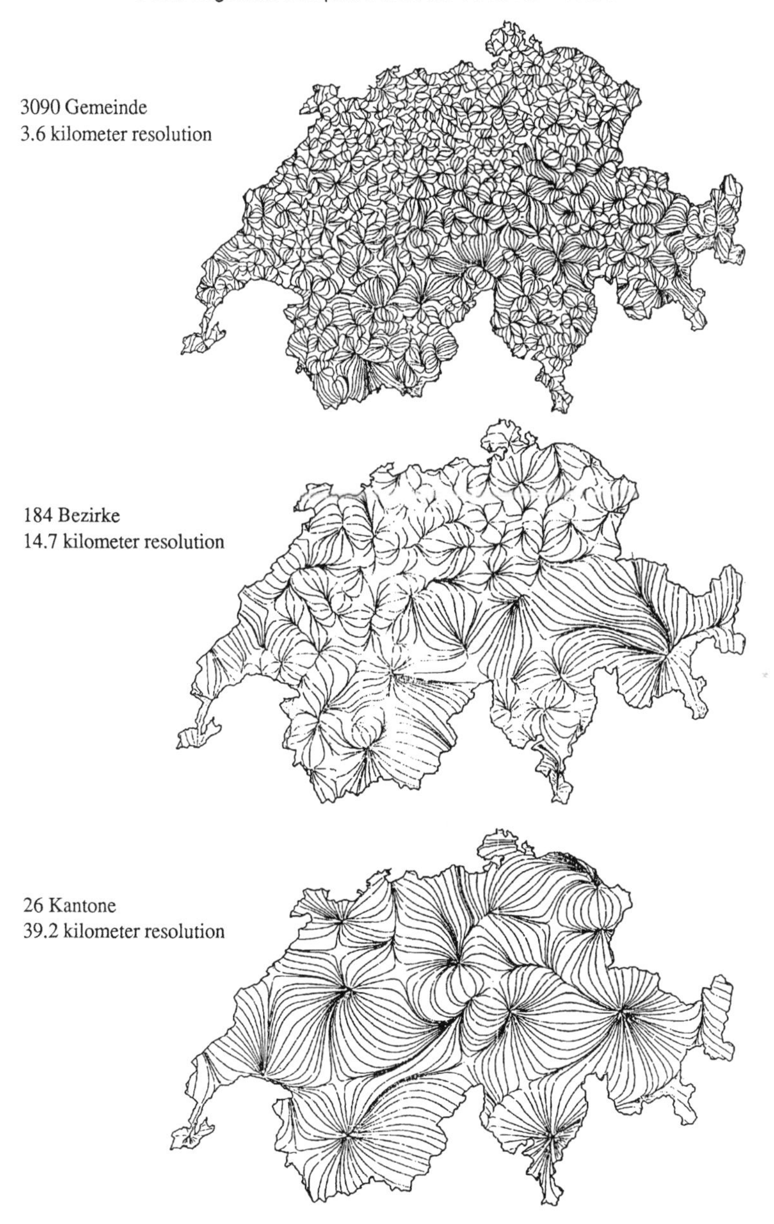

Figure 5 Migration patterns.

Robinson, A., 1950, Ecological correlation and the behaviour of individuals. *Am. Soc. Rev.* **15**, 351-357.
Rylander, G., 1986, Areal data reaggregation, Master's Thesis, Michigan State University, East Lansing, Department of Geography
Silverman, B., 1986, *Density Estimation for Statistics and Data Analysis*. (New York: Chapman & Hall)
Tobler, W., 1969, Geographical filters and their inverses. *Geographical Analysis* **1**, 234-253.
____, 1979, Smooth pycnophylactic interpolation for geographical regions, *Journal, American Statistical Assn*. **74**, 121-127.
Wilson, A., 1967, A statistical theory of spatial distribution models. *Transportation Research* **1**, 253-269.

Section IV

Of all the sections in the book, this third section probably best emphasizes the diversity of approaches and problems which fall under the heading of accuracy in spatial data. If they have anything in common, it is a desire to look carefully at individual, basic issues, rather than to take a broad or applied view over the whole.

Although it is the only paper on this topic in the book, Dutton's paper represents what many have seen to be perhaps the most comprehensive way of resolving the accuracy problem, which is to reduce the spatial resolution of the database and the processes which operate on it to a level which is consistent with the data's accuracy. In a sense this is a return to the days of manual cartography, as it was argued in the preface that such a balance between precision and accuracy was common in traditional methods of map analysis, but has become unfortunately uncommon in the newer digital world. The scheme which Dutton explores for the globe has its planar parallels in quadtree addressing and in raster representations where cell size is determined by data accuracy. In a paper presented at the meeting but not included in this volume, Alan Saalfeld of the Bureau of the Census, US department of Commerce, argued the need for a system for a system of finite-resolution computation in GIS which would complment this approach to database creation.

Dutton's proposal would five every object on the earth's surface an address which reflects its size, or the accuracy of its location, expressed in a single string of base-4 digits. Whether such a suy can ever replace latitude and longitude as a method of spatial addressing on the globe remains to be seen, but its advantages in spatial databases are much more immediate.

Slater's paper is concerned with one of a number of possible methods for simulating spatial data. The Ising model was developed to represent the magnetization of solids, but its results have useful parallels with a range of two-dimensional phenomena. A simple method of simulating of simulating spatial data would allow the performance of GISs and spatial databases to be benchmarked using data with known and controlled properties. With a stochastic element, it might be used as a method of simulating the propagation of error through GIS processes, and of connecting uncertainty in input data with uncertainty in output measures and statistics.

The third paper in the section, by Laskowski, presents a new perspective on an old problem, that of measuring the distortions introduced by use of map projections. These are perhaps the most fundamental geographical errors of all, and Tissot's indicatrix provides a simple and readily understood means of displaying distortion as a function of location. Laskowski provides a novel way of computing distortions, and represented by the dimensions of the indicatrix.

Chapter 12

Modeling locational uncertainty via hierarchical tesselation

Geoffrey Dutton

Abstract

Information about error and uncertainty sometimes accompanies digital cartographic data; when present, it may characterize an entire dataset, a feature class or a spatial object, but rarely is accuracy documented for individual spatial coordinates. In spatial analyses such as map generalization, object coalescing and thematic overlay, where amalgamation of many primitive features may be required, it is useful to know the precision of locational data being processed. It is suggested that maintaining coordinate data in a hierarchical tesselated framework can facilitate documenting the certainty of coordinates and dealing with its consequences. Points thus encoded can identify the precision with which they were measured, and can be retrieved at lower degrees of precision, as appropriate. This scale sensitivity is an inherent aspect of quadtree and pyramid data structures, and one which the literature on GIS data quality has yet to address in detail. A specific hierarchical tesselation of the sphere into triangular facets is proposed as a basis for indexing planetary data; although composed of triangular facets, the tesselation is a quadtree hierarchy. Its geometry is such that its facets are planar, subdivide a sphere naturally and are efficient to address, in comparison with rectangular tesselations. Some general properties of hierarchical tesselations are explored, focussing on one particular structure and its potential as a basis for encoding spatial data. Methods for generating and manipulating hierarchical planetary geocodes are described.

Locational data quality

Whatever else they may convey, all spatial data possess coordinate locations. Each geographic entity recorded in a GIS must have an identifiable spatial signature among its properties. As a GIS must be relied upon to integrate, analyze and display independent collections of spatial data, it should possess means for coping with variability in the quality of coordinate and other information in the features, layers and themes it records, according to their nature, source and purpose. This is usually not possible, hence rarely done.

Without the ability to generate spatial inferences, a GIS is little more than an inventory of digitized geographic facts. In order to draw quantitative conclusions about objects in space and time, one must know or be able to estimate the reliability and certainty of the tools and information employed. All too few GIS tools in common use attempt to utilize the scant quality data that their databases may provide. Much has been written about building data quality information into GIS, but few actual systems deliberately do so, and none seem to take its implications seriously. While this state of affairs is not new, it is even more a cause for alarm today than it was five years ago: "We experience difficulty in articulating the quality of information represented in a database principally because we don't understand how to analyze data based on information about its qualities" (Dutton, 1984b).

One important source of this difficulty is that, however well-registered features may be, without some form of geodetic reference surface to which each can be related, they are are doomed to hover like fuzzy spaceships in search of landing pads. A reasonable rule of thumb to apply might be "the smaller the spaceship, the greater the importance of knowing where it may land". A few hundred-meter errors in describing the shape of a state may be of much less significance and consequence than a few hundred-millimeter mistakes in specifying the envelope of a downtown office building. Where is the World Trade Center? Where is New York State? What's different about these facts? We should insist that a GIS be capable of presenting such questions and of helping to interpret our answers. Having done this, the system might go on to formulate a new set of questions, then attempt to answer some of them. As impractical as this may seem, why should we settle for less?

Accuracy, precision, error and uncertainty

"Accuracy" and "precision" are often used interchangably; the *American Heritage Dictionary* in fact defines both as terms for "correctness" and "not deviating from a standard." I shall try to be a bit more rigorous: in the current context, *accuracy* will mean "conforming to external truth", and *precision* will signify "the degree of detail recorded for measurements." Spatial *error* may pertain to either accuracy or precision. Highly precise coordinates may err in accuracy if they are positioned without good ground control. Imprecise coordinates can still be relatively accurate if they are consistent with a model of ground truth. The more precise coordinates are, the more confidently we can talk about errors in their description. The more accurate they are, the fewer errors should exist for us to talk about. Spatial *uncertainty* occurs when no model of ground truth exists or can be agreed upon in relation to a particular set of measurements. The location of the World Trade Center may be subject to some uncertainty, but not so much that one cannot speak meaningfully of the error in measuring it; the location of New York State is so much of a fuzzy concept that it is quite meaningless to speak of the error associated with measuring it (Centrography notwithstanding; see Neft, 1966).

Mensuration as modeling

Most GIS enforce a distinction between recording locations and modeling features. Locations are denoted by coordinates, which in turn are associated with features (objects in the real world described via some model or abstraction mechanism). The coordinates pin the models to the Earth at one or more locations, but do not specify how features are encoded. Should coordinates change (due to resurvey, editing or recalculation, for example), this normally has no impact on the feature(s) associated with them beyond causing changes in size and shape. Yet, when coordinates change, something important may have happened.

We have been so paraDIMEed into fanatically enforcing a dichotomy between the topology and coordinates of cell complexes that we have come to assume that topology alone supplies structure, and there is no structure worth knowing about in a feature's coordinates. This ignores much of the "deep structure" (Moellering, 1982) that geographic data - even coordinates - may be viewed as having. We believe that the coordinates of features indeed have a "depth" component, that this can be modelled via hierarchical tesselation, and that this approach can better characterize uncertainty about cartographic features. The work reported here stems from a scheme (appropriately known as DEPTH) for storing digital elevation data using polynomials stored as quadtrees (Dutton, 1983); this was subsequently recast into a global hierarchical triangular tesselation for terrain modeling called GEM (Dutton, 1984a); the work being reported here is focussed on addressing, feature coding and error handling in such models, rather than on elevation and other attribute data management methods.

Hierarchical Tesselations

Hierarchical tesselations are recursive subdivisions of space-filling cells on a model surface, or manifold. The most familiar group of hierarchical tesselations is the family of data structures known as *quadtrees*, square lattices of 2-cells that double their resolution as their number multiplies by four, down to some limit of resolution [see Samet (1984) for a detailed review of the quadtree literature; quadtrees are discussed in relation to GIS by Samet (1986) with a rejoinder by Waugh (1986)]. Other geometries and branching schemes more suitable for modeling global distributions have been proposed or developed (Dutton, 1984a; van Roessel, 1988; Mason and Townshend, 1988), but few have gained acceptance in the GIS realm. In reviewing and comparing data models for global GIS applications, Peuquet states:

> ... a regular, hierarchical spherical tesselation would have many advantages as a global data model. First of all, such a model would retain all of the desirable properties of a planar tesselation including implicit spatial relationships; geographic location is implied by location in the database. Multiple scales and a regular structure are also amenable to rapid search. (Peuquet, 1988, pp. 74-75).

Quadtrees were developed to facilitate image processing operations, and for the most part have continued to be oriented toward raster technology. This, as Waugh (1986) notes, can be a drawback for GIS applications, which tend to use vector data. Furthermore, while map sheets can be regarded as images and handled as rasters, it is a mistake to think of a GIS as a catalog of maps; while a GIS may manage map data, it can go much further than maps in representing properties of spatial phenomena. A GIS, after all, should model the world, not just maps. As the Earth is neither flat nor a cube, any scheme that is based on subdividing rectangular map images of a planet will fail to provide consistent global coverage (consider how the UTM grid system contorts itself to cover the globe). Cubic quadtrees have been developed to store global data (Tobler and Chen, 1986; Mark and Lauzon, 1986). These have tended to stress storage and retrieval of map and image data (including segmentation and other data conversion tasks), rather than the modeling of planetary phenomena.

Quadtrees represent a technology in search of applications; planetary modeling is a set of applications in need of technologies. GIS offers an environment where they may connect, provided some basic outstanding issues are addressed. In a brief but well-informed overview of global database issues, Goodchild (1988) describes the need for research on planetary spatial analysis:

> ... there is as yet no (spherical) extension of the Douglas-Peucker line generalization algorithm, and only limited literature on the generation of Thiessen polygons and polygon skeletons. There is no spherical version of point pattern analysis, and no literature on spatially autocorrelated processes. It is clear that much research needs to be done in developing a complete set of spatial analytic techniques for the spherical case.

We feel that geodesic hierarchical tesselation may provide keys to unlock some of these problems, by enabling higher-order data modeling capabilities that vector, raster, quadtree and hybrid data structures can draw upon to handle planetary data in a consistent fashion, as the remainder of this paper will attempt to demonstrate.

Polyhedral tesselations

Rather than starting with a map -- or even a map series -- and developing data structures (either raster or vector) to encode it, one can base one's efforts on the requirement to describe an entire planet, then subdivide the model into tiles of useful size. This will at least assure that (unlike UTM sheets) tiles will fit together regularly and consistently. The most obvious choices for a basis for tesselation are the five platonic solids; other regular polyhedra (such as a cubeoctahedron, rhombic dodecahedron or rhombic tricontahedron) can be used (and have been for map projections), although not all are capable of self-similar, recursive tesselation (the shape of facets may change when subdivided).

Given a basis shape that can be indefinitely subdivided, it is necessary to select one of several alternative tesselation strategies. Triangular facets, for example, may be subdivided into 2, 3, 4 or 9 triangular tiles. In some of these tesselations the shapes of tiles may vary, in others their sizes may vary, or both size and shape may vary. This is the same problem that designers of geodesic domes face; they tend toward solutions in which struts and connectors are as uniform as possible, as this expedites the manufacture and assembly of geodesic structures (Popko, 1968).[1]

A geodesic planetary model

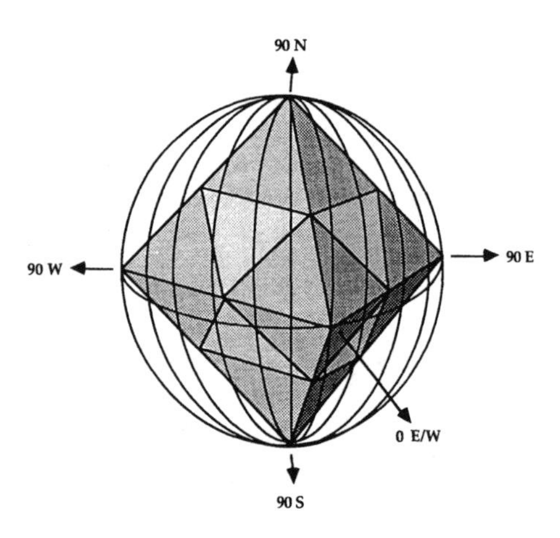

Figure 1 Modeling locational uncertainty via hierarchical tesselation.

We have been investigating a modeling system for planets based on a triangular tesselation rooted in an octahedron, in which each facet divides into four similar ones; this yields successive levels of detail having 8, 32,128, 512, 2048, ... facets overall, or 1, 4,

1 When a sphere is geodesically tesselated, edge members of different lengths have different chord factors; when a geodesic strut is subdivided into two halves, the sum of the lengths of the two will be greater than the length of their parent strut. This does not happen to the edges of triangles subdivided in 2-space, but must be accounted for when computing edges and areas of polyhedral facets.

16, 64, 256, ... facets per octant of the basis shape. Figure 1 illustrates the basic form and orientation of the model, and Figure 2 illustrates the development of this hierarchy, shown in orthographic views. Table 1 lists statistics for this structure and the dimensions its elements would assume if earth-sized.

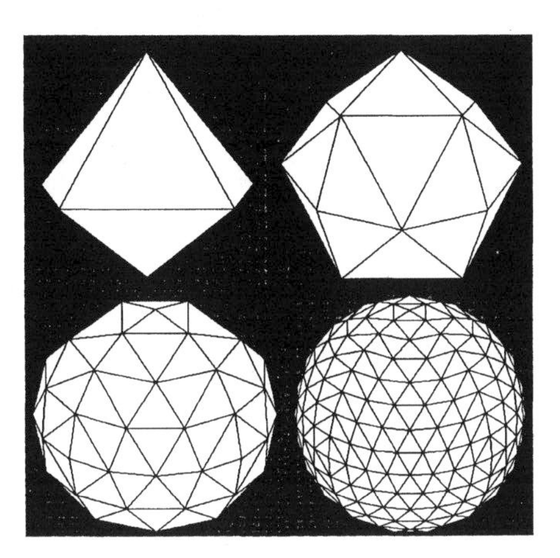

Figure 2 Development of quarternery triangular mesh to level 3 on a basis octahedron.

In Table 1, column 1 indicates the hierarchical level of breakdown, column 2 (= 4^{level}), and column 3 (= 2^{level}), respectively indicate the number of triangular facets and edges that partition an octant at each level. Columns 4 and 5 itemize the linear resolution and unit area each level has on a sphere 6,371 km in radius: approximate distances and areas are given for the spherical wedges defined by the polyhedral facets. Columns 6 and 7 relate to the storage of facet identifiers, to be discussed later.

Level	Facet	Divisions	Resolution	Facet	Area	Code bits	Hex Chars
1	4	2	1444km	15,924,500	kmsq	2	0.5
2	16	4	722km	3,981,125	kmsq	4	1.0
3	64	8	361 km	995,281	kmsq	6	1.5
4	256	16	180 km	248,820	kmsq	8	2.0
5	1,024	32	90km	62,205	kmsq	10	2.5
6	4,096	64	45 km	15,551	kmsq	12	3.0
7	16,384	128	23km	3,888	kmsq	14	3.5
8	65,536	256	11 km	972	kmsq	16	4.0
9	262,144	512	6 km	243	kmsq	18	4.5
10	1,048,576	1,024	3km	61	kmsq	20	5.0
11	4,194,304	2,048	2 km	15	kmsq	22	5.5
12	16,777,216	4,096	705 m	3,796,696	msq	24	6.0
13	67,108,864	8,192	352 m	949,174	msq	26	6.5
14	268,435,456	16,384	176 m	237,294	msq	28	7.0
15	1,073,741,824	32,768	88 m	59,323	msq	30	7.5
16	4,294,967,296	65,536	44 m	14,831	msq	32	8.0
17	17,179,869,184	131,072	22 m	3,708	msq	34	8.5
18	68,719,476,736	262,144	11 m	927	msq	36	9.0
19	274,877,906,944	524,288	6 m	232	msq	38	9.5
20	1,099,511,627,776	1,048,576	3 m	58	msq	40	10.0
21	4,398,046,511,104	2,097,152	1 m	14	msq	42	10.5
22	17,592,186,044,416	4,194,304	69 cm	4	msq	44	11.0
23	70,368,744,177,664	8,388,608	34 cm	9,052	cmsq	46	11.5
24	281,474,976,710,656	16,777,216	17 cm	2,263	cmsq	48	12.0

Table 1 Planetary octahedral triangular quadtree statistics for 1 to 24 hierarchical levels (per octant)

No attempt will be made to justify this particular tesselation as an optimal one; the scheme does appear, however, to strike a balance between geometric uniformity and computational simplicity as a model for the surfaces of spheroids. As its vertices are at right angles, an octahedron readily aligns itself to cardinal points in a geographic world grid; subsequently-introduced vertices are easily computed, as they bifurcate existing edges (as shown in Figure 1). The breakdown generates eight quadtrees of facets; this data structure may be handled as if it were a set of rectangular region quadtrees. However, as their elements are triangular rather than square, many of the geometric algorithms devised for rectangular quadtrees will not work on such datasets without modification.

We call this spatial data model a *Quaternary Triangular Mesh* (QTM). The remainder of this section will explore some of QTM's geometric, infometric and computational properties. The sections to follow will focus on the use of QTM in modeling spatial entities, and how this might address problems of precision, accuracy, error and uncertainty in spatial databases. Throughout, the discussion's context will remain fixed on exploring QTM as a geodesic, hierarchical framework for managing and manipulating planetary data. The tesselation geometry employed for QTM is similar to that proposed by Gomez Sotomayor (1978) for quadtree representation of digital terrain models adaptively split into triangular facets. We use a different numbering scheme (described below), and embed QTM in a spherical manifold rather than a planar one (although for computation and display of QTM data, a projection is normally employed); see section on zenithal orthotriangular projection below.

QTM as geocoding

In a QTM tesselation, any location on a planet has a hierarchical address, or *geocode*, which it shares with all other locations lying within the same facet. As depth in the tree increases, facets grow smaller, geocodes grow longer and tend to become more unique, being shared by fewer entities. A QTM address identifies a particular triangular facet at a specific level of detail; that triangle's vertices are fixed on the QTM grid, covering a definite area fixed on the planet. Any triangle can be subdivided (by connecting its edge midpoints) into four similar ones, numbered 0 through 3, as illustrated by Figure 3; we refer to the four children of each facet as its *tiles*. Each tile thus generated can be identified by a 2-bit binary number, so that *2L* bits (or *L/4* bytes) are needed to specify a QTM address at *L* levels of detail.

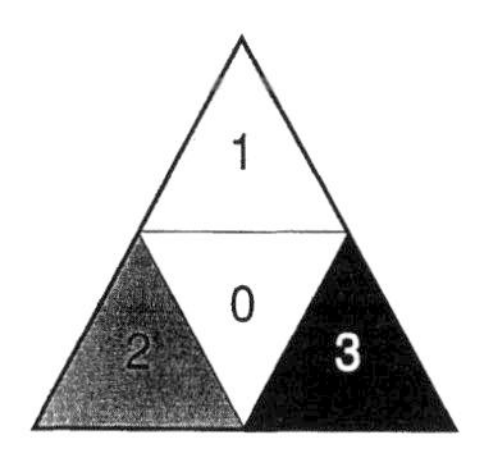

Figure 3 QTM facet breakdown numbering.

QTM addresses therefore consist of variable-length strings of 2-bit numbers, for example *0311021223013032*. Such identifiers lend themselves to being represented by base 16 numbers, having *L/2* hexadecimal digits; the 16-level QTM address *0311021223013032* is, in hex notation, the (32-bit) number *3526B1CE*. To relate this to a more familiar context, QTM addresses at level 16 provide the same order of resolution as LANDSAT pixels. Refer to columns 6 and 7 of Table 1 for the sizes of binary and hex identifiers at various QTM levels of resolution.

QTM as geometry

To identify exactly where on earth QTM hex geocode *3526B1CE* (or any other) lies, one must know the specific method for assigning numbers to QTM facets that was employed to construct the geocode. While there are a number of ways to do this, few of them seem useful. The QTM tesselation always generates a triangle for each vertex of a facet plus one triangle at its center; we always number corner triangles 1, 2 or 3, and designate the central triangle as zero. This scheme has a convenient property: any number of zeros may be appended to a QTM address without affecting its geographic position. While trailing zeros do not modify the location of a measurement, they do signify the precision to which it may be evaluated. We shall return to consider applications of this property later on.

Having fixed the central triangle as facet 0, we must then assign each of the remaining ones as 1, 2 or 3. This presents a fair number of distinct possibilities, giving us the opportunity to influence the way in which geocodes are spatially arrayed. Noting that triangles point either upwards or downwards, we identify the orientation of facets as either *upright* or *inverted*: An *upright facet* has a horizontal base with an apex above it, while an *inverted facet* has a horizontal base with an apex below it. All four octants of the northern hemisphere are upright; all four of the southern hemisphere are inverted. Tesselating an octant generates three outer tiles (numbers 1,2,3) sharing its orientation, and an inner one (tile 0) having opposite orientation. Let us designate the apex of each triangle (regardless of N/S orientation) as *vertex 1*, which locates *tile 1*. Vertices 2 and 3 thus define the endpoints of the octant's equatorial base; we can assign them arbitrarily but consistently, thus defining where *tile 2* and *tile 3* are located within each octant, as Figures 3, 4 and 8 show.

When we do this, we find that the 8 tiles numbered 1 cluster about the north and south poles, and that tiles numbered 2 and 3 lie on the equator. We arbitrarily fix vertex 2 (hence four of the tiles numbered 2) at the equator (0^0 N/S) and the Greenwich Meridian (0^0 E/W), and another (with its four surrounding tiles) at the antipode (180^0 E/W). Finally, the points where the equator intersects longitudes 90^0 E and 90^0 W colocate four octant vertices (and tiles) numbered 3, fully defining the numbering of vertices and facets for the first QTM level. Figure 4 maps this ordering of vertices and facets for a sphere and an octahedron.

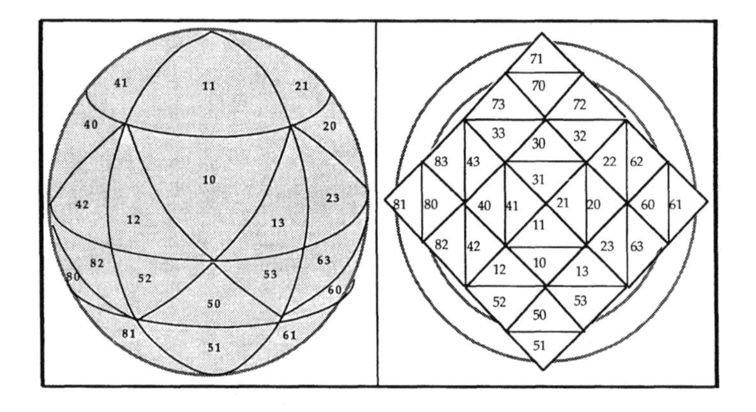

Figure 4a First order QTM tesselation of a sphere, showing facet numbering and 4b First-order QTM tesselation of an octohedron, unfolded from S pole.

QTM as addressing

A depth-first ordering of QTM geocodes traces a specific pattern in the process of enumerating an octant's facets. This pattern represents a memory map of the geocodes, as it delineates the sequence in which facets are ordered in computer storage. The compactness of this arrangement helps one map point coordinates to memory addresses which are close to those of nearby points. Exploiting this property can simplify the problem of spatial search from a 2-dimensional procedure to a 1-dimensional one. The pattern generated by visiting successive QTM addresses is the set of self-similar, self-intersecting curves shown in Figure 5.

Figure 5 Second-order QTM code sequencing (memory map order). ZOT projection.

QTM location encoding is clearly a form of spatial indexing; not only are geocodes systematically ordered into quadtrees, they have the property that numerically similar QTM geocodes tend to lie in close spatial proximity to one another. Furthermore, as a consequence of the numbering pattern described in section 3.2, facets at the same level having QTM codes terminated by the digits 1, 2, and 3 form hexagonal groups of six triangles regularly arrayed across the planet; those ending in 0 are isolated triangles filling gaps in the hexagonal pattern. Figure 6 is an equal-area mapping of this pattern for third-order facets for northern hemisphere octants 1 and 3.

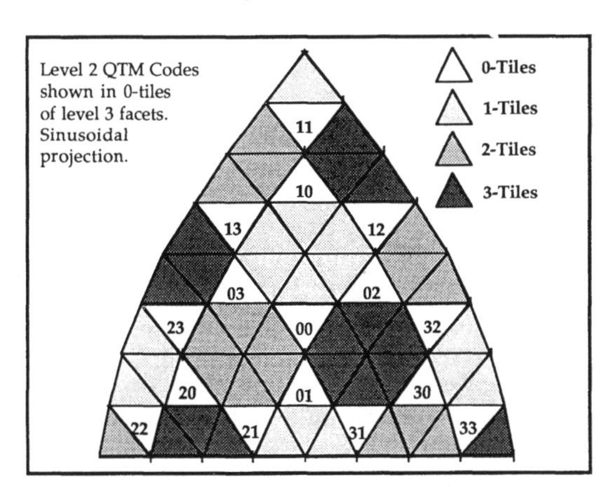

Figure 6 Pattern of least significant digits of QTM codes; note clustering of nonzero codes in hexagonal groups.

This numbering pattern has properties worth noting. The centerpoint of each hexagonal cluster of tiles is a vertex in the QTM grid shared by each tile in that group. This nodal point may be thought of as a locus of attraction, or *attractor*, to which nearby observations gravitate. Only tiles numbered 1,2 or 3 are attracted to such nodes; 0-tiles, which provide no new locational information, are not attracted to any node. That is, points in 0-tiles are attracted to the tile's centerpoint, while points in other tiles are attracted to the closest vertex of their facet. The three interlocking hexagonal grids that result from this arrangement are illustrated in Figure 7; they cover 75 percent of the planet, with 0-tiles occupying the remaining triangular gores.

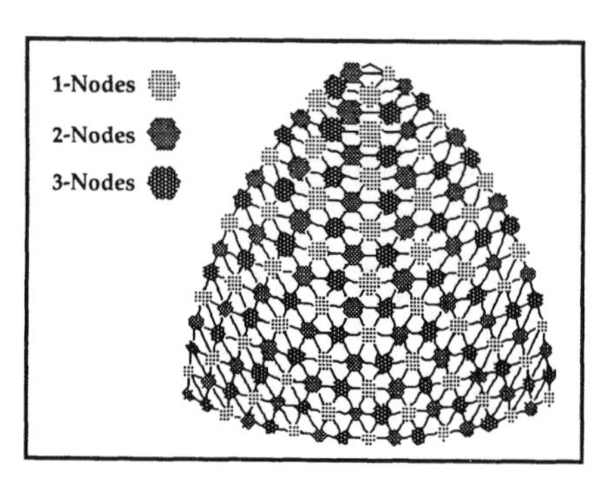

Figure 7 Arrangements of QTM attractors (nodes) across one octant at level 5. Orthographic projection.

Aliasing, attraction and averaging

As an alternative to mapping locations to QTM facets, one may consider QTM grid nodes as their loci. That is, all locations falling within 1-, 2- or 3-tiles can *alias* to their common QTM node. By aliasing tiles to nodes, a higher degree of spatial generalization results. It differs from allocating coordinates into facets in that it averages, rather than partitions, observations into sets. Node aliasing provides a key to dealing with a particularly vexing consequence of many quadtree schemes: the unrelatedness of the attribute encodings of adjacent high-order tiles that share an edge also separating lower-order facets. Each branch of a quadtree partitions each facet (and any values that may be recorded for them) into four subtrees. Whether values are built up from area estimates or obtained from progressive point sampling, discontinuities can occur between sub-branches of the tree simply due to the placement and orientation of the sampling grid. While this can be mitigated by smoothing the resultant grid of values (as demonstrated in Dutton, 1983), this solution is inelegant and should not be necessary.

We can better understand node averaging by conducting the following thought experiment: sample a continuous surface, such as topographic relief, assigning QTM addresses to a set of 3D point observations, aliasing all source locations which happen to fall into the same QTM facet to the same elevation, as there is only one value stored per facet.[2] Let us assign the elevations of the 0-tiles to their centroids, and assign averages of the elevations of proximal 1-, 2- and 3-tiles to their common QTM nodes, as in Figures 6 and 7. We thus obtain a mesh of triangles, the vertices of which have fully-defined latitudes, longitudes and elevations. The edges of the mesh connect QTM nodes to the centroids of their facets. A surface defined by these facets will, in general, be smoother (exhibit less aliasing) than one defined by interconnecting the centers of adjacent atomic tiles. Furthermore, because node elevations are spatially symmetric averages, a surface thus defined is relatively stable under translation and rotation (unlike an unaveraged QTM

2 This can be done by stratifying elevations and recording the changes between strata as attributes of facets, as described in (Dutton, 1983) and (Dutton, 1984a). As it is difficult to represent the input data without severe aliasing, the surface as encoded is excessively quantized.

coverage, or any quadtree for that matter); its contours would not appreciably change were the orientation of the QTM grid to be incrementally shifted.[3]

Evaluating spatial data at QTM grid nodes can simplify spatial analysis tasks. For example, the need to identify and remove slivers following spatial overlay might be lessened by filtering the coordinates of the features of input coverages via QTM tesselation. As all coordinates in the neighborhood of a node are mapped to its location, slight variations in otherwise identical vector strings will tend either to vanish or to alias into structured caricatures of themselves. In a QTM framework, all addresses represent specific triangular areas; each facet, while representing a unit of area, can be mapped to a particular *attractor* (a node in the QTM grid). The ordering of facet identifiers (described in section 3.2) is such that, once an attractor manifests itself, its basis number (1, 2 or 3) will persist in place at all higher frequencies. Space in the vicinity of an attractor is affected as by gravity; the larger an attractor (the shorter its path to its root node), the stronger is its influence. Higher order attractors have smaller ranges of influence than lower order ones, and consequently exhibit less locational uncertainty. Figure 8 illustrates the development of nodes; when an edge is bifurcated, the new node is given a basis number equal to 6 - (a + b), where a and b are the basis numbers of the parent nodes.

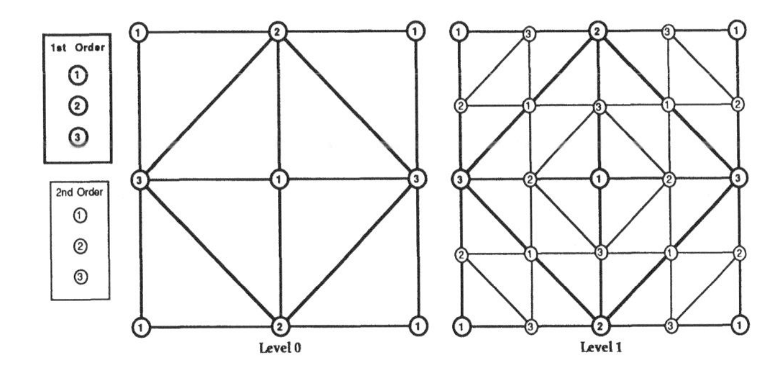

Figure 8 Octa and first level QTM attractor (node) numerology; child nodes are numbered 6 - (a+b), where a and b are the nodes of the parent edge. ZOT projection.

Computational Considerations

There has been interest in using tesselations as ways to partition and index spatial data. The majority of this work seems to be oriented toward decomposing vector and raster databases into tiles of fixed or varying size and content (Weber, 1978; Vanzella and Caby, 1988). Such approaches lead to various hybrid data structures, in which an overview is provided by a tesselated component and details furnished by the vector and raster primitives. Conceptually, this seems little different from storing data as electronic map sheets of equal or differing sizes. We feel that geodesic tesselations have considerably greater modeling power than has been exploited to date.

QTM addresses can replace coordinates in a georeferenced database. When their length is allowed to vary, the accuracy of the coordinates they encode can be conveyed by

3 While I am confident of this, a formal proof (that node averaging results in a more representative sampling of spatial attributes) remains to be constructed.

their precision. Therefore, the number of digits in a QTM geocode may be used as a parameter in processing the coordinates and/or attributes it represents. This permits the precision of coordinate points to be independently specified, and in turn allows analytic procedures to make more informed judgements in modeling the behavior of spatial entities. Describing features at varying precision may or may not result in greater efficiency: the QTM model does *not* specify how spatial entities are defined, how storage for them is structured or how to manipulate QTM elements. Although we understand how to perform certain operations on QTM geocodes, we know little about how to optimize data structures or processing based on QTM's tendency to cluster nearby locations in memory, or how to best take advantage of the facet-node duality that we have called attractors.

QTM in context

There is an increasing amount of literature and interest concerning the properties and computational geometry of hierarchical tesselations. The subject appears to connect many branches of knowledge and goes back many years, involving disciplines as diverse as crystallography, structural engineering, design science, computer science, solid geometry, lattice theory, fractal mathematics, dynamical systems and geography. One particularly relevant source of information concerning the properties of hierarchical tesselations is a group of research fellows and fellow travelers based in Reading, UK. A recent initiative of the British Natural Environment Research Council (NERC) aims to focus inquiry on geometric and computational properties of hierarchical tesselations (Mason and Townshend, 1988). Most of this work is less than five years old, and tends to view the subject matter in general, theoretical fashion.

The NERC papers are solidly in the tradition of fugitive spatial analysis literature that is GIS's birthright;[4] few of the most influential papers have appeared in journals. As befits workers in a field that knows no bounds, the NERC group has coined the adjective *tesseral* to characterize hierarchical tesselations. The term brings to mind the mathematical objects called *tesseracts*, which are in fact multi-dimensional tesselations; it is rooted in the Latin/Greek word *tessera* - the tiles used in making mosaics. QTM is a tesseral construction.

Amalgamated attractors incorporated

In one (published but obscure) tesseral paper (Holroyd, 1983), the subject of attractors is illuminated. Using a combinatorial and group theoretic perspective, Holroyd defines a number of tesseral properties: "Let T be any tiling. Then a tiling S each tile of which is a general tile of T is said to be an *amalgamation of T*." That is, in any tesselation, there will be some number of tilings that can be constructed of groups of its elements which themselves constitute tesselations. What we have termed *attractors* Holroyd defines in a more general way as *amalgamators*. We have yet to decide if QTM attractors qualify as what Holroyd calls *strict amalgamators*, which seem to have special properties.[5] Whether or not attractors involve strict amalgamation, they do have properties that facilitate data handling. As the QTM model involves eight independent

4 Lest we forget: The Michigan geographic community's discussion papers; *Harvard Papers in Theoretical Geography*; David Douglas's subroutine library; ODYSSEY (a fugitive GIS); the Moellering Commission's reports, and multitudes of other government research studies, reports and documents. Evidently, spatial analysis remains an immature discipline.

5 The tesseral literature is not recommended for casual browsing. Holroyd's paper invokes terms such as *isohedral type*, *automorphism*, *dominator* and *hierarchical isohedral sequence*, which may be quite foreign to non-mathematicians (the author included).

quadtrees, there are potential difficulties in handling entities which cross octant edges (as they will be rooted in more than one quadtree). Attractors along octant boundaries should behave like any other attractors, and can be called upon to assist in merging data held in adjacent quadtrees.

In another published article from the NERC group, tesselations are discussed both abstractly and with reference to computational issues. Bell *et al.* (1983) itemize properties of isohedral tilings (81 types of which exist), and categorize the 11 regular ones as *Laves nets*. A Laves net is described by the number of edges in an atomic tile in conjunction with the number of edges which meet at its vertices. Rectangular quadtree tesselations have a Laves index of [4^4]; the index for a QTM triangular quadtree is [6^3]. This is interpreted as three edges per tile, with each vertex linked to six neighboring ones. The paper defines other properties of atomic tilings, one of which is called *democracy*. In a democratic tiling any numbering system may be imposed, and each may require different algorithms for addressing and other manipulations. The QTM tesselation is *undemocratic*, as the address of the central tile is always assigned as zero (the remaining three addresses are democratic, however). In general, democracy is not a desirable property, as "... it is better to avoid the complicated addressing structures which democratic tiles often produce" (Bell *et al.*, 1983). In tilings like QTM which have a central cell with others symmetrically arrayed around it, the arithmetic involved in addressing neighboring tiles is more tractable than when tile zero is off-center or its location is permuted.

One of the more interesting aspects of the tesseral perspective is the possibility of developing special tesseral arithmetics for manipulating hierarchical elements. This was demonstrated for the generalized balanced ternary (GBT) system, a k=7 hexagonal hierarchy developed at Martin Marietta in the 1970's as a spatial indexing mechanism (Lucas, 1979). GBT's undemocratic numbering system allows direct computation of properties such as distances and angles between locations without manipulating coordinates (van Roessel, 1988). Other tesselations have related arithmetics, some of which have been explored in the Tesseral Workshops (Diaz and Bell, 1986). Such an arithmetic could be developed to manipulate the QTM [6^3] tesselation should none already exist.

Polyhedral operations

One common objection to polyhedral data models for GIS is that computations on the sphere are quite cumbersome (in the absence of tesseral arithmetic operators), and that for many applications the spherical/geographic coordinates stored by such models must be converted to and from cartesian coordinates for input and output.

Because planar geometrics are generally much more straightforward than spherical ones, it is almost always easier to compute quantities such as distances, azimuths and point-in-polygon relations on the plane rather than on the sphere. The former may involve square roots and occasional trignometric functions, but rarely to the degree involved in geographic coordinates, where spherical trigonometry must be used unless rather small distances are involved or approximations will suffice. Polyhedral geometry, being faceted, is locally planar but globally spherical. The maximum practical extent of localities varies, both in cartesian and faceted cases, according to the projection employed (for cartesian coordinates) or the type and level of breakdown (for polyhedral hierarchies).

Perhaps the most basic polyhedral operation is the derivation of facet addresses (geocodes) from geographic coordinates (or its inverse). This involves recursive identification of triangular cells occupied by a geographic point, appending each identifier to the location code already derived. This process has been named *trilocation*, and is described in Dutton (1984a). The simplest trilocation algorithm derived to date for [6^3] tiles determines cell identifiers by comparing the squared distance from the test point to the centroid of its facet's 0-tile and each of the three outer ones until the closest one is found (this usually takes 2 or 3 squared distance computations and comparisons for each level). If performed using geographic coordinates, great circle distances are needed, but if performed in the planar domain cartesian distances will suffice (in neither case need square roots be extracted, as we are interested in ordering distances, not in their absolute magnitudes). An even simpler test requiring no multiplication at all is possible under certain circumstances.

Zenithal orthotriangular projection

An extremely simple projection of the QTM tesselation has been developed (Dutton, 1988) to facilitate spatial operations and display. As shown in Figures 4 and 7, the Zenithal Orthotriangular (ZOT) is a doubly periodic map projection having the following properties: it is zenithal (azimuthal) because meridians remain straight and of constant radial spacing; longitudes may be measured directly with a protractor. There is, however, more than one azimuthal origin, as longitudes are only true within a hemisphere. As the South pole occupies four locations, the origin for meridians in the southern hemisphere will be one of the four corners of the projection, as appropriate. ZOT is also an *equidistant* projection, as lines of latitude are both parallel and uniformly spaced. It is termed *orthotriangular* because it maps triangular regions on the sphere to right triangles on the plane.

ZOT meridians and parallels map to straight lines which flex at the equator, due to the piecewise continuous (polyhedral) nature of its basis. In most polar azimuthal projections, parallels map to circles or ellipses. In this one, they map to diamonds (squares). This derives from the fact that distances are computed using a "city block" metric, and that the object being projected is an octahedron rather than a sphere. This orthogonality makes the projection trivial to compute, as it permits all geographic coordinates to be mapped to the plane using only linear arithmetic, without recourse to trigonometric formulae or square roots.

The ZOT projection can be used to convert geographic coordinates to QTM addresses (and *vice versa*) in the planar domain, where trilocation is most efficiently performed. In doing so, it is even possible to use integer coordinates, up to a certain level of detail. When 32-bit integers are used, trilocation can only be accurate to roughly 15 quadtree levels (attempting more would cause either integer overflows or result in aliasing cells beyond the 15th level). Projecting coordinates from longitude and latitude into ZOT space only involves solving two linear equations per point. In short, the QTM model evaluates relationships among spherical distances that are preserved in the ZOT plane, where they can be efficiently computed. This synergy offers much to spatial analysis.

Conclusions

Effective spatial modeling in a GIS environment seems to require detailed information about data quality, not just statistical error summaries. As Chrisman (1983) among others have pointed out, numerical representations of map data - particularly coordinates - can convey the illusion of accuracy simply because numerical digital data tend to be represented at uniform, relatively high precision. There is no GIS in general use which alters the precision of coordinate data (whether real or integer) to reflect its inherent accuracy or precision. As a result, information that could be used to guide spatial analytic and cartographic operations tends to be lacking, complicating those procedures and engendering inconsistent and *ad hoc* solutions. Solving this problem,

Chrisman stresses, is critical and calls for the development of new models of spatial phenomena:

> Space, time and attributes all interact. Quality information forms an additional dimension or glue to tie those components together. Innovative data structures and algorithms are needed to extend our current tools. No geographic information system will be able to handle the demands of long-term routine maintenance without procedures to handle quality information which are currently unavailable.

A recurring problem, and one that we create for ourselves, involves the very idea of coordinates; it is generally assumed that coordinates exist in nature, when in fact they are rather artificial notations for spatial phenomena. Features in a GIS don't actually *have* coordinates, coordinates are in fact *ascribed* to them. Too much of the work in spatial error handling has been devoted to tools that deal with spatial coordinates rather than with spatial entities; too little consideration has been given to exploring alternative models of spatial organization. A polyhedral, tesseral perspective might shed new light on how spatial phenomena can be described, by providing a unified framework for representation, an inherent sensitivity to scale and appropriate mechanisms for dealing with spatial error and uncertainty. It offers the GIS community a rare opportunity to create tools that effectively address some of the multitude of problems, both local and global, now facing us and our planet.

References

Bell, S. M., Diaz, B. M., Holroyd, F., and Jackson, M. J., 1983, Spatially referenced methods of processing vector and raster data. *Image and Vision Computing* **1**, no. 4 (Nov), 211-220.

Chrisman, N. R., 1983, The role of quality information in the long-term functioning of a geographic information system. *Canadian National Committee for the 6th International Symposium on Automated Cartography, Proc. Auto-Carto Six,* Ottawa, pp 303-312.

Diaz, B. M. and Bell, S. B. M., 1986, *Proc. of the Tesseral Workshops*, 13-14 Aug 1984 (Swindon) and 22-23 Sept. 1986 (Reading). (Swindon, Wilts UK: Natural Environment Research Council).

Dutton, G., 1983, *Efficient Encoding of Gridded Surfaces, Spatial Algorithms for Processing Land Data with a Microcomputer.* (Cambridge, MA: Lincoln Institute for Land Policy Monograph).

Dutton, G., 1984a, Geodesic modeling of planetary relief , *Cartographica* **21**, nos. 2 & 3. (Toronto: U. of Toronto Press), pp 188-207.

Dutton, G., 1984b, Truth and its consequences in digital cartography. *Proc. 44th Annual Mtg. of ASP-ACSM*, 11-16 March. Falls Church, VA: American Cong. on Surveying and Mapping, pp 273-283.

Dutton, G., 1988, Zenithal orthotriangular projection, in preparation.

Goodchild, M., 1988, The issue of accuracy in spatial databases. In *Building Databases for Global Science*, Mounsey, H., and Tomlinson, R.(Eds.), pp 31-48. (London: Taylor & Francis).

Holroyd, F., 1983, The geometry of tiling hierarchies. *Ars Combinatoria* **16B**, 211-244.

Lucas, D., 1979, A multiplication in N-space. *Proc. Amer. Math. Soc.* **74**, 1, 1-8.

Mark, D. M., and Lauzon, J. P., 1986, Approaches to quadtree-based geographic information systems at continental and global scales, *Proc. Auto-Carto 7,* Falls Church, VA: ASPRS/ACSM, pp 355-364.

Mason, D. C. and Townshend, J. R. G., 1988, Research related to geographical information systems at the Natural Environment Research Council's Unit for Thematic

Information Systems. *Int. J. of Geographical Information Systems* **2**, no. 2 (April-June), 121-142.
Moellering, H., 1982, The challenge of developing a set of national digital cartographic data standards for the United States. *National Committee for Digital Cartographic Data Standards,* Rpt. no. **1**, pp 1-15.
Neft, D.S., 1966, Statistical analysis for areal distributions , *Monograph Series* no. 2. Philadelphia: Regional Science Research Institute, 172 p.
Peuquet, D., 1988, Issues involved in selecting appropriate data models for global databases. In *Building Databases for Global Science*, Mounsey, H., and Tomlinson, R. (Eds.), pp 66-78, (London: Taylor & Francis).
Popko, E.F., 1968, *Geodesics*. (Detroit: University of Detroit Press).
Samet, H., 1984, The quadtree and related hierarchical data structures, *ACM Computing Surveys* **16**, no. 2 (June), 187-260.
Samet, H., 1986, Recent developments in quadtree-based geographic information systems , *Proc. 2nd International Symposium on Spatial Data Handling*, Seattle, WA, 5-10 July. Williamsville, NY: International Geographical Union, pp. 15-32.
Sotomayor, Gomez, D.L., 1978, Tesselation of triangles of variable precision as an economical representation of DTMs. *Proc. Digital Terrain Models Symposium,* May 9-11, St. Louis, MO. Falls Church, VA: Amer Soc of Photogrammetry, pp 506-515.
Tobler, W. and Zi-tan Chen, 1986, A quadtree for global information storage, *Geographical Analysis* **18**, 360-71.
van Roessel, J.W., 1988, Conversion of cartesian coordinates from and to generalized balanced ternary addresses, *Photogrammetric Engineering and Remote Sensing* **54**, no. 11 (November), pp 1565-1570.
Vanzella, L., and Caby, S., 1988, Hybrid data structures. *Proc. GIS/LIS '88* vol **1**. Falls Church, VA: ASPRS/ACSM, pp 360-372.
Waugh, T.C., 1986, A response to recent papers and articles on the use of quatrees for geographic information systems , *Proc. 2nd International Symposium on Spatial Data Handling*, Seattle, WA, 5-10 July, (Williamsville, NY: International Geographical Union), pp. 33-37.
Weber, W., 1978, Three types of map data structures, their Ands and Nots, and a possible Or, *Harvard Papers on Geographic Information Systems* **4.** (Reading, MA: Addison-Wesley)

Chapter 13

Minimum cross-entropy convex decompositions of pixel-indexed stochastic matrices: a geographic application of the Ising model

Paul B. Slater

Abstract

The problem of modeling class membership (marginal) probabilities over an n x m array of pixels and c classes, taking into account the possibility of spatial autocorrelation in the phenomenon under study, is addressed. A stochastic process that yields the marginal probabilities, as expected values, is constructed. The specific probability distribution that it assigns to the c^{nm} possible realizations of the process (representable by nm x c [0,1] extremal stochastic matrices) is the unique one, among all that achieve the target, that is the closest in minimum information distance (cross-entropy) to the Ising distribution over these realizations for a pre-assigned "temperature" (degree of spatial autocorrelation). An algorithm to find the desired distribution is described and implemented in an example with n = 4, m = 3 and c = 2. Several strategies for incorporating this convergent (but combinatorially explosive) procedure into the analysis of the large arrays encountered in practice, are outlined. The choice of a specific temperature, or more generally, a distribution over temperature, and related thermodynamic, autoregressive and continuous formulations of the problem, are raised. Other geographic applications of the Ising model to population distribution, spatial interaction and aggregation questions, are briefly discussed.

Introduction

Goodchild and Wang (1988) raise the problem of modeling (nominal) class membership probabilities over an array of pixels, so as to incorporate effects of proximity, and present a (heuristic) smoothing approach to its solution. Their general aim is the development of appropriate representations of uncertainty for geographical information systems, and more specifically, the modeling of error distributions in categorical maps. They maintain that this "requires a conceptualization of the manner in which pixel probabilities are realized in a stochastic process".

In this paper, a cross-entropy minimization strategy is adopted with this purpose in mind. The approach taken is strongly analogous to a methodology previously applied (Slater, 1987; Slater, 1989) to express two 5 x 5 nonnegative matrices, one with given row and column sums, and the other, a doubly stochastic version of it, as convex combinations of their extreme points. (Such points, 120 permutation matrices in the 5 x 5 doubly stochastic case and 6,985 transportation matrices in the other, are the ones that cannot be represented nontrivially as convex combinations of other members of the sets.) Through this estimation procedure, a stochastic process is constructed that yields the class membership (marginal) probabilities for the pixels, as its expected values. (Constraints on average values are foreign to non-thermodynamic probability models (Lavenda and Scherer, 1989), so recourse to entropy-based methods seems appropriate.)

In Slater (1987) uniform prior distributions over the extreme points were employed. Here, to incorporate potential spatial autocorrelation (Goodchild, 1988) effects, the probabilities assigned by the two-dimensional Ising model (Ellis, 1985; McCoy and Wu, 1973) of ferromagnetism, which can be considered as a family (parameterized by "temperature") of prior distributions, is utilized to assign initial probabilities to the extreme points ([0,1] stochastic matrices, the rows of which, say, correspond to the pixels, and the columns to the classes). The temperature corresponds (inversely) to the degree of spatial autocorrelation. At infinite temperature, all the extreme stochastic matrices are equally likely (that is, a uniform prior is imposed), and spatial autocorrelation is lacking, while at zero temperature only the matrices having all ones in one of their columns and all zeros in the other(s) have non-zero prior probability, and spatial autocorrelation is at its strongest (most positive).

Initially, temperature is regarded as exogeneously assigned by the investigator, who is presumed able from his subject matter knowledge to fix an estimate of the extent of autocorrelation. The possibility of endogeneously determining a temperature that best expresses the degree of spatial autocorrelation in the data, is also discussed. If one has at least one realization (extreme stochastic matrix) of the process, then the method of maximum-likelihood estimation could be applied. Alternatively, a probability distribution over the full temperature range can be constructed by: (1) assigning the antilogarithm (that is, the likelihood) of minus the minimum discrimination information statistic (computed below) at a specified temperature, as the thermodynamic probability for that temperature; (2) weighting this by a prior distribution for the temperature; and (3) normalizing these values by the integral of this quantity over the temperature range (the positive half-axis). (The [Jeffrey's] assumption that the prior probability of the logarithm of the inverse temperature is uniformly distributed, is usually suitable in a statistical mechanics framework (Lavenda,1988; Slater,1988) .) As temperature is assumed to be positive, then the average, but not most probable (modal), state will display positive spatial autocorrelation.

Statement of the problem

A rectangular n x m array of (nm) pixels is assumed given. To the pixels, typically, by any of several techniques of classifying remotely sensed images (Schowengerdt, 1983) (discriminant analysis, for example), have been assigned binomial parameters (p_{ij} and $1-p_{ij}$; $i=1,...,n$; $j=1,...,$ m) expressing the probabilities of occurrences of two events (say soybeans [1] or corn [0 or possibly, -1], in an agricultural setting). The extension to multinomial probabilities is evident, though computationally more demanding for comparable-sized arrays. Information on source errors [differences between reality and representation input] is not available to geographical information systems from classified pixels (Goodchild and Wang, 1988), so probabilistic methods, such as those presented here, are necessary for error modeling. A realization of a stochastic process generating the nm pairs of binomial probabilities is an nm x 2 matrix, each row of which contains a 1 and a 0. There are 2^{nm} distinct such matrices, which are the extreme points of the convex hull of the nm x 2 stochastic matrices, which, by definition are nonnegative and have row sums equal to 1. The problem addressed in this paper is the placing of a probability distribution over these 2^{nm} matrices that yields the nm x 2 matrix (P) formed by the nm pairs (p_{ij}, $1-p_{ij}$), as its expected value, and is closest in the sense of minimum discrimination information (cross-entropy, Shore and Johnson, 1981) to one of a family (the two-dimensional Ising model) of prior distributions. These express, to varying degrees, the geographic truism that there is a greater tendency for pixels that are near to one another to have similar characteristics than those that are farther apart. The relative probability there of the occurrence of a given (1,-1) n x m array (D) is

$$\exp\left[\frac{\beta}{2} J\,(i,j,k,l) \sum_{i=1}^{n} \sum_{j=1}^{m} \sum_{k=1}^{n} \sum_{l=1}^{m} \sigma_{ij}\sigma_{kl}\right] \qquad (1)$$

where σ is the value (1 or -1) of the corresponding entry of D. In the essentially equivalent lattice gas model, this exponentiation of an energy function (Hamiltonian) is expressed in terms of 1 (occupied) and 0 (empty) cells, rather than 1 (up) and -1 (down) "magnetic spins". J, the coupling, is taken as unity when ij and kl are neighboring sites in the lattice, that is either $i = k$ and $|j-l| = 1$ or $|i-k| = 1$ and $j = l$, and zero for all pairs of more mutually distant sites. (Free, as opposed to periodic boundary conditions are imposed below. Thus, the four corner sites are regarded as having only two neighbors each, and the other edge sites three. In a periodic situation, on the other hand, all sites border four other sites; in effect, the lattice is considered to lie on a torus. Though geographically nonmeaningful, periodic boundary conditions can lead to analytic and computational simplifications.)

The parameter β is the inverse temperature. Configurations in which neighboring lattice sites tend to have similar values, as in positive spatial autocorrelation, are probabilistically favored by this thermodynamic term (1) (as contrasted with the antiferromagnetic model, which, consequently, would appear to have little geographic relevance).

The Ising model has been exhaustively studied, building on landmark work of Onsager in exactly solving the model (asymptotically) in two dimensions (Ellis, 1985; McCoy and Wu, 1973). It provides a suitable family of prior distributions for the problem at hand. If the probabilities (1) alone are assigned to the corresponding 2^{nm} extreme stochastic matrices, the expected value (for any β) is the nm x 2 matrix all the entries of which are, by symmetry, 0.5,which, in general, of course, is different from the entries of the target, P. To achieve the probability distribution over the possible realizations which yields P and is closest to these Ising probabilities in the sense of cross-entropy, iterative scaling (Elfving, 1980) is employed. In effect, a system of nm equations, each of degree nm + 1, needs to be solved for nm + 1 [nonnegative] unknowns; nm probabilities and one normalizing factor. The problem can also be viewed as that of determining the inverse [multivariate] Laplace transform of P (Lavenda and Scherer, 1989; Lavenda,1988; Rothaus, 1968).

Algorithm

An nm x 2 array (Q) of dual binomial probabilities (q_{ij}, $1 - q_{ij}$) is initialized to P. The current weight assigned to any extremal stochastic matrix (E) is the product of nm + 2 terms: (a) the associated Ising probability, given by expression (1); (b) nm entries of Q taken in the natural fashion (that is q_{ij} if e_{ij} is 1, or $1 - q_{ij}$ if it is 0); and (c) a probability normalizing factor (f) the partition function or "permanent" (Rothaus, 1974), initialized here to 2^{nm}. Within each iteration, one cycles through the nm pixels, scaling the corresponding pairs, q_{ij} and $1 - q_{ij}$, in turn, so that the current weighting distribution over the 2^{nm} E's yields p_{ij} and $1 - p_{ij}$. At the end of each iteration of nm such steps, all the rows of Q are normed to sum to 1, and f recomputed from the nm norming factors used. The procedure was terminated in the example below, when the adjustments in the entries of Q over an iteration (measured as the sum of the 2^{nm} squares of the differences between 1 and the ratios of the old and updated entries) fell below 10^{-7}.

If this convergent process were repeated, using Q now, not P, as input, f (which is, in effect, a Lyapunov function in this dynamical system) could not decrease in value (Rothaus, 1974) . P would converge under repeated iterations to an extremal stochastic matrix, a fixed point of this interesting and generalizable mapping, which is the inverse of a gradient transformation. The uniform matrix, with all entries .5, is its only nonextremal fixed point.

The rule or function for determining the minimum cross-entropy representation of P as a convex combination of (pixel-indexed) stochastic matrices can, on a higher level of abstraction, be viewed as a member of the convex set of continuous barycenter functions (Fuglede, 1986; Brondsted, 1986) on the convex polytope formed by the stochastic matrices. (This set consists of all "well-behaved" [continuous] stochastic processes that could realize an arbitrary P.) The function itself would then have a maximum entropy representation as a convex combination of extreme continuous barycenter functions. Such functions are nonzero on no more than nm + 1 extreme stochastic matrices, which is a fact of potential importance in studying large arrays of pixels (see sec. 5), as only nm + 1 (as contrasted with 2^{nm}) weights would need to be computed.

Numerical example

A FORTRAN program to implement the algorithm was written. To test it, twelve pairs of binomial probabilities were generated, using a uniform (0,1) random number generator, for a 4 x 3 array of pixels (Table 1). The upper member of each pair is [arbitrarily] considered to correspond to 1 occurring, and the bottom to 0 or -1.).

.01006	.33221	.96647
.98944	.66779	.03352
.91283	.17129	.74186
.08716	.82870	.25813
.62453	.81930	.32617
.37546	.18069	.67382
.14514	.27622	.71939
.85485	.72377	.28060

Table 1 Input binomial parameters.

The FORTRAN subroutine NEXSUB (Nijenhuis and Wilf, 1978) was used to generate the 4,096 (that is, 2^{12}) 12 x 2 (0,1) extremal stochastic matrices. NEXSUB employs a "Gray code" so that each matrix differs from its predecessor/successor at only one entry, leading to efficient updatings in related computations. The Ising probabilities (1) were computed for each configuration employing β =.4407 (the motivation for selecting this particular value, which is quite near the "critical" one (Ellis, 1985; McCoy and Wu, 1973), is indicated at the end of sec. 5). After nine iterations, the sum of squares of the differences of the entries of Q as shown in Table 2.

.00418	.46568	.97878
.99581	.53431	.02121
.98272	.02599	.80483
.01727	.97400	.19516
.48335	.96684	.07854
.51664	.03315	.92145
.10964	.16485	.87625
.89035	.83514	.12374

Table 2 Dual binomial parameters.

and those at the end of the eighth iteration equaled 4 x 10^{-9}. The normalizing factor f (initialized at 4096) equalled 82002.45, the twelfth root of which is 2.567. The accuracy of this test program (employing double precision arithmetic throughout) was confirmed by regenerating Table 1, to six decimal places, from Table 2, f, and the Ising probabilities (1). The objective function, the (Kullback-Liebler) information distance from the achieved distribution to the prior Ising distribution,was 4.871. Twice the minimum information distance multiplied by the sample size [assumed infinite here] would, under a null hypothesis, be distributed as chi-square with 4,095 degrees of freedom. (For $\beta = 0$, this distance is 2.791, the minimum over nonnegative β .) The entropy of the uniform distribution over 4,096 points is 8.317, while the entropy of the Ising distribution for $\beta = .4407$ is 6.243. The entropy of the binomial distribution (Table 1) is 5.526 (8.317 - 2.791), computed either over the extreme points, or as the average entropy of the 12 binomial distributions in Table 1. The greatest *a posteriori* probability (.10895) was attached to the (maximum likelihood) extremal stochastic matrix, that is, the one which selected in all twelve cells of Table 1 (or 2), the larger of the two probabilities.

The variances/covariances, which can serve as the basis for a multivariate normal approximation to the [Gibbs] distribution over the extreme points (Lavenda and Scherer, 1987) and correlations of the values (0 or 1) assigned to each pixel were taken over the 4,096 configurations, using the achieved (f x dual x Ising) distribution as weights. The twelve variances simply equaled those predicted by the formula for the variance of a binomial distribution, p_{ij} (1 - p_{ij}). All interpixel correlations (Table 3) were positive, confirming the intuition, more formally expressible in mathematical physics as moment inequalities (Ellis, 1985; Percus, 1975) that in a spatially (positively) autocorrelated situation, any pixel has a tendency (declining with distance, a fact also supported by the results) to compel other pixels to assume the same value that it possesses.

	(1,1)	(1,2)	(1,3)	(2,1)	(2,2)	(2,3)	(3,1)	(3,2)	(3,3)	(4,1)	(4,2)
(1,2)	.088										
(1,3)	.009	.103									
(2,1)	.026	.039	.007								
(2,2)	.033	.333	.054	.113							
(2,3)	.008	.081	.176	.029	.203						
(3,1)	.008	.027	.008	.244	.081	.040					
(3,2)	.007	.057	.017	.103	.168	.101	.345				
(3,3)	.003	.035	.049	.031	.096	.283	.089	.243			
(4,1)	.002	.009	.003	.057	.027	.019	.235	.120	.054		
(4,2)	.002	.015	.007	.037	.044	.043	.137	.226	.130	.333	
(4,3)	.001	.013	.013	.017	.037	.086	.060	.129	.298	.094	.274

Table 3 Interpixel correlation matrix.

In the 2-D Ising model itself, intersite correlations decline as $\exp\{-|i-j| / \{\xi(\beta, 0)\}\}$ as the intersite distance $|i-j| \to \infty$ where $\xi(\beta, 0)$ is the correlation length [at zero external field] that becomes infinite at the critical temperature [below which, spontaneous magnetization occurs]. The final distribution, the normalized product of the Ising probabilities (1) and the factors generated from Table 2, can be represented as proportional to

$$\exp\left[\frac{\beta}{2} J(i,j,k,l) \sum_{i=1}^{n}\sum_{j=1}^{m}\sum_{k=1}^{n}\sum_{l=1}^{m} \sigma_{ij}\sigma_{kl} + \sum_{i=1}^{n}\sum_{j=1}^{m}\left(\log q_{ij} - \log(1-q_{ij})\right)\sigma_{ij}\right] \quad (2)$$

Within the framework of the Ising model, the term $(\log q_{ij} - \log(1 - q_{ij}))$ would be regarded as an external magnetic field acting on the lattice sites. The problem addressed in this paper could, thus, be posed as that of determining the strength of a (possibly variable) field at each site, so as to yield expected values (magnetizations) at the sites.

To determine β endogeneously, one might simply consider setting it to the value for which the information distance from the (natural) assignment to the extreme stochastic matrices of the probability distribution based on Table 1 to the assignment to these matrices of the Ising probability distribution (1) (parameterized by β) is a minimum. This occurs (setting the derivative of the cross-entropy to 0) where

$$\beta = \left[\sum_{r=1}^{2^{nm}} \frac{a_r}{2} \sum_{i=1}^{n}\sum_{j=1}^{m}\sum_{k=1}^{n}\sum_{l=1}^{m} \sigma_{ij}\sigma_{kl}\right] \quad (3)$$

where a_r is the (natural) product (transversal) of nm entries of Table 1 corresponding to the r th extremal stochastic matrix (E). The quadruple summation can be viewed as yielding the number of even bonds minus the number of odd bonds, where two neighboring cells with the same value contribute an even bond, and two neighboring cells with dissimilar values, an odd bond. The inverse temperature $\beta = 0$ (that is, the absence of spatial autocorrelation), however, always gives the Ising (equivalently, uniform) distribution closest to the class probabilities. The requirement of non-trivial spatial autocorrelation entails a greater information distance. Given a sample of realizations of a spatially autocorrelated two-class process, one could apply maximum-likelihood estimation to find the β for which the sample has the greatest probability of occurrence.

The ratio exp(-2.791)/exp(-4.871) of the likelihoods is the relative thermodynamic probability, the ratio of the associated number of microstates compatible with the two states,(Ellis, 1985; Snickars and Weibull,1977) of the uniform distribution ($\beta = 0$) versus the distribution for β =.4407. It would appear that such ratios also reflect the relative convergence rates of the iterative scaling algorithm to the minimum cross-entropy solutions (Ellis, 1985). This information can be employed to construct a (proper) probability distribution over β. In particular, if β is restricted to positive values (that is, only positive spatial autocorrelation is deemed possible), the average value of β will be greater than 0, while the most probable would be the limit point for $\beta = 0$. In Table 4 are presented: (1) the minimum discrimination information statistic (m.d.i.s.); (2) the (scaled) likelihood (that is, exp(-m.d.i.s.) times 10^5); and (3) one tenth the value of column (2) multiplied by β^{-1}, for the eleven evenly-spaced values of β from 0 to 1.

β	m.d.i.s.	likelihood	weighted likelihood
.0	2.791	6135	
.1	3.036	4802	4802
.2	3.397	3347	1673
.3	3.892	2040	680
.4	4.550	1056	264
.5	5.399	452	90
.6	6.444	159	26
.7	7.660	47	6
.8	8.997	12	1
.9	10.409	3	0
1.0	11.863	0	0

Table 4 Statistics for inverse temperature (β)

A cursory examination would indicate that β should be estimated to lie somewhere between 0 and 0.2.

Within the Bayesian thermostatistics (Lavenda, 1988) framework, it appears that the problem of estimating β can be cast as that of finding the set of dual probabilities ($Q(\beta^*)$), which when applied to the entire range (grand canonical ensemble) of Ising distributions ($0 < \beta <= \infty$), each weighted by β^{-1}, and integrated over this range (in effect, a Laplace transform), yields P. Alternatively, it might be viewed as the problem of finding a set of dual probabilities that when applied to all possible (not just extremal) tables of binomial parameters (weighted by the likelihood of these tables) and integrated over the nm ranges (0 to 1), yields P. Tables with at least one of their binomial parameters equal to 1 would be assigned zero weight.

The first of these two directions was pursued. The values of the Ising probabilities were found for each of the 4,096 configurations for each of the 25 evenly-spaced (.04) values of β from .02 to .98. This required computation of the [normalizing] partition function for each β. The high-temperature expansion [Parisi, 1988, formula (4.50)], an eighth order approximation, was used, for this purpose. These Ising values were then divided by the corresponding β, using the Jeffrey's principle that log β is uniformly distributed (Lavenda, 1988), and summed (numerically integrated) over the 25 values of β for each configuration, to yield a new prior over the configurations. The minimum cross-entropy algorithm was then implemented, using this prior distribution. The m.d.i.s. obtained was 3.246, suggestive (Table 4) that the mean value of β lies between .1 and .2. (In further support of such a conclusion, the twelfth root of the f found was 2.082, while it was 2.047 for $\beta = .1$ and 2.132 for $\beta = .2$.) Table 5 compares the dual binomial probabilities (of event "1" happening) obtained through this numerical integration (top line) with those previously found (Table 4) for b = .1.

.00837	.34928	.96762
.00891	.36308	.96650
.93466	.12767	.74204
.93410	.12648	.74753
.60419	.85079	.26933
.59014	.85728	.25849
.13979	.25979	.74445
.14700	.25845	.75317

Table 5 Comparison of dual binomial ("1") parameters.

The Gibbs variational formula (Ellis, 1985, sec. IV.7), expressing the Gibbs free energy as the supremum of an energy functional minus an entropy functional, reduces here to the fact that 0 (the logarithm of 1, which is the sum of the distinct products [transversals] of any stochastic matrix, in particular, the 4,096 products from Table 2) equals the sum of each of the 4,096 products times their logarithms, minus the entropy of the products.

The approach presented so far has been based on the standard two-dimensional Ising model, in which the probability of either binomial outcome at each site is .5. However, in the general situation under consideration, this is, of course, not the case. To tailor the prior more to this circumstance, one could modify the summand in expression (1) to be $(s_{ij} + p_{ij} - .5)(s_{kl} + p_{kl} - .5)$, rather than simply $s_{ij}\, s_{kl}$. Expansion of this term leads to the associated probability for a configuration being equal to the product of the standard Ising value with nm "mean-field" (Parisi, 1988) terms, in which each site value (spin) is weighted by the average value of one (by some convention) of the two binomial probabilities in its immediate neighborhood. Possible conditions under which these mean-field terms would equal the dual probabilities (Q) bears investigation, since the mean-field terms can be computed much more simply. (Also, conditions for substituting q_{ij} for p_{ij} and q_{kl} for p_{kl} in the earlier expression in this paragraph seem worth exploring.) The consistency condition for mean-field theories is that the expected value at a site equal the hyperbolic tangent of the product of β with the mean field acting there, which is the resultant of the [at most four] neighboring probabilities, and the external field, if any, acting on the site (Parisi, 1988, eq. [3.9]). If one took as the prior over the extremal stochastic matrices, the (standard) Ising probability multiplied by the nm mean-field terms, the minimum cross-entropy result would be no different than if these mean-field terms had been omitted, due to the multiplicativity (exponential nature) of entropy results.

Modeling of large arrays of pixels

Due to the steep increase in 2^{nm} with n and m (the "curse of dimensionality"), the approach presented above is certainly not strictly implementable in standard-sized arrays of pixels, say configurations of dimension 100 x 100 or 256 x 256. Let us assume, as seems plausible, that 4 x 4 (or possibly, 5 x 4) arrays can be routinely modeled, in the manner indicated above with Tables 1 and 2. One possible approach to studying say a 100 x 100 array, would then be to first separately solve and store the results (the dual probabilities [Q] and probability normalizing factors [f]) of the 625 4 x 4 subproblems embedded in it. Then to generate a 100 x 100 realization (extreme stochastic matrix), one would first randomly sample the 2^{16} 4 x 4 extreme stochastic matrices for some initial 4 x 4 subarray, according to their determined minimum cross-entropy probabilities. Next, one would move to a neighboring submatrix, however, (slightly) temporarily modifying the dual probabilities to reflect the current 0/1 pattern that has been imposed along its border. At the end of 625 such steps, one would have a realization, the (seemingly unbiased) expected value of which would be the target P. It would be appropriate to hold β constant across the 625 subproblems.

Alternatively, these solutions (dual probabilities), aggregated into a 100 x 100 array (or 10,000 x 2 stochastic matrix), could be treated as the true dual probabilities for the 100 x 100 problem. The overall probability normalizing factor would be the geometric mean of the 625 f's. Then, a random (0,1) 100 x 100 matrix, possibly, having the same expected density of zeros and ones as the data,would be generated. The Monte Carlo (importance sampling) (Parisi, 1988) method could then be employed to sample the extreme stochastic matrices in proportion to their probabilities. These are each the product of: (1) f ; (2) 10,000 appropriately chosen entries of the stochastic matrix; and (3) the Ising probability (1) for the configuration, the normalizing factor [partition function] for which, various formulas have been developed (Ellis, 1985; McCoy and Wu, 1973; Morita, 1986). The Monte Carlo process consists of tentatively randomly perturbing the configuration, and accepting the trial alteration only if the negative of the resultant energy change, when multiplied by β and then exponentiated, exceeds a randomly generated number on the unit interval. In the optimization procedure known as "simulated annealing" (Kirkpatrick, Gelatt, and Vecchi, 1983), β is gradually increased, after allowing sufficient steps for equilibration.

Yet another possible approach to utilizing the exact minimum cross-entropy algorithm in modeling large arrays of pixels, could be termed "block renormalization" (Chandler, 1987). The Ising model is exactly renormalizable only at its critical temperature. Consider the case of a 256 x 256 array. Partition it into sixteen 64 x 64 blocks. Represent each block by the mean value of the binomial parameters over its 4,096, or even simply its central four, sites. The sixteen sets of binomial parameters comprise,in a natural fashion, a 4 x 4 array, to which the algorithm in question can be applied. The parameters, f and Q, obtained can then be stored. Proceeding recursively, each 64 x 64 block would then be broken into sixteen 16 x 16 ones, which would then each be compressed to a single site. At the termination of the procedure, the binomial parameters for any site would have entered into three problems. The accumulated parameters need then to be incorporated into an overall consistent (nonrecursive) set.

The inverse temperature (β = .4407) employed above has been set to that for which (Ellis, 1985, p. 153), a famous formula of Onsager and Yang

$$m(\beta, +) = \left[1 - (\sinh 2\beta)^{-4}\right]^{1/8} \qquad (4)$$

where m(β,+), the spontaneous magnetization, is the 2-D Ising model counterpart (.00759 for Table 2, the closeness of which to 0 leads to β being close to the critical inverse temperature) to the average probability of "1" in the data matrix (P). In thermodynamics, parameters are traditionally fitted to average "energy" values [typically, without regard to sampling variance, Lavenda and Scherer, 1989; Lavenda and Scherer, 1987], so such a selection of β would seem to possess theoretical justification, at least in the limit of infinite lattices.

In an another approach (one not based on the minimum cross-entropy algorithm above) to studying large arrays, one could sample from the extreme members of the convex set of probability distributions over the extremal stochastic matrices, that yield the P. The appeal of such an approach is that by Caratheodory's theorem (Grunbaum, 1967), no more than nm + 1 of the probabilities assigned to the 2^{nm} extremal stochastic matrices in any extreme distribution, are nonzero. One possible strategy to follow would be to: (a) generate a (0,1) random number r; (b) generate by the Monte Carlo method a random nm x 2 (0,1) stochastic matrix (R) with respect to the Ising probabilities (1) for a given b; and (c) find the minimum probability (s_{ij}) in P corresponding to the values of 1 in R; and (d) replace P by P - s_{ij} R. Steps (b)-(d) would be reiterated, requiring, however, that the next R generated possess a zero in the ij-cell. The sum of the s's would be accumulated over the iterations (guaranteed not to exceed nm + 1 in number), and when it surpassed r, the process would be stopped, with the current R serving as the realization. This process is unbiased but possibly not minimum cross-entropy, that is, it yields P as its expected value.

As a first step in such a direction, a FORTRAN program to compute the extreme points of the convex decompositions of an arbitrary stochastic matrix was written. The subroutine BACKTR (Nijenhuis and Wilf, 1978) was incorporated. A test was run with a 3 x 2 row-stochastic matrix, the first column of which was (.2, .6, .3). The maximum entropy representation (at β = 0) of this matrix has the weight vector (Y) over the 2^3 or 8 extreme stochastic matrices (generated by NEXSUB, Nijenhuis and Wilf, 1978) of (.224, .056, .084, .336, .144, .036, .024, .096). Ten extreme decompositions of the 3 x 2 matrix were found. Iterative scaling (Elfving, 1980) was then employed to find the weights (and the dual multipliers) to attach to these ten representations to yield the weight vector above. (The uniform distribution over the ten decompositions yields [.17, .06, .05, .42, .09, .04, .05, .12].) The dual multipliers found were (.119, .133, .119, .127, .119, .126, .119, .133) using a prior over the ten representations based on Y. The weight assigned to a representation was proportional to the product of the entries of Y corresponding to the nonzero entries of the representation. If β were positive, then the prior could incorporate, in some fashion, the Ising probabilities (Goodchild and Wang, 1988). The normalizing factor was 8.027.

Autoregressive approach

Another method of generating spatially autocorrelated realizations of multinomial processes in pixels, that is also exactly implementable (that is, reproduces class membership probabilities) in small arrays, but only approximately in large ones, is based on an autoregressive model (Haining, Griffith, and Bennett, 1983) . For an n x m array of pixels, one considers the nm x nm matrix $(I - \rho W)$, where ρ is a parameter akin to temperature, and W is the (0,1) pixel adjacency matrix, an entry of which is 1 if the corresponding pair of pixels are contiguous, and zero otherwise. The inverse of $(I - \rho W)$ is required. This can be suitably approximated by the power series

$$(I - \rho W)^{-1} = I + \rho W + (\rho W)^2 + (\rho W)^3, \dots, \quad (5)$$

where the ij-entry of W^k is the number of distinct paths of length k from i to j. A (1 x nm) vector of standard random normal deviates is then generated. The product of this and $(I-\rho W)^{-1}$, another (1 x nm) vector (U), is the realization of a multivariate normal process with zero mean vector and variance/covariance matrix $[(I - \rho W)^{-1}]^2$. If an entry of U is set to 1 if it falls below the (one-tailed) cumulative probability corresponding to the binomial parameter p_{ij} for that pixel (for example, 1.65 for $p_{ij} = .95$), and 0 if above, the expected value of the (0,1) vector so generated would be the target P. (M. F. Goodchild has also, along the lines of a Gaussian random field, informally suggested: (a) generating standard normal variates at each lattice site; (b) generating a new normal variate at each site by averaging [say] the four nearest-neighbors; and (c) applying the cumulative probabilities, corresponding to the expected pixel value, to this variate.)

Multivariate normal processes are maximum entropy distributions for their mean vectors and covariance matrices. The possibility of a formal relationship between ρ and β would be interesting to explore. Without one, the minimum cross-entropy model appears to be advantageous, in that it yields both a measure of fit, (the minimum discrimination statistic), and an explicit expression (2) for the probability of any particular realization along with the ready capacity to determine conditional distributions, given the occurrence of particular outcomes in some subset of pixels. The autoregressive approach might, possibly however, be used more easily to generate a string of realizations themselves, due to the convenience of normal distribution generators.

The autoregressive procedure is suggestive of a quite direct approach to the problem, based on the Ising model. One would generate, via Monte Carlo methods, a "standard" (.5 probability "up" and .5 probability "down") configuration (at some β). (A FORTRAN program for this is available in Parisi, 1988.) If, say, the "up" spins at a given site (pixel) are, however, expected to be more probable, one would flip enough of the down spins at that site to up spins on the average, while leaving the generated up spins intact, to equal the target probabilities (p_{ij} and $1 - p_{ij}$). This process, depending on the overall degree of imbalance between up and down, should dilute (but not nullify) the spatial autocorrelation associated with β.

Continuous modeling

In mathematical modeling, it is often advantageous (conceptually and/or computationally) to pass from a discrete framework to a continuous one. In the case at hand to pursue this direction, one could smooth, by some spline procedure (Dierckx, 1982), for example, the binomial probabilities (p_{ij} and $1 - p_{ij}$) over the pixels to obtain a smoothly varying field of probabilities. The question then arises what is the appropriate counterpart here of an extremal stochastic matrix. The analogous question has been analyzed for doubly stochastic matrices (Rota and Harper, 1971; Iwanik and R. Shiflett, 1986), and, to a lesser extent for stochastic matrices (Shiflett, 1979) themselves. The difficulty there lies "in guessing the correct characterization of all extremal doubly stochastic measures, in view of the non-existence of a measure-theoretic counterpart to the notion of a permutation matrix. It turns out that extremal doubly stochastic measures can be characterized by an approximation property of the spaces of integrable functions they define." (Rota and Harper, 1971) The extremal stochastic measures for the pixel problem considered here would seem to be the step functions over the pixel surface that assume the values 0 and 1, and are reasonably well-behaved, that is their sets of discontinuities have Lebesgue measure zero. This situation is somewhat similar to the renormalization scheme, in which groups of spins are replaced by blocks in which the spins are uniform, so transitions (discontinuities) are restricted.

Other geographic applications of the Ising model

As a device to express spatial autocorrelation, the Ising model has wide potential throughout geographical analysis, in addition to the stochastic process application described above. Both population distribution and spatial interaction (the movement of people, goods from location to location) can be conceptualized as occurring on a (regular) lattice. In the case of population distribution, one can assume that the coordinates of the residences of each inhabitant of an area are available, and that these coordinates lie at the grid points. There would seem to be much leeway, given locational information, in devising the lattice, the fineness of which and consequent density are important factors. The ratio of occupied (or vacant) lattice sites to all available ones is the counterpart of the magnetization. By formula (4), one can find the associated β. One can then test the optimality (entropy-maximization) of the observed configuration by importance sampling (Parisi, 1988) .

Given the average population density of an area, one can find an equilibrium configuration of a lattice laid over the area, possessing that average density. At the critical temperature, where the density is 1/2, such an arrangement is fractal in nature [hence, renormalizable, with no intinsic scale], possessing Hausdorff dimension 1.86 (Ito and Suzuki,1987). Given several adjoining geographic subdivisions with differing population densities, one can find an equilibrium configuration for each separately, then in some reasonable (smooth) manner piece them together (Tobler, 1979). Other approaches to imputing detailed distributions over such multi-unit spatial aggregations, could be based on the concept of a multifractal (Stanley and Meakin, 1988).

The four-dimensional Ising model (for which, exact solutions, as in the three-dimensional case, are not available), is of potential analogous relevance to spatial interaction. Each interareal mover could be assigned the Cartesian product of his or her (two) origin (O) and (two) destination (D) coordinates, each pair assigned to a two-dimensional lattice (Slater, 1985) . The analog of the external magnetic field in the Ising model can be considered to be the (generalized) cost of travel from O to D. In this way, the model can be regarded as generalizing traditional spatial interaction models (Wilson, 1970; Slater,1989), in which interactions between the movers have not heretofore been incorporated. In the two-dimensional population model, the external field, in all probability, a spatially inhomogeneous one, can be viewed as a reflection of spatial preference.

References

Brondsted, A., 1986, Continuous barycenter functions on convex polytopes. *Expositiones Mathematicae* **4**, 179-187.
Chandler, D., 1987, *Introduction to Modern Statistical Mechanics.* (New York: Oxford)

Dierckx, P., 1982, A fast algorithm for smoothing data on a rectangular grid while using spline functions. *SIAM Journal of Numerical Analysis* **19**, 1286-1304.
Elfving, T., 1980, On some methods for entropy maximization and matrix scaling. *Linear Algebra and Its Applications* **34**, 321-329.
Ellis, R. S., 1985, *Entropy, Large Deviations, and Statistical Mechanics*. (New York: Springer-Verlag)
Fuglede, B., 1986, Continuous selection in a convexity theorem of Minkowski. *Expositiones Mathematicae* **4**, 163-178.
Goodchild, M. F., and Wang, Min-hua, 1988, Modeling error in raster-based spatial data. *Proceedings of the Third International Symposium on Data Handling*, Sydney, 97-106.
Goodchild, M. F., 1986, *Spatial Autocorrelation*. (Norwich: CATMOG).
Grunbaum, B., 1967, *Convex Polytopes*. (New York: Interscience)
Haining, R., Griffith, D. A., and Bennett, R., 1983, Simulating two-dimensional autocorrelated surfaces. *Geographical Analysis* **15**, 247-255.
Ito, N., and Suzuki, M., 1987, Fractal configurations of the two- and three-dimensional Ising models at the critical point. *Progress of Theoretical Physics* **77**, 1391-1401.
Iwanik, A., and Shiflett, R., 1986, The root problem for stochastic and doubly stochastic operators. *Journal of Mathematical Analysis and Its Applications* **113**, 93-112.
Kirkpatrick, S., Gelatt, C. D., and Vecchi, M.P.,1983, Optimization by simulated annealing. *Science* **220**, 671-680.
Lavenda, B. H., and Scherer, C., 1987, The statistical inference approach to generalized thermodynamics: I. Statistics; II. Thermodynamics, *Nuovo Cimento* **100B**, 199-227.
Lavenda, B. H., 1988, Bayesian approach to thermostatistics. *International Journal of Theoretical Physics* **27**, 451-472.
Lavenda, B. H., and Scherer, C., The role of statistical inference in equilibrium and nonequilibrium thermodynamics. Rivista *Nuovo Cimento* in press.
McCoy, B. M., and Wu, T. T., 1973, *The Two-Dimensional Ising Model*. (Cambridge: Harvard University Press)
Morita, T. T., 1986, Partition function of a finite Ising model on a torus. *Journal of Physics A: Mathematical and General* **19**, L1191-L1196.
Nijenhuis, A., and Wilf, T. T., 1978, *Combinatorial Algorithms: For Computers and Calculators*. (New York: Academic Press)
Parisi, G., 1988, *Statistical Field Theory*. (Redwood City: Addison-Wesley)
Percus, J. K., 1975, Correlation inequalities for Ising spin lattices. *Communications in Mathematical Physics* **40**, 283-308.
Rota, C. C., and Harper, L. H., 1971, Matching theory, an introduction. *Advances in Probability* **1**, 169-215.
Rothaus, O. S., 1968, Some properties of Laplace transforms of measures. *Transactions of the American Mathematical Society* **131**, 163-169.
Rothaus, O. S., 1974, Study of the permanent conjecture and some of its generalizations. *Israel Journal of Mathematics* **18**, 75-96.
Schowengerdt, R. A., 1983, *Techniques for Image Processing and Classification in Remote Sensing* (New York: Academic Press)
Shore, J. E., and Johnson, R. W., 1981, Properties of cross-entropy minimization. *IEEE Trans. Inform. Th.* **IT-27**, 472-482.
Shiflett, R. C., 1979, Extreme stochastic measures and Feldman's conjecture. *Journal of Mathematical Analysis and Its Applications* **68**, 111-117.
Slater, P. B., 1988, *Statistical Inference in an Entropy-Maximizing Model of Spatial Interaction*. CORI, University of California, Santa Barbara.
Slater, P. B., 1985, Point-to-point migration functions and gravity model renormalization: approaches to aggregation in spatial interaction modeling. *Environment and Planning* **17A**, 1025-1044.
Slater, P. B., 1987, Maximum entropy convex decompositions of doubly stochastic and nonnegative matrices. *Environment and Planning* **19A**, 403-407.

Slater, P. B., Maximum entropy representations in convex polytopes: applications to spatial interaction, *Environment and Planning A*, in press.
Slater, P. B., A field theory of spatial interaction. *Environment and Planning A*, in press.
Snickars, F., and Weibull, J. W., 1977, A minimum information principle. *Regional Science and Urban Economics* **7**, 137-168.
Stanley, H. E., and Meakin, P., 1988, Multifractal phenomena in physics and chemistry. *Nature* **335**, 405-409.
Tobler, W. R., 1979, Smooth pycnophylactic interpolation for geographical regions. *Journal of the American Statistical Association* **74**, 519-530.
Wilson, A. G., 1970, *Entropy in Urban and Regional Modelling*. (London: Pion)

Chapter 14

The traditional and modern look at Tissot's indicatrix

Piotr H. Laskowski

Abstract

A new method of perception of Tissot's Theory of Distortions and the computation of the parameters of Tissot's Indicatrix, used to analyze map distortions of cartographic projections, is proposed. The new approach is based on the algebraic eigenvalue problem and makes use of the Singular Value Decomposition of the column-scaled Jacobian matrix of the mapping equations. The semiaxes of Tissot's Indicatrix are evaluated directly, and the usage of quadratic forms, which may cause unnecessary loss of numerical precision, is avoided. Several advantages of the new approach, versus the original Tissot's approach, are detailed.
The traditional concept of Tissot's Indicatrix is also examined for comparison. This original approach is presented here from the non-traditional point of view, as the study of a variation of the positive definite quadratic form under the action of a mapping transformation. The complete collection of the computational formulas of the original Tissot's approach (improved for numerical efficiency) is included.

Introduction

In 1881 Nicolas Auguste Tissot published his famous theory of deformation of map projections (Tissot 1881). The main concept of Tissot's theory is the geometric deformation indicator called the Indicatrix. The Indicatrix is the ellipse drawn on the map (it may reduce to a circle or fragments of straight line in limiting cases). This ellipse represents a map projection image of the infinitesimal circle on the surface of the sphere or ellipsoid. For over 100 years the Indicatrix proved very useful in revealing the inherent distortion characteristics of the projection transformation. Nevertheless, the computational formulas of Tissot's theory, to be found in the existing literature, differ in numerically important details between different authors.

A completely new and more systematic approach to map distortion, both conceptually and computationally, is possible on the basis of the modern theory of the algebraic eigenvalue problem (Laskowski 1987). It uses the Singular Value Decomposition with its numerically stable algorithm (Golub and Reinsch 1970) to directly access all parameters of Tissot's Indicatrix.

The algebraic approach offers significant advantages over the traditional approach; some of them are listed here.

Theoretical advantages:

- unified approach based on algebraic eigenvalue problem
- new insight into the nature of Tissot's Indicatrix
- application of modern theory of deformation
- possible generalization to mapping in higher dimensions

Practical advantages:

- numerically stable algorithm (orthogonal matrices)
- direct evaluation of Tissot's parameters which avoids square roots of intermediate quantities
- no need for specialized cartographic code (routines from standard algebraic software packages - such as LINPACK or MATLAB - can be used).

The presentation of map distortion theory will be divided into two major parts. The first part is devoted to the original Tissot's approach. The traditional concept of Tissot's Indicatrix is presented here from a rather non-traditional point of view - as the study of the variation of the symmetric, positive definite quadratic form. As a practical result of the theory, the complete set of the traditional Tissot's computational formulas is given. A self-contained demonstration of the existence of Tissot's Indicatrix (on the basis of Sylvester's Law of Inertia for quadratic forms) is provided in the Appendix B. The necessary theoretical background needed for the development of both traditional and modern approaches to map distortions can be found in the Appendix A.

The second part of the presentation describes the new approach to map distortions. The development of the new concept is self-contained and completely independent of the traditional Tissot's theory of distortions. It is based on a single theorem of linear algebra, with some preliminary use of differential geometry. However, since the results (but not the concepts) of the new method are numerically identical to the results of the traditional method, analogies between the two approaches are frequently made.

In conclusion, the important differences and analogies between the two approaches are summarized in the form of a table (Table 1).

TISSOT'S APPROACH	MODERN APPROACH BASED ON SVD ANALYSIS (DIMENSION=2)
DIFFERENCES	
Theoretical Basis:	
differential geometry plane geometry trigonometry	differential geometry linear algebra
Problem Linearization:	
by using total differential of the mapping equations	by using Taylor expansion of the mapping equations
The Object of Study:	
the quadratic form (B5)	the matrix A, equation (15) and its SVD
Algorithmic Technique:	
based on eigenvalues of quadratic form $A\,A^T$	based on singular values of A
ANALOGIES	
Local Analysis	
Tissot's Indicatrix	Ellipse of Distortion associated with the SVD of matrix A
semiaxes of Tissot's Indicatrix	singular values of A
principal directions on the projection plane	left singular vector of A
scale factors k and h	norms of columns of A

Table 1 The analogies and the differences between the Singular Values Decomposition (SVD) approach and the original Tissot's approach to map distortions

The traditional concept of Tissot's Indicatrix

The concept of the Indicatrix as the ellipse of distortion in cartography was included in his map projection textbook of 1881 by N. A. Tissot, although much of the concept had been introduced in his earlier papers of 1859 and 1878. Tissot's theory of distortions (Tissot 1881) states that an infinitesimally small circle on the spherical or ellipsoidal Earth will be transformed on the projection plane into an infinitesimally small ellipse. This ellipse is called the Indicatrix and describes the local characteristics of a map projection at (and near) the point in question. The orthogonal directions of the semiaxes of the Indicatrix define the principal directions at that point on the map. Tissot's famous proof of existence of orthogonal principal directions uses geometric concepts rather than the sole analytical derivation (Robinson *et al.* 1978, p.430). Tissot's original approach was based on formulas of planar geometry and trigonometry. The planar assumption was made possible by considering only infinitesimally small areas on the surface of the ellipsoid and the corresponding areas on the surface of the map.

The important statements of the original Tissot's theory of deformation can be summarized for cartographic purposes as follows:

- An infinitesimally small area on the surface of the ellipsoid can be treated as a plane figure. This area will remain infinitely small and plane on the projection surface.
- The corresponding infinitely small areas on the ellipsoid and on the projection plane are related by the 2-dimensional affine transformation (so that the rules of the projective Euclidean geometry apply).
- The circle on the ellipsoid with the differential unit radius ds=1 is portrayed as a perfect ellipse with differential semiaxes a and b. This ellipse is called Tissot's Indicatrix.
- The dimensions a and b of the Indicatrix and the directions of its semiaxes (principal directions) are uniquely determined by the equations of the map projection and the intrinsic geometric properties of the ellipsoidal surface at the point of evaluation.
- Tissot's Indicatrix describes the local properties of the mapping transformation, including distortions in lengths, angles, and areas.

Although Tissot's original proof was based on the rules of Euclidean projective geometry, it is also possible to express his ideas in the modern language of quadratic forms. It is shown in Appendix B that both the existence and the parameters of Tissot's Indicatrix can be derived by studying the variation of a certain quadratic form and its eigenvalues. Sylvester's Law of Inertia plays the key role in the proof.

It should be stressed that the quadratic form approach, although (to the author's knowledge) never used in the cartographic literature before, is conceptually fully equivalent to the original Tissot's approach. This means it results in the identical set of computational formulas. In contrast, the new approach (based on Singular Value Decomposition, to be presented in the second part of the paper) produces a completely different set of computational formulas. Of course both approaches theoretically describe one and the same set of Tissot's distortion parameters. However, as will be shown later, the method of singular values may produce more reliable numerical results than the traditional method.

Summary of traditional computational formulas

The traditional computational formulas for the dimension parameters a and b, and the orientation parameters of Tissot's Indicatrix are summarized here for completeness (see Appendix B for details):

At a given point of interest, evaluate the matrix

$$\begin{bmatrix} a_{11} a_{12} \\ a_{21} a_{22} \end{bmatrix} = \begin{bmatrix} 1_\lambda, 1_\phi \end{bmatrix} = A \tag{1}$$

where

$$A = J K^{-1} \tag{2}$$

and J, K are defined by equations (A16) and (A12), Appendix A. Vectors l_λ, l_ϕ in equation (1) denote the columns of A. Then, the semiaxes a and b of Tissot's Indicatrix are determined by the following formulas (identical to those used in algebra for computation of eigenvalues of AA^T).

$$k^2 = a_{11}^2 + a_{21}^2 = l_\lambda^T \, l_\lambda$$
$$h^2 = a_{12}^2 + a_{22}^2 = l_\phi^T \, l_\phi \qquad (3)$$

where angle ϕ is defined on Figure 1.

$$ab = \left| a_{11} a_{22} - a_{12} a_{21} \right| = \left| kh \sin \phi \right| \qquad (4)$$

$$a' = \left(k^2 + h^2 + 2ab\right)^{1/2}$$
$$b' = \left(k^2 + h^2 - 2ab\right)^{1/2} \qquad (5)$$

$$a = a' + b'/2$$
$$b = a' - b'/2 \qquad (6)$$

The orientation β_- (Figure 1) of the semimajor axis, a, of the Indicatrix in the Easting, Northing coordinate system is determined by the following set of formulas:

$$\beta_\phi = \arctan\left(a_{22} / a_{12}\right) \qquad (7)$$

$$\beta_{-\phi} = \arctan\left(1 - h^2/a^2\right)^{1/2} / \left(h^2/b^2 - 1\right)^{1/2} \qquad (8)$$

$$\beta_- = \beta\phi - \beta_{-\phi} \qquad (9)$$

In the actual computations, the equivalent of FORTRAN function ATAN2 should be used in equation (7).

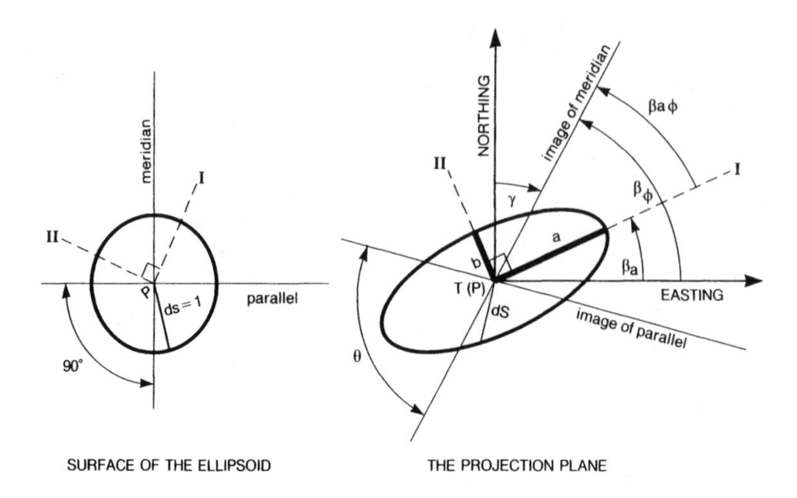

Figure 1 An infinitesimal unit circle on the ellipsoid, and Tissot's indicatrix on the map surface.

Remarks:

- One reason for summarizing the relevant equations in the form ready for computations is that the formulas appearing in different cartographic publications differ significantly in details and occasionally are not clearly stated.
- The traditional Tissot's method is indirect, in that it effectively uses the formulas for the eigenvalues of matrix AA^T (see Appendix B for details) which are precisely the squares of the target parameters a and b of the Indicatrix. As a result, at least two extra square root operations are unavoidably required (as seen in equation (5)) to successfully complete traditional computations.
- The introduction of the determinant (4) represents an improvement in the computational efficiency of the traditional method as compared with the existing published versions known to the author. It replaces the evaluation of cos ϕ followed by sin ϕ (involving square root) in the version published by Maling (1973), and it replaces the direct evaluation of ϕ (at the cost of three trigonometric functions) in the version published by Richardus & Adler (1972).

A modern look at Tissots's Indicatrix

The traditional approach to map distortions proposed by Tissot in 1881 may be classified as a geometric approach, for the reasons explained above. The object of Tissot's original study was (from the point of view of modern algebra) the equation of quadratic form describing (in general) the ellipse, and its alterations under the action of a mapping transformation. In contrast, the new approach, first proposed by Laskowski (1987), could be classified as the algebraic approach. The object of study is the mapping transformation itself. The underlying theory is modern distortion analysis, and the algorithmic apparatus comes from the algebraic eigenvalue problem. The geometric interpretation, so fundamental to Tissot's original theorem, is optional with the new approach. In fact, the algebraic approach is more general and completely independent of Tissot's original theorem. Nevertheless, both methods lead to the identical geometric description of map distortions: both produce the ellipse with identical semiaxes and identical orientation on the projection plane. This ellipse is known as the distortion ellipse in modern distortion analysis, and as the Indicatrix in the traditional Tissot's analysis.

In this section the algebraic approach to map distortions is explained. The development is divided into two major steps. In the first step, the original mapping equations are linearized by Taylor's expansion theorem and the suitable matrix of linear transformation is extracted. In the second step, the distortion analysis is performed on the matrix of linearized transformation by means of the algebraic technique called the Singular Value Decomposition. The singular values and the singular vectors obtained from this analysis reveal the distortion characteristics of the original mapping transformation in the local neighborhood of the point in question.

For completeness, Appendix C summarizes a modern approach to many other important map elements used in the computational cartography.

Linearization of Projection Transformation

Let the map projection T be given by the (usually nonlinear) mapping equations

$$X = X(\lambda, \phi) \text{ and } Y = Y(\lambda, \phi) \tag{10}$$

where (λ , ϕ) represent the orthogonal curvilinear coordinates on the datum surface (the longitude and latitude on the ellipsoid), and (X,Y) represent the Cartesian set of coordinates (Easting,Northing) on the projection plane. Let the datum surface be the ellipsoid of revolution with equatorial radius a (not to be confused with the semimajor axis of Tissot's Indicatrix) and eccentricity e. Taylor's expansion theorem is used to linearize the originally nonlinear transformation (10). The desired linear term of the expansion is constructed by forming the Jacobian matrix based on equation (10), and by evaluating its numerical entries at a given point P.

$$J = \begin{bmatrix} (\partial X / \partial \lambda) & (\partial X / \partial \phi) \\ (\partial Y / \partial \lambda) & (\partial Y / \partial \phi) \end{bmatrix}_P \tag{11}$$

The regularity requirement assures that J exists and is nonsingular. The square matrix J can then be interpreted as a linear approximation of the original mapping transformation (10) in the neighborhood of P. In fact, from Taylor's expansion theorem, the linear transformation represented by J completely describes all local properties, including distortions, of the original mapping equations in the infinitesimal neighborhood of P.

The action of the linearized mapping equations in the neighborhood of P may be described as follows: the differential changes in the curvilinear coordinates, ($d\lambda$, $d\phi$) on the datum surface generate the differential changes in the resultant grid coordinates (dX,dY) on the projection surface, according to the linear formula:

$$[dX, dY]^T = J[d\lambda, d\phi]^T \tag{12}$$

The linear problem (12), with its matrix J, does not form a good basis for the scale distortion analysis of map projection. The reason lies in the fact that the differential change vector, $[d\lambda, d\phi]^T$, in equation (12) is given in the angular units (radians), and cannot be directly compared with the left-hand-side vector $[dX,dY]^T$, given in the units of distance. The scale distortion analysis requires that the changes in the distance on the map be compared with the respective changes in the distance on the ellipsoidal surface. However, by a simple change of variables, equation (12) may be reformulated to fulfill this requirement. In this procedure, the curvilinear coordinate increments $[d\lambda, d\phi]^T$ will be replaced by associated distance increments $[ds_\lambda, ds_\phi]$ on the surface of the ellipsoid. The necessary result is derived in Appendix A (equation (A11)):

$$[d\lambda, d\phi]^T = K^{-1}[ds_\lambda, ds_\phi]^T \tag{13}$$

where K is defined by equation (A12).

Substituting (13) in (12), the linearized transformation (12) takes the final form

$$[dX, dY]^T = A[ds_\lambda, ds_\phi]^T \tag{14}$$

where

$$A = J K^{-1} \tag{15}$$

Notice that the inverse of A is used to build the quadratic form describing Tissot's Indicatrix in the traditional approach (see Appendix B). Equation (14) forms the appropriate basis for further distortion analysis.

Distortion analysis by singular value decomposition

Since the matrix A (equation (15)) is a local linear representation of the original mapping transformation in the infinitesimal neighborhood of P, it is sufficient to analyze distortions of A instead of the original mapping equations, provided the results are considered valid only sufficiently close to P. (This indicates that the new method of analysis is local in nature, as is the traditional method of Tissot's Indicatrix.)

In order to perform the distortion analysis of matrix A, which would produce the parameters of Tissot's Indicatrix, a very useful algebraic theorem will be introduced. This relatively new technique is called the Singular Value Decomposition (SVD). The SVD theorem leads to a very simple interpretation of the action of any linear transformation represented by a given matrix A, which is ideal for the distortion analysis of map projections.

Following the notation presented by Forsythe, Malcolm and Moler (1967), the SVD theorem may be stated as follows: Given any n-by-n real matrix A, there exist two n-by-n real orthogonal matrices U,V, so that U^TAV is a diagonal matrix D. Moreover, U and V may be chosen so that the diagonal elements of D are $d_1 \geq d_2 \geq \ldots \geq d_r \geq d_{r+1} = \ldots = d_n = 0$, where r is the rank of A. In particular, if A is nonsingular then $d_1 \geq d_2 \geq \ldots \geq d_n \geq 0$.

The proof of this theorem can be found in Forsythe and Moler (1967). In the remainder of this paper we will focus on the interpretation of SVD and its application to map distortions. The numbers $d_1, d_2, \ldots, d_n$ are the singular values of A. The column vectors of U and V are respectively the left and right singular vectors.

The use of orthogonal matrices U and V in SVD is of great theoretical importance. Geometrically, any orthogonal matrix can be associated with the rigid rotation of n-space onto itself, possibly preceded by reflection. Therefore, multiplication by orthogonal matrices preserves the length of a vector and the angle between any two vectors. The SVD theorem states that only the diagonal matrix D is responsible for all the distortions in distances and in angles caused by the transformation A, and therefore (locally) by the original mapping equations. The singular values determine the extremal scale distortion factors that occur along the orthogonal principal directions.

The SVD theorem has a clear geometric interpretation which, for a special case of a 2x2 matrix, gives the desired link with Tissot's Indicatrix. In general, an nxn matrix A represents a linear transformation of one n-dimensional linear space, called the datum space (which in cartographic application is tangent to the ellipsoidal surface at the point of analysis), into another n-dimensional linear space, called the projection or the image space. In the initial orthogonal coordinate systems in the datum and in the projection spaces, this linear transformation is represented by matrix A. Matrices U and V define the orthogonal transformations (change of basis) in the projection space and in the datum space respectively. In terms of the transformed coordinate axes, the linearized mapping transformation originally represented by A obtains a new diagonal representation D. The diagonal matrix D contains the singular values of A. The system of the transformed coordinate axes (in the datum and in the projection spaces) in which A obtains this diagonal representation is called the coordinate system of principal directions. Since U and V are orthogonal, the principal directions are necessarily orthogonal.

Let $x=[x_1,..,x_n]^T$ be any vector in the datum space, expressed in the orthogonal basis of principal directions. Let a vector $y=Dx=[y_1,..,y_n]^T$, in the projection space, be an image of x, expressed in the orthogonal basis of principal directions. In the coordinate system of principal directions, transformation A has a diagonal form D. The action of any linear transformation with the diagonal matrix has a simple explanation in terms of normed linear spaces: D projects the n-dimensional unit sphere $S:=\{x : x_1^2 +...+ x_n^2 =1\}$ in the datum space onto the r-dimensional hyperellipsoid in the projection space, composed of all vectors y such that $(y_1/d_1)^2+...+(y_r/d_r)^2=1$ and $y_{r+1}=...=y_n= 0$. This is known as the hyperellipsoid of distortion of the transformation A. The diagonal elements $d_1,...,d_r$ (non-zero singular values) are semiaxes of the ellipsoid which coincide with the orthogonal principal directions. If $r<n$, then D and hence A are singular matrices. If $r=n$, both D and A are nonsingular and have inverses. It is important to mention that the geometric notion of the hyperellipsoid of distortion is only a byproduct of SVD and not its main focus. The distortions are characterized in terms of purely algebraic quantities (singular values - singular vectors), and geometric interpretation is optional.

Let us now apply the Singular Value Decomposition theorem to the analysis of linearized mapping transformation (14). The SVD analysis of a 2x2 matrix A of equation (14) results in

$$A = UDV^T \tag{16}$$

where

$$D = \begin{bmatrix} a & 0 \\ 0 & b \end{bmatrix} \tag{17}$$

and U and V are 2x2 orthogonal matrices. The diagonal elements, a and b, are the singular values of A. The transformation A converts the differential unit circle on the surface of the ellipsoid onto a hyperellipsoid of distortion which, in this two-dimensional (nonsingular) case, reduces to a perfect ellipse on the projection plane, with the semimajor axis, a, and the semiminor axis, b. This agrees precisely with the definiton of Tissot's Indicatrix. Therefore, the singular values a and b determine the semiaxes of the Indicatrix. The mutually orthogonal axes corresponding to a and b define the two principal directions in the projection plane. Along these axes the distortions in length (scale factors) have the extremal values a and b.

The orientation of Tissot's Indicatrix on the projection plane can also be easily determined from the Singular Value Decomposition (16). The SVD theorem states that the orthogonal matrix U represents the change of basis in the projection plane. Specifically, U transforms coordinates of vectors expressed in the basis of principal directions to restore the original coordinates of the same vectors expressed in the standard basis on the projection plane. The standard basis is composed of the usual pair of orthogonal coordinate axes X and Y (grid axes) on the map surface. The change of basis theorem of linear algebra (Strang 1980) gives the desired relationship between the two bases, as implied by the action of the matrix U. From this theorem we conclude that the first column of U contains the representation of the first basis vector of principal directions, written in terms of the standard basis (X,Y). The SVD theorem guarantees that this first basis vector corresponds to the larger singular value: the semimajor axis, a, of Tissot's Indicatrix. Therefore, we have the direct access to the orientation of Tissot's Indicatrix, just by looking at the first column of the matrix U, $[u_{11},u_{21}]^T$. The angle, β- (Figure 1), that the semimajor axis, a, of the Indicatrix makes with the positive X-axis on the projection plane, is determined by the slope of the first column vector of the matrix U (extracted in (16)):

$$\beta- = \arctan\left(u_{21} / u_{11}\right) \tag{18}$$

Summary of computational procedure in modern approach using singular value decomposition

At a given point of interest, evaluate the (2x2) matrix

$$A = J K^{-1} \tag{19}$$

where J and K are defined by equations (11), and (A12), Appendix A, respectively.

Using the Singular Value Decomposition subroutine from any standard linear algebra software package (such as LINPACK), decompose A into the two orthogonal matrices U and V, and the diagonal matrix D:

$$A = UDV^T \tag{20}$$

where

$$D = \begin{bmatrix} d_1 & 0 \\ 0 & d_2 \end{bmatrix} \quad U = \begin{bmatrix} u_{11} & u_{12} \\ u_{21} & u_{22} \end{bmatrix} \tag{21}$$

(Note: V is of no immediate interest in this procedure).

Then, the desired dimensions, a and b, and the orientation angle β– of Tissot's Indicatrix are:

$$a = d_1 \text{ and } b = d_2 \tag{22}$$

$$\beta- = \left(\arctan\ u_{21} / u_{11}\right) \tag{23}$$

Conclusion

The Singular Value Decomposition offers a powerful conceptual and computational tool for analyzing distortions of a projection transformation. It allows for the direct and numerically stable computation of all parameters of Tissot's Indicatrix, avoiding the formation of quadratic forms which may cause a loss of significant digits or unnecessary computer overflows near singular locations on the projection. Computationally, the new method replaces the specialized cartographic code with a subroutine from the standard math library.

The analogies as well as the differences between the new and the traditional methods are best summarized in tabular form. In Table 1 the original Tissot's theory is compared with the new approach based on Singular Value analysis.

Acknowledgements : I thank Lincoln Bragg for helpful comments concerning this approach to Tissot's theory of distortions. The editors are grateful to the ACSM for permission to publish this paper, which also appears in *The American Cartographer.*

References

Dongarra, J., Bunch, J. R., Moler, C.B., and Stewart, G.W., 1978, *LINPACK Users Guide.* (Philadelphia: SIAM Publications)

Forsythe, G., Moler, C. B., 1967, *Computer Solution of Linear Algebraic Systems.* (Englewood Cliffs, NJ: Prentice-Hall)

Forsythe, G., Malcolm, M. A., Moler, C. B., 1977, *Computer Methods for Mathematical Computations.* (Englewood Cliffs, NJ: Prentice-Hall)

Golub, G. H., Reinsch, C., 1970, Singular value decomposition and least squares solutions. *Numer. Math.* **14**, 403-420.

Herring, J., 1965, *Changes in the Geodetic Coordinates Due to a Change in the Reference Ellipsoid.* M. S. Thesis, Ohio State University, Columbus.

Laskowski, P. H., 1987, Map distortions and singular value decomposition. *Technical Papers, ASPRS-ACSM Annual Convention*, Baltimore, March 29-April 3, 1987, Vol.4, pp.42-51.

Maling, D. H., 1973, *Coordinate Systems and Map Projections.* (London: George Philip and Son Limited)

Moler, C. B., 1980, *MATLAB User's Guide*, Technical Report CS81-1, Department of Computer Science, Univerity of New Mexico, 87131.

Richardus, P., and Adler, R. K., 1972, *Map Projections for Geodesists, Cartographers and Geographers.* (Amsterdam: North Holland Publishing Company)

Robinson, A. H., Sale, R. D., and Morrison, J. L., 1978, *Elements of Cartography.* Fourth edition, (New York: John Wiley and Sons)

Strang, G., 1980, *Linear Algebra and Its Applications.* Second Edition, (NJ: Academic Press)

Tissot, A., 1881, *Memoire sur la representation des surfaces et les projections des cartes geographiques.* (Paris: Gauthier Villars)

Appendix A

Map Projections and Differential Geometry

Due to the strictly local nature of Tissot's theory of distortions, the language of differential geometry is traditionally used for a systematic description of Tissot's Indicatrix. This appendix provides a short review of some basic facts from differential geometry that are used in the chapter. It also summarizes the notation traditionally used in the field of mathematical cartography.

Computational cartography deals with the mathematical model of the Earth and with various computational procedures for projecting the curved Earth's surface onto the flat surface of the map. The model of the Earth used in this presentation is the ellipsoid of revolution defined by the equatorial radius a and the eccentricity e. The system of orthogonal curvilinear coordinates, known as longitude λ and latitude ϕ, is established on the surface of the ellipsoid (which is called the datum surface). Curvilinear coordinates on the surface of the ellipsoid and Cartesian coordinates in space are related by

$$\begin{aligned} x &= N \cos\lambda \cos\phi \\ y &= N \sin\lambda \cos\phi \\ z &= N\left(1-e^2\right)\sin\phi \end{aligned} \tag{A1}$$

where

$$N = a / \left(1-e^2 \sin^2\phi\right)^{1/2} \tag{A2}$$

The origin of the (x,y,z) system is at the center of the ellipsoid. The X-axis is at the intersection of the plane of Equator with the plane of the Zero-Meridian in the direction towards the Zero-Meridian. The Z-axis is along the semiminor axis of the ellipsoid in the direction towards the North Pole. The Y-axis is in the plane of Equator in the direction towards the 90 E - Meridian, and completes the definition of the orthogonal, right-handed system.

The differential element of distance is, in general, given by the fundamental law of differential geometry:

$$ds^2 = dx^2 + dy^2 + dz^2 \tag{A3}$$

Differentiating equations (A1), we have in general

$$\begin{aligned} dx &= (\partial x / \partial\lambda)\, d\lambda + (\partial x / \partial\phi)\, d\phi \\ dy &= (\partial y / \partial\lambda)\, d\lambda + (\partial y / \partial\phi)\, d\phi \\ dz &= (\partial z / \partial\lambda)\, d\lambda + (\partial z / \partial\phi)\, d\phi \end{aligned} \qquad \text{(A4)}$$

Specific partial derivatives required in (A4) can be evaluated from equation (A1) (Herring 1965):

$$\begin{aligned} (\partial x / \partial\lambda) &= -N \sin\lambda \cos\phi \\ (\partial y / \partial\lambda) &= N \cos\lambda \cos\phi \\ (\partial z / \partial\lambda) &= 0 \\ (\partial x / \partial\phi) &= -M \cos\lambda \sin\phi \\ (\partial y / \partial\phi) &= -M \sin\lambda \sin\phi \\ (\partial z / \partial\phi) &= M \cos\phi \end{aligned} \qquad \text{(A5)}$$

where

$$M = a(1 - e^2) / (1 - e^2 \sin^2\phi)^{3/2} \qquad \text{(A6)}$$

Using (A5), the general equations (A4) take the form:

$$\begin{aligned} dx &= (-N \sin\lambda \cos\phi)\, d\lambda + (-M \cos\lambda \sin\phi)\, d\phi \\ dy &= (N \cos\lambda \cos\phi)\, d\lambda + (-M \sin\lambda \sin\phi)\, d\phi \\ dz &= (M \cos\phi)\, d\phi \end{aligned} \qquad \text{(A7)}$$

Finally, using equation (A7), the element of distance (A3) on the ellipsoidal surface is written as:

$$ds^2 = (N \cos\phi)^2 d\lambda^2 + (M)^2 d\phi^2 \qquad \text{(A8)}$$

As a special case of (A8), by setting $d\phi = 0$, an element of length of the parallel of latitude is given by

$$ds_\lambda = (N \cos\phi)\, d\lambda \qquad \text{(A9)}$$

Similarly, by setting $d\lambda = 0$, an element of length of the meridian is given by

$$ds_\phi = (M)\, d\phi \qquad \text{(A10)}$$

In matrix notation, (A9) and (A10) may be written as

$$\begin{bmatrix} ds_\lambda \\ ds_\phi \end{bmatrix} = K \begin{bmatrix} d_\lambda \\ d_\phi \end{bmatrix} \qquad \text{(A11)}$$

where

$$K = \begin{bmatrix} N\cos\phi & 0 \\ 0 & M \end{bmatrix} = \begin{bmatrix} r_\lambda & 0 \\ 0 & r_\phi \end{bmatrix} \qquad \text{(A12)}$$

is a nonsingular matrix. The diagonal elements, $r_\lambda = N\cos\phi$ and $r_\phi = M$, are the radii of curvature of the parallel and the meridian at a given point on the ellipsoid. The inverse matrix, K^{-1},(which may be called the curvature matrix) is essential in the linearization step of the modern approach to map distortions (see equation (14) in text).

On the projection plane, the position of a point is specified by means of the rectangular coordinates X and Y, also called the Easting and Northing system.

Let T denote a one-to-one and differentiable transformation (the mapping equations) between the curvilinear coordinates λ , ϕ on the surface of the ellipsoid, and the rectangular coordinates X,Y on the projection plane: $T : (\lambda , \phi) \dashrightarrow (X,Y)$. The (usually nonlinear) mapping equations of the transformation T will be written as:

$$X = X(\lambda,\phi)$$
$$Y = Y(\lambda,\phi) \qquad \text{(A13)}$$

and its differential version, valid in the local neighborhood of the point in question, is

$$dX = (\partial X / \partial\lambda)\, d\lambda + (\partial X / \partial\phi)\, d\phi$$
$$dY = (\partial Y / \partial\lambda)\, d\lambda + (\partial Y / \partial\phi)\, d\phi \qquad \text{(A14)}$$

In matrix notation, (A14) may be written as

$$\begin{bmatrix} dX \\ dY \end{bmatrix} = J \begin{bmatrix} d\lambda \\ d\phi \end{bmatrix} \qquad \text{(A15)}$$

where

$$J = \begin{bmatrix} \partial X/\partial\lambda & \partial X/\partial\phi \\ \partial Y/\partial\lambda & \partial Y/\partial\phi \end{bmatrix} \qquad \text{(A16)}$$

is known as the Jacobian matrix.

T is assumed to be regular in the differential neighborhood under consideration, so that formula (A15) may be inverted for $d\lambda$, $d\phi$:

$$\begin{bmatrix} d\lambda \\ d\phi \end{bmatrix} = J^{-1} \begin{bmatrix} dX \\ dY \end{bmatrix} \tag{A17}$$

On the projection plane, the element of distance (dS) is defined by

$$dS^2 = dX^2 + dY^2 \tag{A18}$$

By substituting (A14) in (A18), the important formula results which shows that the differential distance change dS on the surface of the map is a result of the corresponding differential changes $d\lambda$, $d\phi$ in the curvilinear coordinates on the surface of the ellipsoid, propagated through the mapping equations T:

$$\begin{aligned} dS^2 = & \left[(\partial X / \partial \phi)^2 + (\partial Y / \partial \phi)^2 \right] (d\phi)^2 \\ & + 2 (\partial X / \partial \phi)(\partial X / \partial \lambda) + (\partial Y / \partial \phi)(\partial Y / \partial \lambda)\, d\phi\, d\lambda \\ & + \left[(\partial X / \partial \lambda)^2 + (\partial Y / \partial \lambda)^2 \right] (d\lambda)^2 \end{aligned} \tag{A19}$$

The coefficients in equation (A19) are the First Gaussian Fundamental Quantities, traditionally denoted by E, F, and G respectively:

$$\begin{aligned} E &= (\partial X / \partial \phi)^2 + (\partial Y / \partial \phi)^2 \\ F &= (\partial X / \partial \phi)(\partial X / \partial \lambda) + (\partial Y / \partial \phi)(\partial Y / \partial \lambda) \\ G &= (\partial X / \partial \lambda)^2 + (\partial Y / \partial \lambda)^2 \end{aligned} \tag{A20}$$

As a special case of (A19), by putting $d\phi$ and $d\lambda$ in succession equal to zero, the effect of the changes in just one curvilinear coordinate on the surface of the ellipsoid can be studied separately. Specifically, the element of distance in the direction of the projected parallel on the map surface is given by

$$dS_\lambda = G^{1/2} d\lambda \tag{A21}$$

Similarly, the element of distance along the image of the meridian on the map surface is given by

$$dS_\phi = E^{1/2} d\phi \tag{A22}$$

In general, the ratio of the element of distance dS on the map to the corresponding element of distance ds on the ellipsoid defines the point scale implied by a given mapping transformation T. Usually, the point scale changes as the element of distance ds changes its location or its direction on the surface of the ellipsoid. Specifically, the scale along the parallel is given by

$$k = dS_\lambda / ds_\lambda = G^{1/2} (N \cos\phi) \tag{A23}$$

Similarily, the scale along the meridian is given by

$$h = dS_\phi / ds_\phi = E^{1/2} / M \tag{A24}$$

Appendix B

The traditional theory of Tissot's Indicatrix in the language of quadratic forms

Consider an infinitesimally small circle of the unit radius on the surface of the ellipsoid. In the (planar) local coordinate system formed by the two orthogonal vectors which are tangent respectively to the parallel and to the meridian on the surface of the ellipsoid, the equation of that circle may be written as

$$dS_\lambda^2 + dS_\phi^2 = 1 \tag{B1}$$

This may also be written as a quadratic form with the symmetric positive definite matrix (the unit matrix):

$$[dS_\lambda, dS_\phi] \begin{bmatrix} 1 & 0 \\ 0 & 1 \end{bmatrix} \begin{bmatrix} dS_\lambda \\ dS_\phi \end{bmatrix} = 1 \tag{B2}$$

Our goal is to show that this circular quadratic form on the datum surface transforms into another quadratic form describing the perfect ellipse on the projection plane. The transition between the distance elements on the ellipsoid (composing equation (B2)) and the corresponding distance elements on the projection plane will be increments in the ellipsoidal curvilinear coordinates $[d\lambda, d\phi]^T$.

In the second step, the mapping transformation T is utilized to complete the desired change of variable procedure. More specifically, equation (A17), Appendix A, is used to further change variables from the increments in the ellipsoidal curvilinear coordinates $[d\lambda, d\phi]^T$ to the desired distance increments $[dX, dY]^T$ on the projection plane. Combination of step 1 and step 2 gives the desired formula

$$\begin{bmatrix} dS_\lambda \\ dS_\phi \end{bmatrix} = C \begin{bmatrix} dX \\ dY \end{bmatrix} \tag{B3}$$

where

$$C = K J^{-1} \tag{B4}$$

is nonsingular. Notice that $C=A^{-1}$, where A (equation (15)) is a matrix of linearized mapping transformation used as a basis for singular value analysis in the modern approach.

Equation (B3) establishes the desired total change of variables between the chosen local coordinates on the ellipsoidal surface and the corresponding local coordinates, parallel to the Easting and Northing axes, on the map surface (at the point of interest). Under the change of coordinates (B3), the initial circular quadratic form (B2) transforms into

$$[\, dX, dY]\; C^T \begin{bmatrix} 1 & 0 \\ 0 & 1 \end{bmatrix} C \begin{bmatrix} dX \\ dY \end{bmatrix} = [\, dX, dY]\; C^T\; C \begin{bmatrix} dX \\ dY \end{bmatrix} = 1 \tag{B5}$$

where

$$C^T\, C = \left(J^{-1}\right)^T K^T K\; J^{-1} \tag{B6}$$

Equation (B5) represents another quadratic form in which the unit matrix of the initial circular form (B2) is replaced by a symmetric positive definite matrix C^TC, through a process known in linear algebra as the congruence transformation (Strang 1980,p.259).

The geometry of this new quadratic form depends on the properties of matrix C^TC. In particular, if eigenvalues of C^TC are all positive, then equation (B5) describes a perfect ellipse on the projection surface (Strang 1980,p.256). If this is the case, the semiaxes a and b of the ellipse are related to the (necessary positive) eigenvalues (s_1^2) and (s_2^2) of matrix C^TC by the following (Strang 1980,p.256):

$$a = 1/\left(s_1^2\right)^{1/2} = 1/|s_1|,\; b = 1/\left(s_2^2\right)^{1/2} = 1/|s_2 \tag{B7}$$

The unit matrix of the initial circular quadratic form (B2) has obviously two positive eigenvalues. The application of Sylvester's Law of Inertia (Strang 1980,p.259) to the matrices of quadratic forms (B2) and (B5) guarantees that the matrix C^TC has also two positive eigenvalues. Therefore, the transformed quadratic form (B5) indeed describes the perfect ellipse on the projection surface. This confirms the original Tissot's claim that an infinitesimal circle on the datum surface transforms into a perfect ellipse on the map plane. The only prerequisite is that C, and therefore J, be nonsingular.

The relation between computational formulas of the Tissot's original approach and computation of eigenvalues

It was shown in equation (B7) that the traditional computational formulas used in Tissot's theory, although originally derived from a few simple geometric relationships using rules of Euclidean projective geometry, are identical to the algebraic formulas for the computation of the eigenvalues (d_1^2), (d_2^2) of the symmetric, positive definite matrix $(C^TC)^{-1}$ (where C^TC is defined by (B6)). Remark: it is a well-known fact in the algebraic eigenvalue problem (Strang 1980,p.188), that the eigenvalues (s_1^2), (s_2^2) of C^TC are the inverses of the eigenvalues (d_1^2), (d_2^2) of the matrix $(C^TC)^{-1}$:

$$d_1^2 = 1 / s_1^2 \quad \text{and} \quad d_2^2 = 1 / s_2^2 \tag{B8}$$

Therefore, the parameters a and b of Tissot's Indicatrix are finally expressed as (combining (B7) and (B8)):

$$a = d_1 , \; b = d_2 \tag{B9}$$

where d_1 and d_2 are positive square roots of the two positive eigenvalues of the symmetric, positive definite matrix $(C^TC)^{-1}$. This matrix may be obtained by inverting (B6):

$$\left(C^TC\right)^{-1} = \left(J\,K^{-1}\right)\left(J\,K^{-1}\right)^T = AA^T \tag{B10}$$

where A is again the familiar matrix of linearized mapping transformation, defined by equation (14).

Appendix C

This Appendix will demonstrate how all the important map elements used in computational cartography are easily expressed in terms of the square matrix A of linearized mapping equations (see equation (14)). Only elementary linear algebra is needed. All the formulas will be derived from the fundamental properties of the matrix A of linear transformation (14) in a given (local) coordinate system.

Equation (14) represents a linear transformation which describes all the properties of the original mapping equations within some infinitely small area on the surface of the ellipsoid. Matrix A is the matrix representing this linear transformation. From linear algebra, (Strang 1980), it is a fundamental property of a matrix transformation that the two columns of A are the respective projection images of the two orthogonal basis vectors on the datum surface: the differential unit vector along the parallel, ds_λ, and the differential unit vector along the meridian, ds_λ. Let l_ϕ (the first column-vector of A) be the projection image of the first unit vector, ds_λ.

$$1_\lambda = \left(1/\,r_\lambda\right)\left[\left(\partial X / \partial\lambda\right), \left(\partial Y / \partial\lambda\right)\right]^T \tag{C1}$$

Similarly, let the second column of A, denoted by l_ϕ, represent an image of the unit vector ds_λ along the meridian:

$$1_\phi = (1/r_\phi)\left[(\partial X/\partial\phi), (\partial Y/\partial\phi)\right]^T \tag{C2}$$

where r_λ and r_ϕ are defined by (A12), Appendix A. The ratio of the length of the image vector, l, on the projection plane to the length of the corresponding vector, ds, on the surface of the ellipsoid determines the scale in the direction of ds. Since the differential basis vectors on the surface of the ellipsoid are chosen to have unit lengths, the lengths of their respective images on the projection plane (columns of A) determine the scales along the parallel and along the meridian through P. Traditionally, these scales are called the scale factors and are denoted by k and h respectively.

Specifically, the scale factor k along the parallel is given by the Euclidean length of the first column of A:

$$k = |1_\lambda| = \left(1_\lambda^T 1_\lambda\right)^{1/2}$$
$$= (1/r_\lambda)\left[(\partial X/\partial\lambda)^2 + (\partial Y/\partial\lambda)^2\right]^{1/2} \tag{C3}$$

The scale factor h along the meridian is given by the Euclidean length of the second column of A:

$$h = |1_\phi| = \left(1_\phi^T 1_\phi\right)^{1/2}$$
$$= (1/r_\phi)\left[(\partial X/\partial\phi)^2 + (\partial Y/\partial\phi)^2\right]^{1/2} \tag{C4}$$

(compare (A23) and (A24), Appendix A.)

The columns of matrix A can also be used to determine the slopes of the projected meridians and parallels (graticule lines) on a given projection. It is sufficient to notice that, since the basis vectors ds_λ , ds_ϕ on the ellipsoid have been chosen tangent to the meridians and parallels, the respective image vectors l_λ, l_ϕ (represented by columns of A) are tangent respectively to the projected parallel and the projected meridian at the point of interest. Specifically, the slope, β_λ, of the projected parallel (measured counterclockwise from the X grid axis (Easting) on the projection surface) is defined by the slope of vector l_λ :

$$\tan\beta_\lambda = (\partial Y/\partial\lambda)/(\partial X/\partial\lambda) \tag{C5}$$

Similarly, the slope, β_ϕ , of the projected meridian is defined by the slope of vector l_ϕ:

$$\tan\beta_\phi = (\partial X/\partial\phi)/(\partial Y/\partial\phi) \tag{C6}$$

The angle of the intersection of the projected meridian and the projected parallel at that point is defined as

$$\theta = \beta_\phi - \beta_\lambda \tag{C7}$$

Also, meridian convergence τ can be expressed in terms of b_ϕ to indicate the angle between the image of the meridian (true north) and the positive Y axis (grid north):

$$\tau = \beta_\phi - 90 \tag{C8}$$

Section V

Although many of the earlier applications were in natural resource management and land use planning, GIS are being increasingly applied to market research, economic and regional forecasting, transportation planning, and health. Adoption of GIS technology in these fields has been rapid because of the availability of much of the basic data in digital form, avoiding the need to incur the high startup costs of database creation.

Some of the accuracy problems of socioeconomic data are well-known - the difficulties of undercounting in census returns have been studied extensively, for example. The use of GIS technology introduces a new set of problems, however, precisely because of the power it offers for sophisticated spatial analysis. Of particular concern to the authors of the chapters in this section are inaccuracies due to aggregation and disaggregation of data, and estimation from one set of reporting zones to another. The Modifiable Areal Unit problem is one particular case of a much wider set of accuracy issues related to the geographical basis of socioeconomic data and its analysis.

Brusegard and Menger open the section by discussing some of the practical problems encountered in the operation of a small data-providing agency, which is obliged to maintain certain standards of data quality, and to document uncertainty in ways which are meaningful to its clients. GIS is increasingly important in coping with changing reporting zones, different spatial bases for data and different classes of spatial objects.

Kennedy examines a problem which plays a key role in the uncertainty associated with analysis of health statistics - the small numbers problem. While GIS offers the ability to correlate spatially disaggregated mortality and morbidity data with a host of possible causal factors, it can be a very dangerous tool in the wrong hands, particularly if data are not interpreted with adequate caution and care. Small numbers are one such liability - others include the ease with which false causal inferences can be drawn from correlations which have no causal basis, but which can be detected quickly with a GIS. Whether GIS technology will be used responsibly or not as an epidemiological tool remains to be seen.

Location analysis is one area where uncertainty in the underlying data has poorly understood effects, and yet uncertainty is almost always present because of the need to aggregate, and because no database of the spatial distribution of demand for a service can be perfectly current. Instead the use of representative points for dispersed demand, the uncertainty in any demand forecast and changes in population distribution all contribute to uncertainty in the database on which any decision about the location of a central facility must be made. Batta's paper looks at the effects of aggregation of demand points on a location model.

The fourth paper in the section, by Herzog, discusses the use of a process of polygon filtering to model the reliability of statistical surfaces based on socioeconomic data. Polygon filtering is one possible way of producing the kind of frame-free analysis which Tobler called for in Chapter 11.

Chapter 15

Real data and real problems: dealing with large spatial databases

David Brusegard and Gary Menger

Abstract

The Institute for Market and Social Analysis (IMSA) develops and maintains large geographic databases for use in market and social research applications. This paper details a number of conceptual and practical problems which are encountered in developing consistent and accurate spatial databases. In particular, problems regarding the use of data derived from numerous sources at various levels of spatial aggregation are discussed.

Introduction

The field of market analysis has always relied heavily on georeferenced data. While the universe of observation may be postal zones, census enumerator areas, sales territories, marketsheds, and custom trade areas, rather than parcels, utility lines, or watersheds, many of the field's methods and spatial data handling concerns parallel those of municipalities, government agencies, and universities. Burgeoning interest in GIS systems and spatial analysis has been felt in this field and has given rise to numerous efforts to use and adapt this technology to the goals of market analysis.

The Institute for Market and Social Analysis is a research and development company providing applications tools and techniques to the market research industry, municipalities, and groups involved in the analysis of demographic, economic, and spatial phenomena. This paper details a number of conceptual and practical problems which are encountered in developing consistent and accurate spatial databases. In particular, problems regarding the use of data of varying quality derived from numerous sources at various levels of spatial aggregation are discussed.

IMSA likely processes and analyses more information than any other private market research and social research agency in Canada. In any discussion of error in large-scale spatial databases, the Institute can provide first hand experience regarding the practical problems of dealing with and attempting to quantify and minimize error. This paper outlines the problem set of any like group dealing with such datasets in an applications context. Accordingly, it raises questions without necessarily attempting to answer them. The objective of this discussion is to provide an overview of the major and immediate accuracy problems facing private sector users of GIS technology. In so doing, a short list is developed of priorities for development and research which are of direct concern to the private sector. The emphasis of private sector firms is without doubt on the side of short term solutions rather than long term and broad approaches.

Error in the market research context

As a developer of large spatially oriented databases and analytical methods, IMSA must continually re-assess the quality and the impact of error in these databases. While some might regard the purposes and ethics of market research as being less than caring with respect to error margins, this is a major misconception for a number of reasons. First, the information products derived from these databases are used in support of what are often major business decisions, resulting in considerable capital investments. If investment decisions are based on inaccurate data or inappropriate methodologies, the results will generally be evident in short order. In order to build and sustain a stable clientele, a market research firm must continually strive to improve the quality of its data holdings, to provide quantitative and defensible levels of accuracy, and to provide guidance on the interpretation of such information. Second, the potential for litigation forces market research firms to maintain and explain their standards for data and models. It is not sufficient to indicate that data are 'estimates'. Market research and other data-dissemination firms must be able to:

- identify sources of error;
- utilize available statistical procedures for spatial data;
- provide sound quantitative error estimates; and
- fully and accurately communicate the implications of those analyses to their clients.

Third, clientele in the public sector have continually demanded levels of accuracy in geographic information and in models using that information which are quantifiable and testable. Public sector clients typically require results which precisely match those of government or other authoritative agencies. In cases where the analyses and outputs of market analysis firms prove more accurate than governmental sources on objective grounds, the case for divergence from authoritative sources must be iron clad.

It is clear that the rapid development of geographical analysis tools has resulted in a situation where the user can very rapidly generate multiplicative and cascading errors in ways which are not fully understood at the outset. This represents a major challenge for groups such as IMSA who provide both data and analytical tools to clients.

Database scale and content

The geographic scale for market analysis

As its primary function, IMSA maintains and updates a set of large databases containing numerous demographic, social, and consumer behaviour attributes for a wide variety of geographical units, ranging in scale from individual neighborhoods to the national level. Two main geographic classification systems serve as the basis for the databases, each defined for distinct purposes.

Statistics Canada has defined an extensive geographic classification system, primarily for the administration and reporting of the national census, which is carried out every five years. The census classification consists of a number of nested hierarchies, defined for both administrative and statistical purposes. The base data collection unit is the Enumeration Area (EA), which is typically compact in shape and consists of areas of approximately two hundred households. For the 1986 census, there were slightly over 44,000 of these units, providing complete coverage of the national territory. Census data is typically reported at the EA, and all higher levels, while the multitude of special-purpose surveys undertaken by the agency are typically reported only at higher levels (although the sampling framework is derived at the EA level).

Enumeration Areas serve as the base building block for the census geographical classification which reflects a mixture of statistical and administrative goals. The census classification system consists of a number of independent and semi-independent 'threads', all having the EA as the base and most completely covering the national territory. Within any thread, units are guaranteed to be hierarchically nested. Between these base points, however, the spatial units between threads typically do not nest cleanly. For example, census tracts often straddle census subdivisions (municipalities) and metropolitan areas can straddle provincial boundaries.

The second major system is based on the national postal system, which is designed purely for postal collection and distribution purposes. The postal system consists of a two-tiered collection and distribution hierarchy of approximately 1300 large Forward Sortation Areas (FSA), each of which is divided into Local Distribution Units (LDU). There are approximately 780,000 such units nationally. Postal distribution is carried out at a third level, the Postal Walk (or carrier route), composed of geographically linked strings of LDU's.

The postal distribution system generally does not match with any of the census units, as there is little regard in such a system for political and administrative boundaries, a problem which will be discussed in a later section.

Temporal change in geographic units

The classification systems in use are subject to revision either on a periodic basis (the census) or on an ad-hoc basis (the postal system). Many of the purely administrative and political geographic units are reasonably stable in spatial extent over time. For example, there have been few changes to provincial or county definitions during this century, especially in rural areas of the country.

On the other hand, a number of the census geographical units are subject to change from one census to another. In some cases, these changes are a result of the growth of metropolitan areas and subsequent territorial annexations. In the areas surrounding the largest metropolitan areas, large scale reorganizations have occurred, typically reducing the number of distinct municipalities surrounding these cities. Often, neighbouring rural counties have gained adjacent territory which is not urbanizing.

Census statistical units are subject to even greater changes over time. The concept of a metropolitan area, and therefore the territory it encompasses, has changed with almost every census. Infilling or major urban redevelopment projects may result in the creation of new census tracts or splitting of existing ones in many urban areas, as these are purely statistical regions. At the most basic level, the number and precise spatial definition of individual enumeration areas or neighborhoods change with each census, causing considerable problems in the comparison of census data at this scale. Fully fifty percent of EA units were modified between the 1981 and 1986 censuses.

The postal system, unlike the census, is subjected to continual revision. Postal walks are updated quarterly but change continually, resulting in significant miscoding of these units. Approximately 16,000 LDU units are added to the system annually. Many of these are temporary and may not be in the parent FSA territorial extent. Significant numbers of LDU units are retired (ie. no longer valid) but are retained on Canada Post administrative files.

At the Forward Sortation Area (FSA) level, regions are often split or retired, and their precise boundaries changed from time to time. In some cases, the boundary may be imperfectly specified or arbitrarily drawn on a generalized map. In some cases, the reporting of changes occurs months after the administrative changes are made.

IMSA maintains digital mapping files for many standard geographical units, both census and postal. Digital files have been created for most of the larger census units and for the postal FSA system. In addition, work has begun on the digitization of street networks for major centers.

Database attributes

IMSA maintains approximately 4000 attributes for census geography units. These are also available for units in the postal distribution system through an enumeration area to LDU conversion table, but must be estimated or derived for all other geographical units. In some cases, data are only available for higher order units.

Attributes encompass a wide variety of themes, which can be classified into the following broad groupings:

- historical and current population and household counts and IMSA generated estimates and projections;
- household and family characteristics including household size, composition, income distributions, and employment characteristics;
- population characteristics including age, sex, occupation, education, marital status, and incomes;
- estimates of household consumer expenditures on approximately 800 categories, e.g. rent, retirement savings plans, gasoline, flowers;
- detailed food expenditure estimates, ie. frozen peas, chicken, restaurant meals, etc. ;
- household financial data, including assets, debts, and income;
- household ownership patterns, i. e. microwave ovens, downhill skis, etc. ;
- employee counts, firms by size and revenues, firms by SIC group;
- new and total vehicle registrations by manufacturer, model and year;
- municipal statistics such as labour force activity, land use, tax rates, elected officials, and facilities (ie. airports, ports, railway access), maintained for most municipalities nationwide.

Linkage of classification units

In order to make effective use of the postal hierarchy, a critical linkage must be maintained between the census EA and the postal LDU. In urban areas, the LDU is typically smaller than the EA and can be geocoded to the block face in most instances. In rural areas, LDU's are actually rural post offices and may serve a number of enumeration areas. To complicate matters, rural LDU's are only loosely defined and may have 'fuzzy' or even unknown boundaries. Current public linkage files are inadequate in newly developing areas.

Detecting and correcting error

The embrace of GIS technology and procedures has led to greatly increased interest in base data, be it in map, coordinate, or attribute form. In many cases these datasets emanate from government or para-governmental sources which generate them primarily as tools for their own administrative or mandated purposes. The rush of GIS users to purchase and install such datasets in their own systems raises two issues of major importance:

(a) what is the responsibility of major base data set providers such as census bureaux, national topographic mapping agencies, and postal systems to provide correctly georeferenced data, or to alert users to the potential for error which occurs in datasets created for administrative purposes when they are applied to the diverse opportunities provided by a GIS environment; and

(b) how can error be detected and handled in such basic datasets by GIS applications users and developers.

The first issue begs the question of what is an error. For example, postal systems will often locate postal codes in rural areas not where the recipient lives, but where the nearest postal station is located, sometimes as much as fifty miles away. For administrative purposes this is appropriate, but for the purposes of using these files for locating mail recipients it is extremely inaccurate.

The minimal requirement of government and related agencies releasing large georeferenced files should be to indicate the impact of their administrative procedures on what is essentially ground truth from a GIS perspective. For example, if postal codes are essentially regarded as block face coordinates from a GIS perspective, then it is undoubtedly incumbent upon providers of this dataset to indicate the potential pitfalls of this perspective. Generalizing this example suggests that the emergence of GIS technology places a requirement upon major dataset providers to establish standards and information practices which permit the datasets to be used with appropriate understanding of their limitations in non-standard environments.

The second issue of simple error detection (not to mention correction) is a difficult one to address. Noted below are some simple examples of the need for very basic tools within or associated with GIS systems in a market analysis context. This phase of error detection and correction is usually left to the inventiveness of the user, but within a GIS context there exist routines which can handle basic error detection processes, or be used to build such processes, such as:

- calculation of distances between point data with known relations in the database;
- determination of location within/without enclosing polygon;
- Thiessen and triangulation routines;
- polygon overlay routines;
- detection of multiple nodes and features in the same location;
- tolerance setting with thresholds;
- attribute transfer routines, etc.

Beyond pointing out the need for GIS systems to provide basic tools for error detection on standard georeferenced files there is an additional and common requirement of cross file sharing of data which emerges often in correction procedures. It is not uncommon in handling street network files to discover a wide variety of levels of accuracy depending upon the source. For example, Canadian National Topographic System street files are (on some measures) more accurate than the government Area Master Files of street networks (the equivalent of U.S. DIME files). However, the NTS files lack the attributes of the AMF files. Correction here requires transfer of attributes from one street network file to another to build a resultant file with the best of both source files united together.

Another example of error correction arises when differing address parsers, or worse, no address parsers, are used when street address data is handled by GIS systems. Knowledge of the limitations of, and differences between, address parsers is a basic requirement of market analysis, and to that extent is central to error detection and correction procedures and practices that attach to error in GIS databases. The use of two official languages in Canada complicates this process somewhat.

This brief discussion can be summarized in noting that from the perspective of the private sector using GIS systems for market or demographic purposes, there is a need for a basic toolkit of error detection and correction methods and routines. Any research agenda looking to provide assistance to this sector should take this into consideration.

Error in data estimation

Attribute data are often made available at only one or more levels of either the census or postal hierarchy. Since IMSA works not only with predefined census, postal, and other administrative boundaries (municipal wards, polling districts), but also with client defined areas and market defined trade areas, considerable efforts are made to disaggregate all information to the smallest feasible unit of analysis which in the case of Canada is the Enumeration Area (EA).

Point-point estimation

The most commonly occurring estimation problem involves the estimation of data for postal distribution units (LDU), using census EA units as source data. The LDU (and the parent FSA level) are the preferred units for reporting for a number of market research applications, such as targeted direct mail and retail sales analysis.

The estimation of EA data at the LDU level is error-laden, since there are no consistent relationships between these units. In urban areas, an LDU is typically smaller in size than the EA while the opposite is often true in rural areas. To compound problems, only the centroid geographic coordinates are currently available for both levels of geography.

There are two primary methodologies employed. The simplistic approach effectively assigns the demographics of the nearest EA to each LDU, splitting its households and population proportionately according to post office reported household counts by LDU. The more complex approach involves the estimation of areas of each unit by computing Thiessen polygons for each area, then using polygon overlay techniques to assign demographics to each LDU. The major sources of error in the latter strategy are known to arise from the fact that LDU units are generally linear, rather than polygonal, in nature and that it is assumed that the distribution of all attributes is constant across an EA.

The second main usage of point-point estimation techniques occurs between EA units at each new census. In order to maintain historical, as well as current, information, the data from previous census periods must be estimated for current EA definitions. Unlike the EA-LDU problem, where there are few higher order units which can be used to control the reallocation, the census does provide historical counts of population only at higher levels of geography, accounting for any adjustments in areal units at those scales which may have occurred in the intervening period.

Point-area estimation

Point-area estimation problems are also frequently encountered, especially at the applications level when dealing with client-defined districts, non-standard administrative districts, and retail trading area analyses. Typically, areal estimates are derived from point estimates using standard point-in-polygon techniques. Errors will occur at the fringes of any such defined area, as border points must be declared as either lying fully within or outside the target area.

Some improvement may be possible by first using Thiessen polygon techniques to convert the point estimates to areas, then using polygon overlay techniques to apportion the estimated areas to the target polygon. It is clear that the quality of these estimates is improved somewhat if the Thiessen polygons can be bounded to an appropriate higher order areal unit into which the points are known to cleanly nest.

Area-area estimation

In some cases, it is necessary to estimate an attribute for one set of areal units based on a different data collection unit. For example, if municipality level data are to be disaggregated to, or re-created from, large postal area data (about 4000-6000) households, then the areas under analysis are overlain and estimates created (up or down) based on either the proportion overlain or based on some associated variable such as population or household density.

Errors arising from database applications

The following are some typical examples of the types of problems for which the Institute and other market analysis firms are asked to provide solutions and methodologies:

Using Institute information and estimates, take a client defined area, provide an estimate of spending on restaurant meals for that area, compare it with the spending experienced by a client restaurant, and determine whether the client restaurant is reaching its estimated potential in that market.

Using Institute information on expenditures on mail order books and gifts, select the areas with the greatest estimated expenditure, select the routes walked by the postman in these areas (Postal Walks) and determine the demographics of these individuals as well as their postal (ZIP) codes. Calculate the number of houses the postman must deliver to and determine what the potential success rate will be on a direct mail campaign.

It should be relatively easy for readers to imagine the spiral of increasingly complex uses of geographically linked data, ranging from simple location-allocation routines to site location modelling, spatial interaction and gravity models, full market planning studies, and specialized studies of government defined target groups for policy creation.

As market analysis firms move to the use of GIS techniques for the creation of reports, maps, and custom analyses for clients, methods for tracking and estimating error will need to become an integral part of the system. In short, standard estimates of error are required after each operation, and must be carried forth through multiple operations with appropriately accessible statistical software for calculating and assessing that error.

Aggregation and disaggregation problems

The aggregation and disaggregation of units requires the modelling of error during these processes, and during the processes of estimation. The problem is especially critical when both processes occur, for example, when census tract data are disaggregated to neighborhood (enumeration area - EA) level data, and the neighborhood data re-aggregated to conform to a client specified boundary composed of neighborhood (EA) centroids.

Errors also occur when rendering point data as polygon data, using polygon based techniques to estimate other polygon data, for example, using EA aggregations (points) to estimate client areas (polygons) and then using this information to aggregate client areas to higher levels of client specified geography. The following questions become important to aggregation and disaggregation exercises:

How does one model error in either direction?

How does one model and take account of multiple stage estimations in either or both directions?

What error minimization techniques can be applied, and to what extent can these be done within a GIS system as opposed to a separate or connected statistical environment?

Assume for a moment the following case. Estimates of the population of a metropolitan area are created with estimated error. The metropolitan area is essentially a polygon with a potential centroid. These estimates are disaggregated in order to develop estimates of component EAs with resulting populations and error. The population estimates of EA units are then used to derive estimates of a client specified area using a digitized polygon with an unknown accuracy, but assumed to be 3 to 10 meters, and the estimates are simply a sum of the data on the EA centroids. EAs do not fit exactly within the client area, resulting in the use of a variety of possible techniques to arrive at the final estimate:

- all centroids which fall into the client area;
- a statistical smoothing technique for parcelling out portions of the estimated EA population based on the size of component Enumeration Areas;
- some form of estimation of partial populations of border EAs which inserts an estimate of the population only in the portion of the EA which sits inside the client specified area.

In this case the main questions in need of answer are:

- How is the error of the population of the client defined area modelled/calculated?
- How are error estimates stored in a GIS system and utilized in additional operations such as creating a density surface or calculating a sub-component of this client area such as a store trading area within the client specified area?
- Can GIS systems handle data with confidence intervals and can they show or calculate variance for the area?

It is misleading to believe that because questions can be posed regarding how error is handled, that there indeed exist methods for handling these types of error in GIS systems as presently available. The methods do not exist. The questions posed here beg the existence of such thinking among GIS developers and researchers.

Discussion

When undertaking all of the above activities, the Institute is constantly required to provide some measure of accuracy of the information provided. For example, how accurate is the estimated count of elderly individuals based on an initial point estimate which is then aggregated using a point in polygon procedure? For example, if the accuracy of a digitized boundary is 10 meters rather than 3 meters, what is the effect on the accuracy of the information gathered using one level of digitizing accuracy versus another? For example, if an overlay routine is used to move data from one set of boundaries to another, how is the error calculated and handled in the analysis process, and if those areas are then aggregated, how is the further error calculated and handled in the analysis process?

A summary of the current problem set faced by IMSA and other market research firms is as follows, bearing in mind that most client interest is understandably focussed on areas which are undergoing change, either in terms of areal definition or of characteristic (attribute) change (e.g. population growth). It is this problem set which gives rise to many of the concerns voiced here.

1. Estimation of attributes of non-standard areal units (e.g. circular areas, custom trading districts) arising from point in polygon estimation techniques.
2. Estimation of overlapping boundaries over time, such as estimating 1981 data on 1986 geographical units. Depending upon the nature of the geographic areas, estimation is point-point, point-area, and area-area in nature. In some cases, areas are estimated using Thiessen polygon techniques.
3. Estimation of attribute data for missing units, such as new areas, areas with unreliable data.
4. Forecasting for a single geographic unit (e.g. municipalities, Census Tracts), confounded by the inconsistencies of areal units over time.
5. Forecasting from one set of data to another and from one set of areal units to another. For example, forecasting EA growth from time-series data on FSA units.
6. The incorporation of newly created neighbourhoods, involving the splitting of existing EA units (for which only the geographic centroid coordinates are known).
7. The estimation of demographic and related attributes for Postal Walks (which are effectively paths defined by LDU units). Each LDU is only loosely attached to the EA base level.

The ability to minimize error in all or some of these activities is dependent upon an ability to effectively measure the error in each component operation in the chain.

Despite the current focus on error induced during the analysis of spatial data, it is far too easy to search for the classic technological fix to problems in datasets. One thing the use of GIS systems should provide us is the realisation that increased attention to the source data themselves will alleviate more problems than all the combined technological toolkits and error correction algorithms built into commercial GIS packages. That is to say, that the revolution in GIS should portend a similar upheaval in the preparation and design of standard datasets which feed the analysis of spatial data by a variety of users with an unpredictable range of uses.

In a document which purports to do no more than raise questions, the only conclusion can be that the set of questions is formidable. Looking forward, the research agenda of import to the private sector market analysis community should to some extent derive from the problem set and the interests of the group describing them.

Using these few examples and describing some of the basic services of market analysis firms, we have attempted to create a comprehension of the nature and problems inherent in this type of enterprise. Since it is our perception that businesses will increasingly become major purchasers and users of both GIS systems, and GIS based services offered by firms such a IMSA, we suggest that the impact of overlooking or neglecting issues of error detection, handling, and correction, is considerable. The investment required by researchers, users, vendors, and developers is high, and the needs of the private sector are viewed in short term horizons. The very capabilities which attract users to GIS systems are those which permit major levels of error propagation when they are used. Any research agenda which hopes to assist the GIS user and developer community in North America must not ignore the very elementary problems touched upon in this paper.

Chapter 16

The small number problem and the accuracy of spatial databases

Susan Kennedy

Introduction

Often when the topic of the accuracy of spatial databases arises one thinks in terms of the positional accuracy of the punctiform, linear, areal or volumetric features of the earth's surface. In many GIS applications, this type of focus is appropriate and necessary for the task at hand which may involve map overlay operations or terrain modelling in a natural resources context or may have legal implications in a cadastral database context. For most GIS applications in human geography, the research context involves the use of census data or data tabulated by census enumeration units. For these applications, which may involve dozens or even hundreds of data layers with the same polygonal boundaries (choropleth maps), the concept of the accuracy of the spatial database is quite different. This paper will briefly review the types of error on choropleth maps which are commonly held to be important, and introduce an overlooked component of choropleth map accuracy as the Small Number Problem. A short history of the Small Number Problem, and a review of various attempts to minimize it or solve it will be given with specific examples from epidemiologic, health care delivery and health care resource allocation contexts.

MacEachren (1985) defines accuracy for any quantitative thematic map as a function of the following four factors: 1) errors in map production (planimetric errors), 2) errors in data collection methods, 3) errors in data classification strategies, and 4) errors of symbolization techniques. Jenks and Caspall (1971) identify three types of error specific to choropleth maps: tabular, boundary, and overview error. Tabular error refers to the error which occurs when a user observes the shaded pattern of an enumeration unit and reads the intensity value for that shading from the map legend. Boundary error is that which occurs when boundaries between choropleth shadings on the map are not breaks in the statistical surface. Jenks and Caspall describe overview error as the error between the choropleth map classification and reality, where the observed data value for each areal unit is considered to be "reality". Goodchild and Dubuc (1987) make a distinction between error in feature location and attribute error on choropleth maps. I propose the following five factor classification for attribute error on choropleth maps:

1) errors in data collection,
2) errors in data classification,
3) errors in data analysis,
4) errors in perception of map content (visualization), and
5) errors in representing reality (random sampling errors).

Errors in data collection and recording are well known and will not be addressed here. Errors in data classification are addressed by Jenks and Caspall (1971) and Jenks (1977). Errors in data analysis occur when inappropriate data analyses are undertaken (either the method chosen is not appropriate for the problem at hand or the assumptions of the method are not met) and displayed on a choropleth map in a geographic information systems context. Errors in perception of map content have been addressed in a limited context elsewhere (Muller, 1979; Petersen, 1979; and Cartensen, 1982), but see Olson (1975) and Costanzo (1985) for a related discussion on visual map complexity. Errors in representing reality will always occur even if it were possible to eliminate all other sources of error, because every phenomenon that the scientist tries to observe or measure is subject to sampling error due solely to chance. This sampling error will occur even if the sample being measured is completely enumerated. Let us assume, for example, that we can accurately count all deaths in an enumeration district, and that there are no recording errors and no deaths are left uncounted. There is sampling error in this complete enumeration because what we are actually interested in is the unknown probability of death and not the observed probability of death (deaths/total population) which varies with the size of the enumeration district.

A major source of inaccuracy, above and beyond data collection errors and bias, in social and economic spatial databases is the variable size of the enumeration districts, which leads to what Jones and Kirby (1980) call the "small-number problem". While such problems as the Modifiable Areal Unit Problem (Openshaw and Taylor, 1981) and the Scale Problem (Haggett, 1965; Harvey, 1968) have received much attention in geography, the Small Number Problem is an equally important although less well-known problem. The Small Number Problem occurs whenever one uses a percentage, ratio, or rate calculated for any geographic area for which the population of interest (denominator) is sparse or the numerator is a rare event. If this occurs, small random fluctuations in the variable of interest (numerator) may cause large fluctuations in the resulting percentage, ratio, or rate. This problem is exacerbated whenever the phenomenon of interest is a relatively rare event, especially for disease mortality or morbidity rates. Indicators of this type are often used for epidemiologic surveillance and in public health planning for resource allocation. Indicators used in other types of resource and fiscal planning are also susceptible to the small number problem when the denominator is small, as is often the case at the local scale.

Related to the Small Number Problem is the problem of missing denominators for rates. The denominator problem has several facets, one of which is related to scale discordance in which the denominators are not available for the set of areas or at the scale one wishes to investigate. Another problem with denominators is what to do in intercensal years, particularly if we require projection into the future. The discordance of scale problem would be solved if data were available by address instead of by enumeration districts. This is not a problem which is likely to be resolved soon because of archaic confidentiality constraints and because address is not always a mandated data item in many information gathering systems. For many health applications, it is necessary to go back to the paper copy of the death certificate and abstract the address and information by hand in order to obtain data at the address level. I will not discuss the denominator problem in detail here, since it is complicated enough to deserve its own treatment, except to mention that it may interact with the small number problem.

The problem

The Small Number Problem poses three interrelated problems: the problem of map pattern analysis, modelling with unreliable spatial series, and the problem of rank-ordering and identifying extreme values.

The first is the problem of map pattern analysis in which the extreme percentages which result from the Small Number Problem dominate the display on the map. As an example, consider the analysis of mortality rates in a small geographic area which has a population of ten people, one of whom dies of cancer. The percentage of people dying from cancer is 10%, or as more commonly specified, as a mortality rate of 10,000 deaths per 100,000 people. Although, the example given here is synthetic, it is nonetheless an example of the problems inherent in the analysis of percentages, ratios, rates and proportions in small geographic populations. This type of occurrence is counter-intuitive because we would like those areas of the map which have higher absolute numbers and presumably greater statistical reliability to dominate the map.

Maps of rare events are often used in setting priorities for public health policy. In the United States, the National Cancer Institute uses maps of standardized cancer mortality rates to target analytical epidemiologic studies to geographic areas at high risk for cancer. An example of the power of this technique is the National Cancer Institute's *Atlas of Cancer Mortality for U.S. Counties: 1950-1969* (Mason *et al.*, 1975) which has sparked a multitude of epidemiological studies of cancer, including work on geographic patterns of lung cancer (Blot and Fraumeni, 1973, 1982), large bowel cancer (Blot, Fraumeni, Stone and McKay, 1976), breast cancer (Blot, Fraumeni and Stone, 1977), prostate cancer (Blair and Fraumeni, 1978) and pancreas cancer (Blot, Fraumeni and Stone, 1978). The maps of lung cancer by county published in the atlas revealed excessive rates of lung cancer among males along the Gulf Coast of the United States from Texas to Florida. Upon further investigation, researchers at the National Cancer Institute, in collaboration with the Center for Disease Control, discovered a relative summary risk of 1.6 associated with employment in area shipyards during World War II (Blot, Stone, Fraumeni and Morris, 1979).

Although a cartographic/geographic or computer-assisted analysis is most often undertaken when it is suspected that the pattern of a disease is influenced by environmental factors, the same techniques can be used to examine the distribution and utilization of health care resources. Computer-assisted cartography has been used as an aid for targeting federal resources to rural areas in the planning and implementation of the U.S. Public Health Service's Rural Health Initiative (RHI) program (Baldi, 1980). The Rural Health Initiative was established by the Public Health Service's Bureau of Community Health Services (BCHS) in 1975 to help medically underserved geographic areas establish primary care health clinics. The BCHS and RHI staff agreed on the following four criteria to identify the severity of need: 1) medically underserved areas, 2) health manpower shortage areas, 3) high migrant impact areas and high impact areas, and 4) high infant mortality areas. All the factors above, with the possible exception of factor 3, rely on percentages, rates, or ratios for presumably small populations in rural areas. It would be interesting to analyze the standard errors of the four indices to see which geographic areas, whose indices were not significantly different from those chosen as priority areas to receive RHI funding, did not receive funding.

The second problem associated with the Small Number Problem occurs when modelling is attempted with spatial data sets of varying degrees of reliability. Difficulties arise when these numbers are treated as ordinary numbers and correlation and regression analyses are applied to them. Wilson (1978) suggested that the "now quite standard application of correlation and regression techniques to arrays of small area mortality rates or ratios may be productive of erroneous results." Wilson found a large range in infant mortality rates for the shires and municipalities of metropolitan Sydney, with a metropolitan average of 17.83 per thousand and a low of 11.04 per thousand and a high of 32.36 per thousand. Although Wilson found that six of the LGA's had infant mortality rates significantly above and five of the LGA's had rates significantly below the metropolitan average, and these rates did tend to occur towards each end of the ranked array, he concluded that even the most extreme rates in any array derived from small numbers of deaths for an area would be unlikely to represent important differences from the mean because the standard error of the rate is likely to be quite high.

Numerous studies based on data such as this do not stop with the descriptive analysis that suggests that further statistical analysis of the data might be inappropriate. Wilson cites himself (1976) as an example of a study which ignored the shortcomings of death rates for small areas in a search for statistical associations between death rates and a variety of social and environmental variables.

Wilson (1978) plots infant mortality rates and their associated confidence interval 'envelope' against a measure of social disorganization and concludes that other sets of infant mortality rates randomly selected from within the confidence interval envelope might generate different results. Wilson ran four regressions using each one of the following, in turn, as the dependent variable: infant mortality rates calculated from the raw data; infant mortality rates representing the outer extremes of the confidence interval around each rate; infant mortality rates representing the inner extremes of the confidence interval for each rate; and infant mortality rates derived by randomly adding to or subtracting from each rate an amount equal to one standard error of the rate. His results indicate that there is substantial variability in the explanatory power of the regression and in each of the regression coefficients, and in one regression, both the slope coefficient and the regression have values which are not statistically significant. Another reason that applying the standard OLS regression to arrays of small area mortality rates will be very likely to produce erroneous results is because with OLS it is possible to produce nonsensical predicted values. Although this result is not strictly part of the small number problem, it is worth mentioning here because most geographers ignore Wrigley's (1973, 1976) suggestion to perform logit or probit transformations on rates before applying OLS regression.

The third problem occurs when it is desired to rank a set of geographic areas in order to identify the n best or worst or the top n percent or bottom n percent using some indicator which is based on percentages, proportions, rates, or ratios. In epidemiology, the studies which attempt to address this problem are known as small area analysis studies. These studies analyze variations in the amounts of health services consumed by inhabitants of small geographic areas and almost unilaterally conclude that any variation observed in the supply of health services (usually surgical procedures) is "bad" (Lewis, 1969; Wennberg and Gittelsohn, 1973, 1982; Roos, 1981). They make this conclusion because they assume that the etiology of disease which leads to the need for a specific medical procedure is the same in all populations and geographic areas and that any observed variation must be the result of physician practice style. The usual analytic method is to calculate the utilization rates for surgical procedures, and observe that the difference between the highest and lowest rate is quite large and to attempt to explain the difference as a function of other variables such as physician practice style or treatment option availability. Diehr (1984) in a paper appropriately titled "Small Area Variation: Large Statistical Problems", estimated that if one assumes a normal distribution and the same underlying mean rate, that just by chance alone the ratio of the largest rate to the smallest will be 10.9 if twenty small areas are compared.

Aggregation solutions

Traditionally, the problem of small numbers has been overcome by aggregation over time, space or disease type. All these methods of aggregation have disadvantages. Aggregation over time can conceal time trends. Aggregation over geographic areas conceals geographic variation and poses the question of the modifiable areal unit problem. Aggregation over disease type will cause epidemiologic confounding when disease types of differing etiology are aggregated. Modal threshold aggregation is an aggregation method in which small census units are aggregated with neighboring units in order to bring them up to the modal value of the data set. Kirby and Tarn (1976) used this method to aggregate census tracts to a level of 10,000 households as an approach to addressing the Small Number Problem.

Traditional cartographic solution

There are several cartographic solutions to the small number problem: map classification, data suppression, proportional symbol maps, and cartograms. It is a commonly held belief among cartographers that the judicious choice of appropriate classification intervals will reduce bias and error in map interpretation. This belief is simply not true when the variables being mapped are subject to the Small Number Problem, because the amount of variation caused by chance can result in an area changing not just one but several map class intervals.

The data suppression method for addressing the Small Number Problem uses a minimum cut-off threshold method and ignores the proportions which fall beneath it. The difficulty with this method is that symbols representing suppressed data may dominate the map display, as in the case of the example given by Jones and Kirby (1980).

The size of the population at risk may be displayed with proportional symbol maps with the size of the symbol proportional to the size of the population at risk and the rate represented by the shading or pattern of the symbol. Examples of this method are given by Cole (1964) and Howe (1970,1972).

In the cartogram solution, each areal unit is drawn in proportion to its denominator. Forster (1966) presented the development of a demographic base map (cartogram) as an alternative to the usual geographic base map for the display of epidemiologic data. His use of the cartogram is one way to include the information about the size of the population at risk on the map, thus providing a way to relate disease rates to local at-risk populations in addition to the more common geographic position. Other cartogram solutions are provided by Tobler (1973,1976), Levison and Haddon(1965), and Hunter and Young (1971).

Analytical cartographic solutions

Jones and Moon (1987) suggest using a chi-square index of chance effect to decide whether or not an index differs sufficiently from 100 for a chance effect to be unlikely. Their index follows:

$$\text{Index of chance effect} = \frac{(O - E)^2}{E} \qquad (1)$$

where O is the observed number of deaths and E the expected number, for five or more observations. The Poisson generating function (White, 1972) can be used to calculate the probability of a given observation occurring by chance if there are fewer than five observations.

Chi-square maps have been used in the analysis of census data by Jones and Kirby (1980). The chi-square solution has the useful property that for a given proportion, chi-square variables are smaller for small values than for larger ones. A chi-square value appropriate for mapping is given using Mantel's formula (Gilliam and MacMahon, 1960;

Jones and Kirby, 1980):

$$\chi^2 = \frac{O_t(O-E)^2}{EO+E^2} \tag{2}$$

where O_t is the total observed count. If the observed value is greater than the expected value the computed chi-square quantity is given a positive sign and if the observed value is less than the expected value then the computed chi-square quantity is given a negative sign. This formula takes into account the absolute size of the observed and expected values. This use of the chi-square statistic for mapping is a deviation from the exact formal manner in which a chi-square statistic is usually used, and it is not intended to demonstrate that a particular observed value in a geographic area is significantly different from the expected value at a certain significance level. The repeated use of the method, the problem of spatial autocorrelation (Cliff, Martin and Ord, 1975), and the small size of the expected values, prevent attaching an exact probability level to the statistic.

Visvalingham (1976,1978) used a chi-square formulation, which differs from the formulation given above by not including the observed total, and found that gridded census data maps display noticeably different map patterns when ratio and chi-square maps are compared. Clark (1980) demonstrated that chi-square and ratio maps of house foundation collapse during a drought were markedly different from one based on proportions.

Other solutions

Wilson presents an alternative approach to using conventional correlation and regression analysis in which he categorizes the mortality rates in a manner which combines the order of magnitude information with measures of significance. He then does a preliminary reduction of the data matrix by using principal components analysis because there are many relevant independent variables and they are not statistically independent. He suggested applying the Kruskal-Wallis One-Way Analysis of Variance and did a step-wise regression on a limited number of categories. As an example, he used three categories: infant mortality rate significantly higher than the metropolitan mean, infant mortality rate not significantly different from the metropolitan mean, and infant mortality rate significantly below the metropolitan mean.

Williams (1979) estimated the component of variation in perinatal mortality rates that was due to random error and then used these estimates as weights in a GLS regression analysis to measure the effectiveness of perinatal care. Williams estimated the extent to which hospital characteristics influenced perinatal outcomes and found (with one exception) that all of his correlations had the expected sign. His success is in great contrast to other attempts to demonstrate a connection between process measures of quality of health care and outcome. Several other studies have produced contradictory results in which observed correlations between process and outcome variables are the opposite of those expected (Ashford *et al.*, 1973; Tokuhata *et al.*, 1973).

Manton *et al.* (1987) present a statistical method to address the problems of small population size and rareness of events in rate estimation for small local areas. They suggest a composite rate estimator, which is a combination of the local rate estimate and the total population estimate, to increase the stability of the local area rate estimates. Manton *et al.*, use a negative binomial regression model in an "empirical Bayes" procedure to determine the best weighting of the local rate estimates for the composite rate estimate. They assume that the local area mortality count is a Poisson variable with an unknown mortality rate which is gamma distributed. If these assumptions are satisfied

(Manton *et al.*. 1987) then the observed mortality count can be described by the negative binomial distribution and the conditional distribution of the unknown mortality rate will also be gamma distributed. The conjugate prior is the most robust choice (Morris, 1983; Manton *et al.* 1987) for estimating the mortality rates in the sense that the expected value of the maximum squared-error loss is minimized, and this property still holds approximately when the mean and variance are estimated from observed data.

The model used by Manton *et al.* (1987) has two stages. In the first stage, a negative binomial regression model is used to produce an estimate of both the local area mortality rate and the super-Poisson variability. In the second stage, the estimates of the local area mortality rates and the super-Poisson variability are used to calculate the local area composite estimates. They calculate the composite rates using the empirical Bayes point estimator

$$h_i = \frac{w_i y_i}{n_i} + (1 - w_i)\,\mu_1 \qquad (3)$$

where the weights w_i are calculated as the ratio of the standardized coefficient of super-Poisson variation to the standardized coefficient of total variation and y_i/n_i is the observed rate and μ_1 is the model based rate.

Manton *et al.* (1987) applied the procedure outlined above to U.S. county age and sex-specific mortality rates for lung and bladder cancer and found that the composite rate procedure produced estimates that were more stable over time than the observed rates for the small areas.

Conclusions

The accuracy and reliability of spatial databases will suffer unless the Small Number Problem is recognized and addressed by GIS users and researchers. This paper has suggested several ways that this problem can be addressed - with aggregation, visual display, and modelling approaches. All three of these approaches, separately or in combination are useful in GIS applications and modelling. For example, it will often be appropriate to aggregate the data over both time and space, and to explicitly model the random variation in the variable of interest and to create maps which display the reliability of the data. Explicit modelling of the random component is the only approach which addresses all three problems which result from the Small Number Problem: difficulties in map pattern analysis, modelling with data of varying degrees of reliability, and rank-ordering and identification of extreme values. The most promising avenue of future research appears to lie in the Bayesian and empirical Bayesian approaches which can be used minimize the maximum squared-error loss over the entire spatial database.

References

Ashford, J. R., Read, K. L. Q., and Riley, V. C., 1973, An analysis of variations in perinatal mortality among local authorities in England and Wales. *International Journal of Epidemiology* **2**, 31-46.

Baldi, J. M., 1980, Cartography as an aid for targeting federal resources to rural areas. *Proceedings of the International Symposium on Cartography and Computing, Applications in Health and Environment*, Volume I. Auto-Carto IV. American Congress on Surveying and Mapping and American Society of Photogrammetry, pp. 26-32.

Blair, A., and Fraumeni, J., 1978, Geographic patterns of prostate cancer in the United States.*J. Natl. Cancer Inst.* **61**, 1379-1384.

Blot, W., and Fraumeni, J. F., 1973, Geographical patterns of lungcancer, industrial correlations. *Am. J. Epidemiol.*. 539-550.

Blot, W., and Fraumeni, J. F., 1982, Changing patterns of lung cancer in the United States. *Am. J. Epidemiol.* **115**(5), pp. 664-673.

Blot, W., Fraumeni, J., Stone, B.J., and McKay, F., 1976, Geographic patterns of large bowel cancer in the United States. *J. Natl. Can. Inst.* **57**, 1225-1231.

Blot, W., Fraumeni, J. F., and Stone, B. J., 1977, Geographic patterns of breast cancer in the United States. *J. Natl. Can. Inst.* **59**(5), 1407-1411.

Blot, W., Fraumeni, J., and Stone, B., 1978, Geographic correlates of pancreas cancer in the United States. *Cancer* **42**, 373-80.

Blot, W., Mason, T. J., Hoover, R., and Fraumeni, J. F., 1977, Cancer bycounty, etiologic implications. In *Proceedings of the Cold Spring Harbor Conferences on Cell Proliferation: Origins of Human Cancer*, Hiatt, H , Watson, J.D., and Winsten, J.A., (Eds.), pp 21-32, Cold Spring Harbor, New York.

Blot, W., Stone, B. J., Fraumeni, J. F., and Morris, L.E.,1979, Cancer mortality in U.S. counties with shipyard industries during World War II. *Environ. Res.* **18**(2), 281-90.

Cartensen, L. W., 1982, A continuous shading scheme for two-variable mapping. *Cartographica* **10**, 53-70.

Clark, M. J., 1980, Property damage by foundation failure. In *Atlas of Drought in Britain 1975-6,* Doornkemp, J. C., and Gregory K. J. (Eds.), (London: Institute of British Geographers Special Publication).

Cliff, A. D., Martin, R. L., and Ord, J. K., 1975, A test for spatial autocorrelation in choropleth maps based upon a modified χ^2 statistic. *Transactions of the Institute of British Geographers* **65**, 109-131.

Cole, J. P., 1964, *Italy*. (London: Chatto and Windus)

Costanzo, C. M., 1985, *Spatial association and visual comparison of choropleth maps*, Unpublished Ph.D. dissertation. University of California, Santa Barbara.

Diehr, P., 1984, Small area statistics, large statistical problems. *Am. J. Public Health* **74**(4),313-314.

Evans, I. S., and Jones, K., 1981, Ratios and closed number systems. In *Quantitative geography*, Wrigley, N., and Bennett , R. J., (Eds.), pp. 123-134.(London: Routledge & Kegan Paul)

Forster, F., 1966, The area of a demographic base map for the presentation of area data in epidemiology. *British Journal of Preventative and Social Medicine* **20**, 165-71.

Gilliam, A. G., and MacMahon, B., 1960, Geographic distribution and trends of leukaemia in the United States. *Acta. Un. int. Canc.* **16**,1623-28.

Goodchild, M. F., and Dubuc, O., 1987, A model of error for choropleth maps, with application to geographic information systems. *Proceedings, Auto-Carto 8*, pp. 165-174.

Haggett, P., 1965, Scale components in geographical problems. In *Frontiers in geographical teaching*, Chorley, R. J., & Haggett, P., (ed), pp. 164-185.(London: Methuen)

Harvey, D. W., 1968, Pattern, process and the scale problem in geographical research. *Transactions of the Institute of British Geographers* **48**,71-78.

Howe, G. M., 1970, *A National Atlas of Disease Mortality in the United Kingdom*, 2nd edition. (London: ??)

Howe, G. M., 1972, *Man, Environment and Disease in Britain* . (New York: Barnes & Noble Books)

Hunter, J. M., and Young, J., 1971, Diffusion of influenza in England and Wales. *Annals of the Association of American Geographers* **61**(2), 637-653.

Jenks, G. F., 1977, *Optimal data classification for choropleth maps*. Department of Geography Occasional Paper 2, The University of Kansas, Lawrence, Kansas.

Jenks, G. F., and Caspall, F. C., 1971, Error on choropleth maps: definition, measurement, reduction. *Annals of the Association of American Geographers* **61**(2), 217-244.

Jones, K., and Kirby, A., 1980, The use of chi-square maps in the analysis of census data. *Geoforum* **11**, 409-417.

Jones, K., and Moon, G., 1987, *Health, Disease and Society: An Introduction to Medical Geography*, pp 59. (London: Routledge and Kegan Paul)

Kirby, A. M., and Tarn, D., 1976, Some problems of mapping the 1971 census by computer. *Environ. Plan. A.* **8**, 507-513.

Levinson, M.E., and Haddon, W. , 1965, The area-adjusted map: an epidemiologic device. *Publ. Hlth. Rep.* **80**, 55-59.

Lewis, C. E., 1969, Variations in the incidence of surgery. *N. Engl. J. Med.* **281**,880-885.

MacEachren, A. M., 1985, Accuracy of thematic maps: implications of choropleth symbolization. *Cartographica* **22**(1), 38-58.

Manton, K. G., Stallard, E., Wodbury, M. A., Riggan, W. B., Creason, J. P., and Mason, T. J., 1987, Statistically adjusted estimates of geographic mortality profiles. *Journal of National Cancer Institute* **78**(5), 805-815.

Mason, T. J., McKay, F. W., Hoover, R., Blot, W. J., and Fraumeni, J. F., Jr., 1975, *Atlas of Cancer Mortality for U.S. Counties: 1950-1969,* DHEW. Publ. No. (NIH) 75-780. U.S. Govt. Printing Office, Washington, D.C.

Morris, C. N., 1983, Parametric empirical Bayes inference: theory and applications. *J. Amer. Stat. Assoc.* **78**(381), 47-65.

Muller, J. C., 1979, Perception of continuously shaded maps. *Annals of the Association of American Geographers* **69**, 240-249.

Olson, J. M., 1975, Autocorrelation and visual map complexity. *Annals of the Association of American Geographers* **65**(2), 189-204.

Openshaw, S., and Taylor, P. J., 1981, The modifiable areal unit problem. In *Quantitative geography*, Wrigley, N and Bennett, R. J., (Eds.), pp. 61-69. (London: Routledge & Kegan Paul)

Petersen, M. P., 1979, An evaluation of unclassed crossed-line choropleth mapping. *The American Cartographer* **6**, 21-37.

Roos, N., 1981, High and low surgical rates: risk factors for area residents. *Am. J. Public Health* **71**, 591-600.

Tobler, W. R.,1973, A continuous transformation useful for districting. *New York Academy of Science Annals* **219**, 215- 220.

Tobler,W. R., 1976, Cartograms and cartosplines. *Proceedings of the 1976 Workshop on Automated Cartography and Epidemiology*, U.S. Department of Health, Education and Welfare, Public Health Service, Office of Health Research, Statistics, and Technology, National Center for Health Statistics. pp. 53- 58.

Tokuhata, G. K., Colfesh, V. G., Mann, K., and Digon, E., 1973, Hospital and related characteristics associated with perinatal mortality. *Am. J. Pub. Health* **63**, 163-185.

Visvalingham, M., 1976, Chi-square as an alternative to ratios for statistical mapping and analysis. Census Research Unit. *Working Paper* **8**, Department of Geography, University ofDurham, 1976.

Visvalingham, M., 1978, The signed chi-square measure for mapping. *Cartographic Journal* **15**, 93-98.

Wennberg, J., and Gittelsohn, A., 1973, Small area variations in health care delivery. *Science* **18**, 1102-1108.

Wennberg, J., and Gittelsohn, A., 1982. Variations in medical care among small areas. *Scientific American* **246**, 120-134.

White, R. R., 1972. Probability maps of leukemia mortalities in England and Wales. In: McGlashan, N.D., (Ed.) *Medical Geography: Techniques and Field Studies,* pp. 173-85. (London: Methuen)
Williams, R. L., 1979, Measuring the effectiveness of perinatal care. *Medical Care* **17**(2), 95-110.
Wilson, M. G. A., 1976, Infant death in metropolitan Australia 1970-73: dimensions and determinants. Paper presented to the IGU population commission symposium, Minsk, USSR, July 1976.
Wilson, M. G. A., 1978, The geographical analysis of small area/population death rates, a methodological problem. *Australian Geographical Studies* **16**, 149-60.
Wrigley, N., 1973, The use of percentages in geographical research. *Area* **5**, 183-6.
Wrigley, N., 1976, *An Introduction to the Use of Logit Models in Geography*, CATMOG **10**. (Norwich, England: Geo. Abstracts)

Chapter 17

Demand point approximations for location problems

Rajan Batta

Abstract

In this position paper we discuss the quality of demand point approximations for location problems. There are three reasons why this research is important: (i) urban planners frequently make the simplifying assumption of restricting facility location to demand points, perhaps because they feel that demand points are the only points that *control* the eventual locations of the facilities and hence are a *good* subset of feasible points for facility location; (ii) reducing the number of feasible solutions enhances our ability to solve larger location problems, which is important from an applications standpoint; and (iii) demand point approximations lead to heuristic solutions for the actual problem. We summarize previous and ongoing research in this area and also provide directions for future research.

Introduction and review of previous work

Consider the p-median problem on a two-dimensional plane; the objective of the p-median problem is to locate p facilities so as to minimize the average distance of a demand point to its closest facility. When using the Manhattan travel metric and in the absence of impenetrable barriers to travel, the set of *intersection* or crossing points comprise a solution set for the p-median problem -- see the location theory text by Francis and White (1974). If we have n demand points, each with unique X and Y values, we generate n^2 intersection points. Yet only a small fraction (4 % for n=25) of these are actual demand points that *control* the eventual locations. The other (96% for n=25) intersection points are *zero-demand* locations. As discussed in Larson and Sadiq (1983), the number of intersection points increases quite dramatically with the addition of barriers. Thus the demand points constitute an even smaller fraction of the total number of intersection points when barriers are present.

From the above statements about the p-median problem on a two-dimensional Manhattan plane, it is intuitively clear that restricting facility location to demand points would probably provide a good approximation to the actual problem. Perhaps it is for this reason that urban planners frequently make this assumption when solving planar facility location problems. The potential savings in computational effort by restricting the solution space to just the demand points enhances the ability to obtain reasonable solutions to large problems -- this increases the ability to effectively tackle real-world location problems, which typically have hundreds of spatially scattered demand points and dozens of facilities to be located. The above discussion also applies when other travel metrics like the Euclidean metric, the l_p -norm, and the Block norm are used instead of the Manhattan metric; see the paper by Thisse, Ward and Wendell (1984) for an excellent discussion on travel metrics on a two-dimensional plane. However, we note that, except in a very special case (see Katz and Cooper, 1981), there is no known analytical solution method for the planar p-median problem with impenetrable barriers when the travel metric is not Manhattan. The above discussion also applies to other location problems like the p-center problem and the Stochastic Queue Median (SQM) problem; the p-center problem seeks to locate p facilities so as to minimize the maximum distance to a demand point from its closest facility (see Hakimi, 1964), and the SQM problem's objective is to locate a facility so as to minimize the average response time to a call, given that the service system is operating as an M/G/1 queue -- see Berman, Larson and Chiu (1985) for an explanation of the SQM problem.

The only work known to the author which addresses the issue of demand point approximations to location problems is a recent paper by Batta and Leifer (1988). In their paper, Batta and Leifer study the accuracy of demand point solutions to the planar, Manhattan metric, p-median problem, both with and without impenetrable barriers to travel. In the absence of barriers to travel, they show that tight worst case bounds are achieved by moving the facilities to their closest demand point. In the presence of barriers, they show that similar worst case bounds are achieved when the set of demand points is augmented with the set of barrier intersection candidate points. They report computational experiments for : (a) random locations and random weights; (b) clustered locations and spatially autocorrelated weights; (c) clustered locations and spatially autocorrelated weights with impenetrable barriers to travel. Their empirical findings are gathered from : (a) census tract data from Buffalo, New York; and (b) census tract data from Buffalo, with natural barriers to travel. Their conclusions show that : (a) restricting facility location to demand points is an excellent approximation in the absence of barriers to travel; and (b) the set of demand points should be augmented with the set of barrier intersection candidate points to preserve the overall quality of the solution in the presence of barriers to travel.

To give the readers a flavour for this type of research, we cite two studies on insensitivities in urban systems. Larson and Stevenson (1972) perform calculations with spatially homogenous demands which suggest that the mean travel time resulting from totally random distribution of facilities in the region served is reduced by only 25% when the facilities are optimally distributed. Larson and Odoni (1981) perform calculations to show that ignoring the size of a city block in computing travel distances induces an average error of no greater than 1/3 the size of the block.

The rest of this paper is organized as follows: Section 2 discusses the status of our current research and section 3 discusses our future research plans.

Status of current research

V.Viswanathan, a graduate student in the Department of Industrial Engineering at SUNY at Buffalo, is currently researching the accuracy of demand point approximations to the planar p-median problem, while using the Euclidean travel metric and with and without polygonal barriers to travel. His work is being supervised by Dr. Rajan Batta and is being assisted by Dr. Wayne F. Bialas, both of the Industrial Engineering Department at SUNY at Buffalo. In this section we will describe the status of Viswanathan's dissertation work.

As mentioned earlier, there is no known analytical solution method for the planar p-median problem while using the Euclidean metric when there are impenetrable barriers to travel, except in the case of one circular impenetrable barrier (see Katz and Cooper). The first step in researching the accuracy of the demand point approximation is therefore to develop an analytical procedure to solve the case with impenetrable barriers to travel;

we assume that the barriers are polygonal. Towards achieving this initial goal, we first develop a method to find the shortest path between two points on a Euclidean plane in the presence of polygonal barriers.

Consider the planar p-median problem with existing facilities at (a_i, b_i), i = 1,...,m. Let B_1 ,..., B_t be vertices of the polygonal barriers. The following theoretical result is useful in identifying the shortest path between two points when employing the Euclidean metric.

Theorem 1 *A shortest path between two points that are not visible from each other must pass through barrier vertices.*

Proof (by contradiction): Let the two points which are not visible to each other be A and C. It is to be shown that the shortest path from A to C will pass through the barrier vertices. Let the shortest path from A to C be A - V_1 - ... - V_k - C, where V_1 is visible from A, V_k is visible from C, V_2 is not visible from A, V_{k-1} is not visible from C, V_l is visible from V_{l+1} for l=1,..., k-1, and V_l is not visible from V_{l+2} for l = 1,... , k - 2 . Let us assume that V_i is not a barrier vertex, where $1 \le i \le k$. We show that there exists a barrier vertex B_j such that the sum of the distance from V_{i-1} (note that $V_0 \equiv A$) to the barrier vertex B_j and the distance from Bj to Vi+1 (note that $V_{k+1} \equiv$ C) is shorter than the distance from V_{i-1} to V_i plus the distance from V_i to V_{i+1} . Refer to Figure 1 and let d(A,B) denote the Euclidean distance between two visible points A and B. We have to show that

$$d(V_{i-1}, V_i) + d(V_i, V_{i+1}) \ge d(V_{i-1}, B_j) + d(B_j, V_{i+1}) \tag{1}$$

Using the triangle inequality it can be shown that :

$$\begin{aligned} &d(V_{i-1}, V_i) > d(V_{i-1}, P); \\ &d(V_i, Q) > d(P, B_j); \text{ and} \\ &d(Q, V_{i+1}) > d(B_j, V_{i+1}). \end{aligned} \tag{2}$$

From these we get

$$\begin{aligned} &d(V_{i-1}, V_i) + d(V_i, Q) + d(Q, V_{i+1}) > \\ &\quad d(V_{i-1}, P) + d(P, B_j) + d(B_j, V_{i+1}), \end{aligned} \tag{3}$$

and hence the result. ■

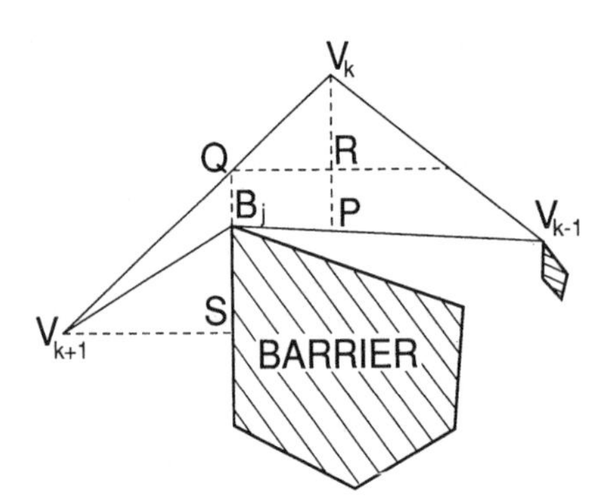

Figure 1. Illustration for Theorem 1.

The above theoretical result can be used to obtain an algorithm to find the shortest path between two points A and B on a Euclidean plane with polygonal barriers. To do this we construct a network as follows: the nodes of the network are points A and B and the barrier vertices $B_1, \ldots, B_t$. Two nodes are connected via an arc if they are visible from each other. The length of the arc is the Euclidean distance between the points. We can now use any well known shortest path algorithm (for example, see Dijkstra, 1959) to find the shortest path between nodes A and B. The length of this shortest path on the constructed network is equal to the length of the shortest path between the two points A and B on the Euclidean plane which does not penetrate a polygonal barrier.

We now develop an exact, albeit inefficient, method to solve the 1-median problem on a Euclidean plane with polygonal barriers. The exact method involves two steps. The first is allocation of the demand points to form pseudo demand points. The second step is to solve the resulting single facility Euclidean distance location problem.

Each demand point can either be allocated to itself or to one of the barrier vertices. In other words we create pseudo demand points. The weight of these pseudo demand points is the sum of the weights of the demand points allocated to it. The algorithm to find the optimal location is given below.

> Step 1: Consider an allocation of demand points (note that a demand point can either be allocated to itself or to one of the barrier vertices). The points to which the demand points are allocated will be referred to as pseudo demand points.
> Step 2: For the above allocation of demand points, calculate the weights of the pseudo demand points. The weight of the pseudo demand point is equal to the sum of the weights of the demand points assigned to it.
> Step 3: Identify the regions on the plane which are visible from all the pseudo demand points. These regions are either bounded or unbounded polygons, since the barriers are assumed to be polygons.
> Step 4: Solve the constrained single facility location problems that arise in the regions commonly visible from these pseudo demand points. The distances to be used are the distances from the candidate facility location site to the demand point via the point to which the demand point has been allocated. The distance of the candidate location site to the pseudo demand point is simply the Euclidean distance between the two points. The distance between a demand point and the pseudo demand point to which it is allocated, however, is not necessarily Euclidean. These distances can be precalculated by using the previously mentioned algorithm to find the length of the shortest path between points A and B on a Euclidean plane with

polygonal barriers. The quantities to be precalculated are the distances between every demand point and every barrier vertex.
Step 5: Repeat the entire procedure (step 1 through step 4) for another allocation of demand points, until all allocations have been exhausted. The constrained single facility location problem (in step 4) with the smallest objective function value is the optimal solution to the problem. The above procedure finds an optimal solution to the 1-median problem on a Euclidean plane with polygonal barriers, due to the following reason: a shortest path from the optimal location to a demand point is either a straight-line path or a sequence of straight-line paths which pass through barrier vertices. If it is a straight-line path, then the demand point can be considered allocated to itself. Otherwise, the demand point is allocated to the first barrier vertex encountered on the sequence of straight-line paths. Thus by considering all possible allocations of demand points to barrier vertices and themselves we are assured of finding the optimal solution. We note that this method of allocating the demand points removes the barriers from the problem, in the sense that the resulting constrained single facility location problems use the Euclidean metric for distance calculations.

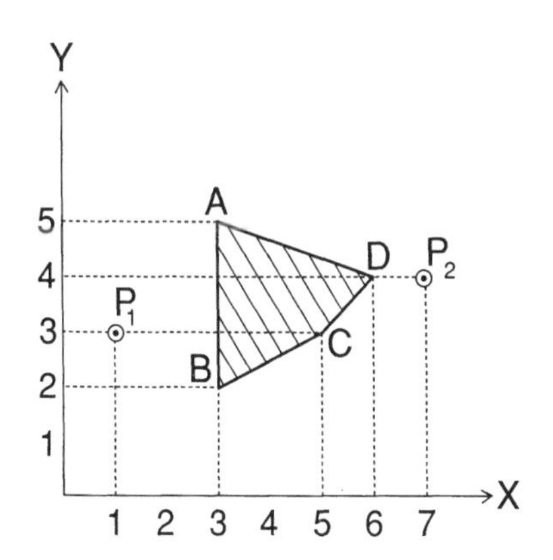

Figure 2 A numerical example.

We now consider a simple numerical example. Let m (number of demand points) =2, and K (number of barrier vertices) = 4, and refer to Figure 2. The number of possible allocations for the above problem is $(K + 1)^m = 25$. The possible allocations of the demand points are listed in Table 1. The exact method is obviously computationally prohibitive, and we therefore seek methods to improve its efficiency.

Allocation Number	Demand point 1 (P_1)allocation	Demand point 2 (P_2) allocation
1	P_1	A
2	P_1	B
3	P_1	C
4	P_1	D
5	P_1	P_2
6	A	A
7	A	B
8	A	C
9	A	D
10	A	P_2
11	B	A
12	B	B
13	B	C
14	B	D
15	B	P_2
16	C	A
17	C	B
18	C	C
19	C	D
20	C	P_2
21	D	A
22	D	B
23	D	C
24	D	D
25	D	P_2

Table 1 Allocations for the Example in Figure 2.

We now focus our attention on dominance properties, which are used to reduce the computational effort of the exact method. There are two kinds of dominance properties. They are:

(i) Dominance of one allocation over another.
(ii) For any particular allocation, dominance of certain regions over other regions.

The first type of dominance enables us to reduce the number of allocations to be considered. The second type allows us to reduce the feasible region over which the problem is to be solved for any particular allocation. The second type is applicable only to disjoint feasible regions. Below we discuss three dominance results. Our first dominance result is of the first kind, establishing conditions when one allocation dominates another allocation.

Theorem 2 *let $K = \{ 1,2,...,k \}$ be the set of barrier vertices.*
Let $N = \{k+1,... , k+m\}$ be the set of demand points. Let

$$f_i = \begin{cases} k+i \text{ if demand point i is assigned to itself} \\ l \text{ if demand point i is assigned to barrier vertex l.} \end{cases}$$

Let $(f_1, f_2, \ldots, f_m)$ be an allocation.

Let $(f'_1, f'_2, \ldots, f'_m)$ be some other allocation.

Then $(f_1, f_2, \ldots, f_m)$ dominates $(f'_1, f'_2, \ldots, f'_m)$ if

$$(f_1 \cup f_2 \cup \ldots \cup f_m) = (f'_1 \cup f'_2 \cup \ldots \cup f'_m) \text{ and}$$
$$d(k+i, f_i \leq d(k+i, f'_i) \; \forall i = 1,2,\ldots,m. \tag{4}$$

Proof: Since the feasible regions are the same for both allocations, the optimal location of the new facility will be the same for both allocations. However, while calculating the objective function value, the distances are measured from the new facility location to the demand points through the allocated points. Hence it is clear that if the distances of all demand points from the new facility location for allocation $(f'_1, f'_2, \ldots, f'_m)$ are more than their corresponding distances for allocation $(f_1, f_2, \ldots, f_m)$, the objective function value for allocation $(f_1', f'_2, \ldots, f'_m)$ is more than that for allocation $(f_1, f_2, \ldots, f_m)$. Thus allocation $(f_1, f_2, \ldots, f_m)$ dominates allocation $(f'_1, f'_2, \ldots, f'_m)$, and hence the result. ■

Our second dominance property applies to visible regions within any particular allocation. By using this dominance property, it is possible to eliminate some feasible regions from consideration.

Theorem 3 *Let $N = 1, 2, \ldots, m$ be the set of demand points and let R_1, R_2, R_k be the set of regions that are visible from the pseudo demand points. Define the following terms:*

$$\delta_{ij} = \max_{x \in R_j} \; d(x,i), \text{ and}$$

$$\gamma_{ij} = \min_{x \in R_j} \; d(x,i), \tag{5}$$

where $d(x,i)$ is the shortest Euclidean distance that does not penetrate a barrier from point x in region R_j to the demand point via the point that i is allocated to.

Region R_j dominates region R_k if

$$\delta_{ij} \leq \gamma_{ik} \; \forall i = 1, 2, \ldots, m \tag{6}$$

Proof: Let us consider regions R_j and R_k which are feasible for any particular allocation. Let x_1 denote an optimal solution to the constrained single facility Euclidean distance location problem over region R_j. Similarly, let the optimal location in region R_k be x_2. If the minimum value of the shortest distance from any point in region R_k to any of the demand points via the pseudo demand points is more than the maximum value of the shortest distance from any point in region R_j, then it is clear that the objective function value obtained by using location x_2 will be greater than the objective function associated with x_1. In other words, since

$$d(i,x_1) \le d(i,x_2) \quad \forall i = 1,2,\ldots,m, \tag{7}$$

it is clear that

$$\sum_{i=1}^{m} w_i d(i,x_1) \le \sum_{i=1}^{m} w_i d(i,x_2). \tag{8}$$

Therefore, no matter where you locate the new facility in region R_k, the objective function value is going to be greater than the objective function value obtained by locating the new facility anywhere in region R_j. So it is not necessary to solve the problem over region R_k, even though the region is feasible. Thus, region R_j dominates region R_k, and hence the result. ■

It is possible that a portion of some region may be dominated by another region. In other words, it is not necessary for a whole region to be dominated. The portion of the region which is dominated can be removed from the feasible space while solving over that region. An example of such a dominance is given in Figure 3. In this example R_1 is a feasible region. However, a portion of the region R_1, R'_1, is dominated by another region R_2. In other words, the minimum value of the shortest distance from any point in region R'_1 to any of the demand points via the pseudo demand points is more than the maximum value of the shortest distance from any point in region R_2. Therefore, we need to consider only $R_1 - R'_1$ while solving over the region R_1.

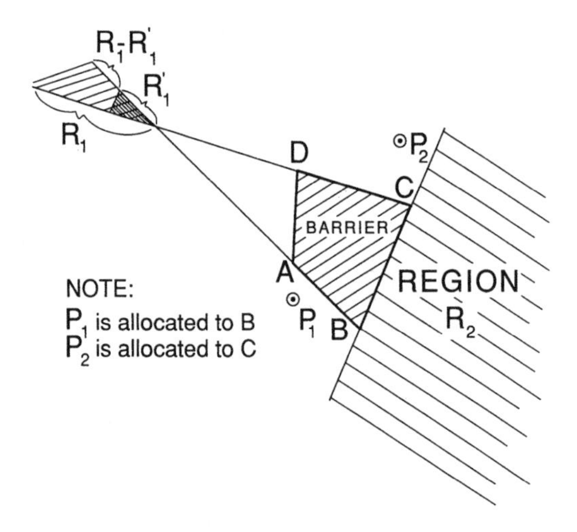

Figure 3 Illustration for partial dominance of a region.

Our third dominance property enables us to identify dominance of portions of regions over other portions of the same region.

<u>Theorem 4</u> *Let $N = \{1, 2, \ldots, m\}$ be the set of demand points on the plane. Let $y_1, y_2, \ldots y_l$ be the set of pseudo demand points and let Y be the convex*

hull of these points. Let Rj be the region visible from all these allocated points. The following results hold:

(i) If $R_j \cap Y \neq \phi$, then we can replace R_j by $R_j \cap Y$.

(ii) If $R_j \cap Y \neq \phi$, then we can replace R_j by a set of points X, such that for each $Z \in R_j - X$ there exists an $x \in X$, such that

$$d(z, y_i) \geq d(x, y_i)\ \forall i = 1,2,\ldots,l \tag{9}$$

Proof: The proof for part(i) is trivial, since the optimal objective function with the new facility located in the convex hull of the allocated points is better than the objective function with the new facility located outside the convex hull of the allocated points. Therefore, we can replace the region R_j by the intersection of the region and the convex hull of the allocated points.

We now focus on the proof for part (ii). Let us consider any point in the interior of the region R_j not belonging to the set X. Clearly, since the distance of any point in the interior region (ie., $x \in R_j - X$) to the allocated points is greater than the distance from any point Z in X to the allocated points, it is obvious that the objective function value with the new facility located at any point $x \in R_j - X$ will be greater than the objective function value with the new facility located at any point in X. Therefore, the set of points X dominates any point in the region $(R_j - X)$. Therefore, we can replace R_j by X. The result follows. ■

It is to be noted that any point in region R_j, such that $R_j \cap Y = \phi$ is always dominated by a set of points on the edge. This leads to the fact that X will be a union of edges of finite length. Therefore, it is sufficient to solve the single facility location problem over these edges of finite length. The dominance property is explained with a simple example shown in Figure 4.

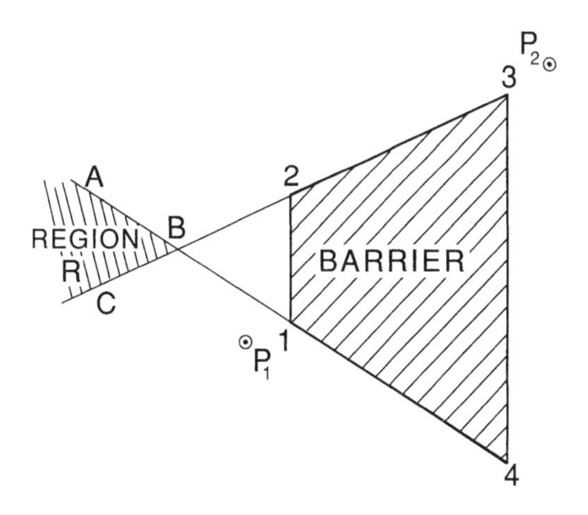

Figure 4 Illustration of dominance result in Theorem 4 .

Let us consider a region R_1 feasible to some particular allocation of demand points to the pseudo demand points. The region R_1 is not in the convex hull of the allocated points P and Q. Any point in the interior of the region R_1 is dominated by points along the edge ABC, since the distance of any point on the edge ABC to the pseudo demand points is less than the distance from any point in the interior of region R_j to the allocated points. Hence, we can replace region R_j by a set of points along the edge ABC and solve the single facility Euclidean distance problem over these edges of finite length. The dominance properties discussed above can be used to reduce the computational effort of the exact solution method.

As we observed earlier, a complete enumeration method is computationally prohibitive. Even with the application of the dominance properties, the number of allocations to be considered would probably still be too large; computational effort grows at an exponential rate with respect to an increase in the number of demand points. It is therefore desirable to have a method to solve the problem which involves less computations and hope to get a *near optimal solution.* Of course, an optimal solution is not guaranteed by this *short cut* approach. However, an intelligent approach can lead us very close to an optimal solution. One such heuristic approach for solving the 1-median problem on a Euclidean plane with polygonal barriers is explained below.

The proposed heuristic is similar to the exact procedure in the allocation of demand points to barrier vertices to form pseudo demand points. In other words, each demand point can either be assigned to itself or to one of the barrier vertices. By doing this we create pseudo demand points. The weight of these pseudo demand points is the sum of the weights of the demand points allocated to it. The heuristic algorithm is given below.

Step 1: Identify all barrier vertices.

Step 2: Start with any arbitrary point as the trial location.

Step 3: Check if the trial location is visible from demand point i. If the demand point is visible from the trial location assign the demand point to itself. The distance to be used in the next iteration is the straight-line distance from the demand point to the trial location. If the demand point is not visible from the trial location, find the barrier vertex which is visible from the trial location through which the shortest path from the demand point i to the trial location passes. Thus, create pseudo demand points with the weight of these pseudo demand points being equal to the sum of the weights of the demand points allocated to them. Also find the distances of the demand points from the trial location via the allocated point. This step is repeated for all demand points.

Step 4: Solve the single facility location problem with these new pseudo demand points' weights and distances. Find the objective function value at the end of each iteration. Proceed to step 5.

Step 5: If the objective function value at the current iteration is not significantly different from the previous iteration, then stop. The current trial location is the optimal location of the 1-median. Else, make the current location point as the trial location, and return to Step 3.

The exact method and the heuristic discussed above apply to the 1-median problem. The solution to the p-median problem is obtained by embedding these exact and heuristic methods for the 1-median within a location-allocation framework.

To summarize, we have so far developed the theory and methodology needed to solve the p-median problem on a Euclidean plane with polygonal barriers. We have also developed a computer code to implement the proposed exact method (with the dominance properties) and the proposed heuristic method.

Future research plans

Our next step is to complete the development and testing of our computer code for solving the p-median problem on a Euclidean plane with barriers to travel. We will then develop theoretical results for the quality of demand point approximations for the p-median problem on a Euclidean plane with barriers. We will also study theoretical results for the quality of the augmented demand point approximation, obtained by restricting facility location to the demand points and the barrier vertices. We then intend to test these approximations both computationally and empirically. We will generate both random and autocorrelated demand points for our computational experiments. For our empirical results we will use census tract data from the City of Buffalo.

After the above tasks have been fully researched, we will research the following two questions: (a) For a given integer k, what is the best set of k discrete points to restrict facility location; and (b) How sensitive is the answer to question (a) to the quantity m, where m is the minimum number of demand points assigned to a facility. We will first address these questions for the Manhattan metric problem, both with and without barriers to travel. We will later address these questions for the Euclidean metric problem, both with and without barriers to travel. We intend to widen our scope from the p-median problem to study the p-center problem and the SQM problem. Theoretical bounding procedures, computational experiments, and experiments with empirical data will be performed. It is envisioned that these future efforts can be completed by the end of August 1990. These efforts will be undertaken jointly with Professor Richard C. Church of the Department of Geography at the University of California at Santa Barbara.

References

Batta, R., and Leifer, L. F., 1988, On the accuracy of demand point solutions to the planar, Manhattan metric, p-Median problem, with and without barriers to travel.*Computers and Operations Research* **15**, 253 - 262.

Berman, O., Larson, R. C., and Chiu, S. S., 1985, Optimal server location on a network operating as an M/G/1 queue. *Operations Research* **33**, 746 - 771.

Dijkstra, E. W., 1959, A note on two problems in connection with graphs. *Numer. Math.* **1**, 269.

Francis, R. L., and White, J. A., 1974, *Facility Layout and Location: An Analytical Approach.*. (Englewood Cliffs, NJ: Prentice - Hall)

Hakimi, S. L., 1964, Optimum distribution of switching centers and the absolute centers and medians of a graph. *Operations Research* **12**, 450 - 459.

Katz, I. N.,and Cooper, L., 1981, Facility location in the presence of forbidden regions; I. formulation and the case of the Euclidean distance with one forbidden circle. *European Journal of Operational Research* **6**, 166 - 173.

Larson, R. C., and Odoni, A. R., 1981, *Urban Operations Research.* (Englewood Cliffs, N.J: Prentice-Hall)

Larson, R. C., and Sadiq, G., 1983, Facility locations with the Manhattan metric in the presence of barriers to travel. *Operations Research* **31**, 652 - 659.

Larson, R. C., and Stevenson, K. A., 1972, On insensitivities in urban redistricting and facility location. *Operations Research* **20**, 595 - 61.

Thisse, J. F., Ward, J. E., and Wendell, R. E., 1984, Some properties of location problems with block and round norms. *Operations Research* **6**, 1309 - 1327.

Chapter 18

Modeling reliability on statistical surfaces by polygon filtering

Adrian Herzog

Abstract

The advances in GIS technology allow us to store spatial information in increasing detail. On the other hand there is an additional demand for analysis functions. Growing demands for information about accuracy in spatial databases will lead to data with integrated error estimation. Visualization in form of maps and cartographic generalization can make use of this additional information. For this purpose a generalization method based on polygon filtering proposed and exemplified for the case of statistical surfaces. The improved reliability of the resulting maps will help the user to make adequate decisions based on the information provided by the GIS.

Introduction

The high geometrical precision of spatial objects stored in a Geographical Information System (GIS) is often in disproportion to the attributive accuracy of these spatial objects. Work has been done to identify different sources of error (MacEachren 1985, Burrough 1986, Goodchild and Dubuc 1987, etc.), additional efforts are necessary to deal with the related problems.

Discussions on error in spatial databases have to cover the whole sequence of processing steps in a GIS. Based on the usual structure of a GIS we could classify errors from the following scheme:

- Data collection: Errors by careless work, error in measurement, inadequate sampling techniques etc.
- Data recording: Errors due to classification, implied by the numerical and spatial resolution, restricting characteristics of the data structure etc.
- Data manipulation and analysis: At this level further abstractions from reality come into the system and thus the amount of inaccuracy can considerably increase (by problematic analysis techniques, inadequately defined variables, by cartographic generalization steps such as classification and aggregation, change of measurement scale, arithmetic combination of variables with different accuracy etc.).
- Data visualization: insufficient technical means (less important in the future), wrong use of cartographic concepts, loss of accuracy by the inevitable use of

various generalization techniques, and perception problems on the part of the map reader.

Most variables stored in a GIS are subjected as well to random errors: "If we repeatedly measure the same element, we would probably obtain a slightly different score each time" (Clark and Hosking 1986, p. 14). The existence of random error leads to a situation, where the "true" value is not known explicitly and, therefore, a general error definition as "difference between a measured quantity and the true value for that quantity" (Muller 1987, p. 1) cannot suffice.

On one hand it is obvious that in all cases where the data accuracy can directly be influenced by the GIS user all efforts should be undertaken to eliminate these sources of error. Errors once introduced in the system can hardly be eliminated, and they will affect all subsequent processing steps. While errors based on data collection and geometrical manipulation can be reduced with diligent work, random errors will remain: they cannot be overcome either by more sophisticated methods or by a more extensive data sampling.

In the following sections we concentrate on modeling and random errors; the following considerations will be the starting point of our analysis:

- The demand for a better integration of spatial analysis methods will substantially enlarge the amount of errors on the data modeling level. Even very simple (nongeometrical) modeling operations on accurate data can yield results with a considerable amount of uncertainty.
- This effect will be intensified by the fact that advances in GIS technology allow us to accumulate information in an increasing spatial and attributive detail. The finer the spatial resolution, the greater is the probability that a particular data value is significantly affected by random error.
- The errors as mentioned are inherent to any spatial variable and therefore to any GIS: there is no alternative to trying to process such error-burdened, "noisy" information by effective methods. This is especially true in the output context, where negative effects have to be minimized to secure correct communication. It would be a pity for all preceding laborious and sophisticated processing steps, if at the very end their result could not be communicated in an adequate manner to the GIS user.

GIS and Cartography

Human access to data stored in a GIS will remain in the form of maps. This way the accuracy of spatial databases is directly limited and related to the quality of the visualization of their contents. Thus, error analysis implies the analysis of the map as a major communication tool for spatial information. The statement by Harvey has still its validity: "Given the great esteem in which geographers hold the map as means of description, analysis, and communication, it is rather surprising to find that many aspects of cartographic representation remain unanalysed" (1969, p. 369).

Cartography as a discipline has its own tradition and not all its concepts and achievements have become know-how of common GIS instruments. Beyond the analysis of cartographic tools, we need to conduct further research in the cartographic area. In the future, especially, the communication efficiency and the accuracy aspect of maps including the development of new representation models will have to be studied. Particularly new technical possibilities will have to be evaluated with respect to the development of new mapping concepts.

The major problem of map quality, produced through GIS or not, is well known, and especially in the recent past many appeals for better maps have been made (Weibel and Buttenfield 1988, Muller 1985, Jupe 1987). Weibel and Buttenfield (1988) mention two ways to overcome the production of poorly designed maps: a better training of map makers (who today often have too little cartographic training) and the hope for expert systems. The group of map makers with little cartographic knowhow will in the near future increase: tools out of the desktop publishing area allow the production of maps to nearly everyone. As long as expert systems in the cartographic environment are scarcely operational we have no choice but to sharpen traditional instruments for the production of better maps.

Decision making in the context of GIS often substantially relies on "communication maps" which "are small scale, single purpose, highly generalized representations. (...) The function of communication maps is not to represent accurately the real world. Rather, it is to show a simplified model of reality, an abstraction which helps to separate the relevant message from unwanted details" (Muller, 1987, p. 5).

Producing communication maps, it is essential not to communicate the information accurately in the sense of presenting information down to the last detail, but to convey the fundamental structure and the areal pattern relationships of spatial information. Generalization is absolutely necessary to prevent the user from being disturbed by eyecatching but unimportant information. In most cases the generalized form will actually be the most accurate form of map, but not the cartographic one-to-one representation of the original and accurate data. The information to be portrayed has to be modeled in order to separate important from marginal information. Thus, the term "accuracy" has to be extended by a nuance of "appropriateness", "reliability", and "meaningfulness". Nevertheless, cartographic generalization and accuracy are not conflicting goals. We come to the paradoxical situation where modified and manipulated data give the map reader the more correct impression rather than the communication of the raw data. This necessary extension of accuracy by the reliability aspect will put a GIS user into a position to take adequate decisions based on the information provided by the GIS.

As we have pointed out, especially the modeling and analysis steps can be the source of great amounts of fuzziness. It can be expected that in the future GIS will have to deal with an increasing data uncertainty. On the other hand, it can be supposed that more information on error estimation will be available not only for data layers as a whole but for each spatial unit. Visualization and cartographic generalization techniques have to make use of this crucial additional information which could be a key to more reliable maps.

While in traditional cartography generalization is an common procedure, today's GIS offer few capabilities of the sort. Basic approaches sometimes are present in the form of mechanisms for the elimination and dissolution of e.g. small polygons, their use is optional and not supported by appropriate assistance. Instead of promoting sophisticated methods which are understandable only to a small community of specialists, we propose to develop a generalization procedure, which is based on simple principles, which is stable, and which can be directed by few parameters.

Statistical surfaces and their visualization

Our discussion is based on the treatment of statistical surfaces: They are "one of the most fundamental of geographical phenomena" and among major problems there are aspects such as "inferences about spatial continuity properties (e.g. spatial autocorrelation), attempts to smooth and generalize complex surfaces, and methods of surface interpolation" (Wilson and Bennett, 1985, p. 83).

Statistical surfaces in a very general sense are often the result of modeling operations in a GIS (e.g. combining information from different information layers for a feasibility study etc.). We will concentrate on the analysis of the easily controllable case of statistical surfaces based on calculated proportions for each map unit (e.g., the percentage of one subgroup in relation to the whole population in that area): this is an elementary and widely used model of abstraction. While under ideal circumstances we are starting with totally accurate data, the estimated height of the statistical surface of interest may be subjected to random variations, specially in case of very detailed data. Even such a single modeling step as the calculation of proportions can therefore lead to a result, where additional treatment is recommended to master different kinds of uncertainty.

In addition, we use proportion data because statistical theory is applicable in a straightforward manner, and so processing can be done under conditions where information is completely associated with uncertainty estimation. From a statistical point of view using a binomial distribution the width of the error band of proportions can exactly be defined (e.g. Yamane 1973, p. 724; tables by Clopper and Pearson 1934). In the case of small absolute values the width of the confidence interval nearly can reach 100%.

With the treatment of statistical surfaces based on proportion data we especially exemplify the higher measurement levels. However, most of the concepts presented here can easily be adapted to other measurement levels.

Like topographic surfaces, statistical surfaces can be visualized by various methods (see Robinson *et al.* 1978, p. 217). We choose here choropleth mapping, one of the most widely used techniques in thematic cartography. The choropleth map has its main advantage in its very simple conceptual foundation, and various problems related to this method, however, are well known (e.g. as the usual reference of population data to areal symbolization (Muller 1985); the use of cartograms, aside from perception problems, could overcome this problem). In spite of the various efforts to improve the effectiveness of choroplethic maps it has been difficult to overcome the traditional concept of the rigidly classified and regionalized (and thereby overgeneralized) map (Board 1987, p. 417). Especially the introduction of unclassed symbolization (Tobler 1973) has sharpened this cartographic method. After a first skepticism on the side of traditional cartographers the unclassed choroplethic map today seems to be a well established tool for the visualization of statistical surfaces. New technological means (such as the widespread of laser printers in the desktop publishing environment) guarantee sufficient quality in most situations, especially with small scales.

The unclassed map has the advantage of showing the smallest changes on the statistical surface. But the more spatially detailed the information, the greater will be the uncertainty of a single value: the map suggests an accuracy which is not justified by the data. This can easily be demonstrated by the confidence limits of proportions calculated from small numbers. Thus, from a cartographic point of view it is absolutely necessary to generalize such a map. In any case, it is not possible to go directly from the unclassed to the unaggregated map (respectively unregionalized map). In the case of noisy data additional generalization steps are required.

Because in many contexts much relevant data is related to polygonally defined zoning systems, we will concentrate on polygon data structures. Statistical surfaces built on a polygonal data structure are discretely stepped by nature and have the handicap that they badly represent the underlying continuous phenomena or generating processes: each reference polygon is represented by a single value. The polygon border inevitably marks changes between different heights of the surface and therefore it is difficult to show the possible fuzziness of boundaries. "These aggregates are often not the most natural or convenient units to use for analysis. They are frequently, however, the only units for which a data base is available" (Cliff *et al.* 1975, p. 146). It is obvious that in some cases the continuous nature of such surfaces can better be represented by other data structures (e.g. a raster). In order not to introduce additional error by the conversion of data structure, we refrain from it. Moreover, the fine resolution of most data such as census data permits us to stick to the polygon environment.

Map processing by polygon filtering

Traditionally, the generalization of statistical surfaces has been mainly achieved by aggregation to higher administrative units. In contrast, we will adapt here the quite simple concept of lowpass filtering of polygons. With this approach we intend to achieve a spatially more homogeneous generalization and to produce maps of substantially greater amount of information.

The idea of polygon filtering (or "resel processing" in a general sense) in this context was proposed by Tobler (1975, 1984): "We need to develop methods of map enhancement, simplification, deblurring, and so on, when the data are arrayed in the form of resels" (1984, p. 142). At the first glance, the use of image processing techniques in the area of map processing seems to falsify correct information. An impression could be, that any modification of "accurate" information has to be omitted. But we agree with Muller (1987, p. 2): "Emphasizing the communication aspect of maps, they (the cartographers) found that in order to tell the truth one often had to lie a little." We will try to show that these map processing techniques are well suited, because the degree of uncertainty of each single observation can be taken into consideration.

Assuming the existence of spatial autocorrelation, we develop an adaptive filtering process based on the following principles:

- The value of a filtered area is normally calculated by a weighted mean of its preceding value and the values of all its neighbors. The relative weights of the center resel and its neighbors or the strength of the influence of the neighborhood are set by a parameter. Thus, the filter used has lowpass properties.
- The neighborhood definition of an area can be set in different ways as known from autocorrelation discussions: topological neighbors (of first, second or higher order), polygons (or polygon centers) in a certain distance, only adjacent polygons having a traffic link (specially in the case of socioeconomic data) etc.. In our situation we usually choose a weight proportional to the length of the common boundary. These boundary lengths can artificially be modified: Topographical, historical, administrative and other boundaries can lead to a change of influence or even to the introduction of barriers. In this view aggregation to regions is just a special case of our filtering approach: within the defined regions the regionalized case is the limit of an infinite application of a lowpass polygon filter: intraregional variability is completely smoothed out or eliminated, respectively.
- Additional restriction: it has to be prevented that the filtering process shifts any data value out of its initial confidence interval. In the case of proportion data the width of this interval is determined by the required reliability (e.g. the 95% level) which defines the lower and upper confidence limits. Depending on the confidence level used, individual data values may change by a variable amount from their original value.
- From a cartographic point of view it would be undesirable to filter over boundaries which represent sharp and significant cliffs in the statistical surface. We therefore modify the initial weighting scheme by excluding filtering operations between neighbors of significantly different data value. Statistical tests for the significance of the difference of two proportions are well known (we analyze contingency tables, and for very small samples we use the exact method of Fisher).
- The filtering process is iteratively applied. Because of the increasing number of areas with a resulting value at one end of the confidence interval, there will be a decreasing number of polygons having a changed value after each step.

- At the end of the filtering process we will usually have a different value for each polygon; in the case of unclassed choropleth maps we will have different symbols for each area. The smoothing effect increases the spatial autocorrelation. Thus, a lot of neighboring resels have just marginally different values. Such a map will pretend a pseudo accuracy and probably give a noisy impression. Therefore an eventual final *ad hoc* regionalization process can be applied to combine areas with nearly equal values. This graphical generalization process can be stopped by setting parameters or under visual control.

For each areal unit the following data items are required: data value, weight, confidence interval, and a list of neighbors with their corresponding weights.

An example

In order to exemplify the proposed generalization and visualization concept we choose a statistical surface of fine resolution. The underlying statistical population consists of the students in Switzerland; investigating the regional pattern in the choice of a university by a student, we treat raw material on the commune level. On this aggregation level in the study area more than 3000 spatial units have to be processed; they have an mean spatial resolution (Tobler 1987) of about 3.7 km with considerable variation in the different parts of the country caused by historical and topographical reasons (regional aggregations often used have resolutions of 15.2 km, cantons are at about 40 km). On this resolution level we find a great variability of absolute values (from communes without students to towns with more than a thousand students).

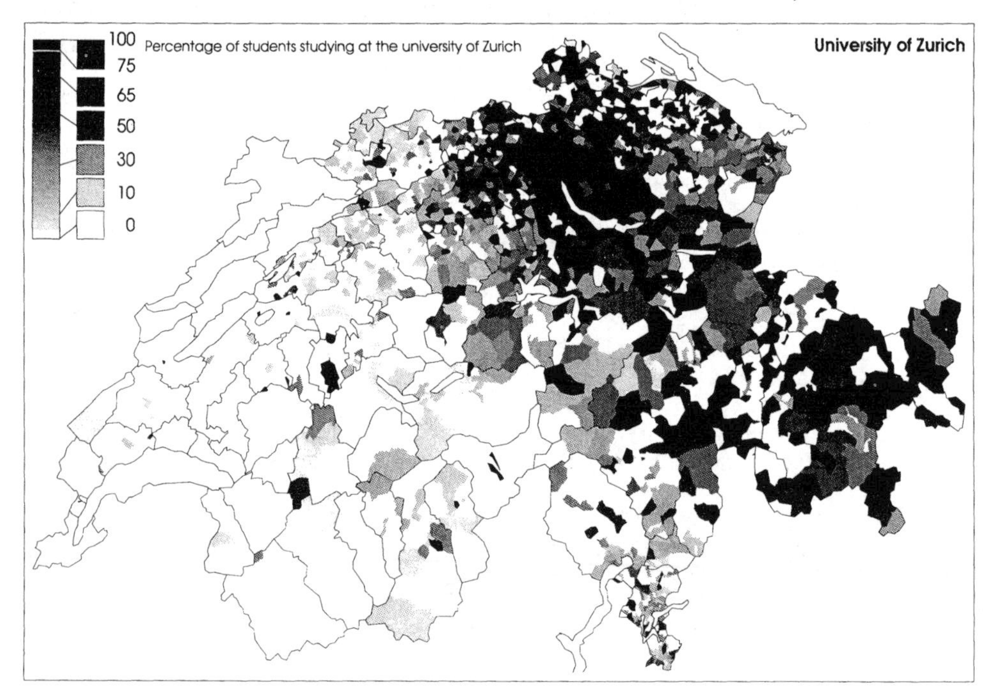

Figure 1 Raw data

The mapping program used here (PSCopam, Herzog 1988) works on a Macintosh computer and generates PostScript page description language, which can be interpreted on various raster output devices.

As an example with relatively high autocorrelation we work with a statistical surface constructed by the proportion of students who enrolled in the university of Zurich.

A great amount of the "roughness" of the visualized statistical surface has its origin in a virtual effect; in the mountainous regions with low population (e.g. in the southeastern part of the country) and in cantons with very small communes we find quite small absolute numbers of individuals and for that reason percentages easily can take extreme values.

After an repeated application of the filter of the same statistical surface a lot of the random effects have been smoothed out (Figure 2: 10 iterations, weight of the central resel: 0.667, total weight of its neighbors: 0.333, width of the confidence interval based on a 95%confidence coefficient , and an exclusion of the filtering over boundaries significant on a 0.99 level).

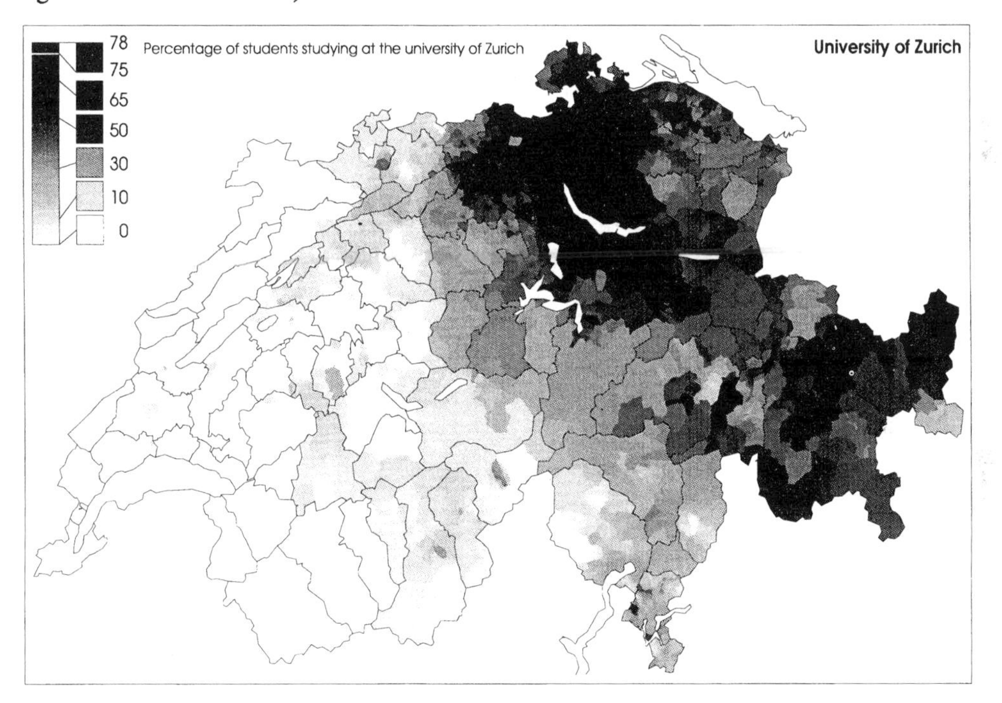

Figure 2 Statistical surface after five iterations

Evaluation

In a first stage, visual evaluation shows the validity of the approach. At the small scales used, the fine spatial resolution leads to close to dasymetric cartographic representations.

The consideration of information reliability of each area renders results which are minimally influenced by individual random values (especially in comparison to some coarse interpolation techniques). The intended generalization affects smooth noisy areas but does not change well confirmed values. At the same time, the continuous nature of the statistical surfaces is emphasized.

The proposed method can be considered as an interpolation process, which especially tries to estimate the underlying spatial phenomena in regions of uncertain information (or even missing values). Areal units where data are missing get estimations based on resels in their neighborhood.

The adaptive treatment of spatial units of different reliability supports the reduction of disturbing side effects of the variable reporting zones.Thus, in comparison to the use of known routine classification and regionalization techniques we display more detailed information about spatial variation. Especially in these cases where the intraregional variance predominates over interregional variance the proposed method has advantages over regionalization techniques. In a first stage the number, weight, and area of elementary areal units, whose modeled values have considerably changed (in relation to the original confidence limits), can give an additional indication for the degree of generalization. Just in these cases we change the observed values in a dangerous way.

Of course we have to proceed and to look for quality measures. Jenks and Caspall (1971) developed some indices for the evaluation of different classification schemes: overview, tabular, and boundary accuracy indices consider the deviations between actual (original) and derived (classed) values (unweighted, weighted by area, and the preserving of boundaries). In this context, these indices can give a first impression; however, they compare different generalizations to a completely accurate starting situation (the true values). This assumption is contrary to our concept in that we start from unreliable values: results of our filtering process are compared to "true" values representing a considerable amount of uncertainty. The presentation of these indices can thus be a supplementary indicator to compare the results with conventional generalization methods. In the same way the development of autocorrelation can give indications of the effect of the filtering process.

The search for more adequate accuracy measures must consider the variable reliability of a single item of information. In a first attempt we use measures based on chisquare statistics (a comparison of the modified to the original values). From a communication point of view the greatest problems arise from areas, which are both visually striking (large polygons with extreme values) and uncertain: therefore further indices will specially emphasize this aspect.

Conclusions

The amount of uncertain attribute information in GIS will grow due to the increasing importance of spatial modeling operations. It is absolutely necessary to integrate information about reliability for each spatial unit and to make use of this information in any further processing steps. GIS will have to include robust instruments for the automatic handling of different kinds of errors. In order to get more reliable GIS, we have focused on the output stage, where, regarding accuracy aspects, cartographic communication often is one important bottleneck. Especially for data of fine spatial resolution we propose an adaptive generalization method for cartographic presentation based on lowpass filtering of polygons. This method is iteratively applied and makes use of associated error information; noisy data is smoothed and missing values are estimated. Compared to traditional map procedures we get more detailed maps with a more homogeneous spatial reliability.

Future work has to be done to extend the method to data of different measurement scales and to develop additional measures for the accuracy and reliability of cartographic representations.

References

Board, C.,1987, The need to integrate map use and map design in a single system for sommunicating seographical information. *Proc. 13th. Int. Cartographic Conference,* Vol. 1, Morelia, 1987, 414-419.

Burrough, P. A. ,1986, *Principles of Geographical Information Systems for Land Resources Assessment.* (Oxford: Oxford University Press)

Chen, ZiTan, 1986, Spatial filtering of polygon data. *Proc. 2nd Int. Symposium on Spatial Data Handling*, Seattle, pp 86-101.

Clark, W. A. V., and Hosking, P. L., 1986, *Statistical Methods for Geographers.* (New York: John Wiley)

Cliff, A. D., Haggett, P., Ord, J. K., Bassett, K. A., and Davies, R. B., 1975, *Elements of Spatial Structures: A Quantitative Approach.* (Cambridge: Cambridge University Press)

Clopper, C.J ., and Pearson, E. S.,1934, The use of confidence or fiducial limits illustrated in the case of the binomial. *Biometrika* **26**, 404-413.

Goodchild, M. F., and Dubuc, O., 1987, A model of error for choropleth maps, with applications to geographic information systems. *Proc. AutoCarto 8*, 165-174.

Goodchild, M. F., 1987, A spatial analytical perspective on geographical information systems. *Int. J. Geographical Information Systems* **1** (4), 327-334.

Goodchild, M F., and Wang, Minhua ,1988, Modeling error in rasterbased spatial data. *Proc. 3rd Int. Symposium on Spatial Data Handling*, Sydney, pp 97-106.

Harvey, D., 1969, *Explanation in Geography.* (London: Methuen)

Herzog, A, 1988, Desktop mapping. Desktop publishing in der kartographie, ein Anwendungsbeispiel. *Geographica Helvetica* **43** (1), 21-26.

Jenks, G. F., and Caspall, F. C., 1971, Error on choroplethic maps: definition, measurement, reduction . *Annals of the AAG* **61**(2), 217-244.

Jenks, G. F., 1976, Contemporary statistical maps: evidence of spatial and graphic ignorance. *The American Cartographer* **3**(1), 11-19.

Jupe, D., 1987, The new technology: will cartography need the cartographer? *The Canadian Surveyor* **41**(3), 341-346.

MacEachren, A. M., 1985, Accuracy of thematic maps / implications of choropleth symbolization. *Cartographica* **22**(1), 38-58.

Muller, JeanClaude, 1983, Ignorance graphique ou cartographie de l'ignorance. *Cartographica* 20 (3), 17-30.

Muller, JeanClaude, 1985, Wahrheit und Lfge in thematischen karten. zur problematik der darstellung statistischer sachverhalte. *Kartographische Nachrichten*, **2/85**, 44-52.

Muller, JeanClaude, 1987, The concept of error in cartography. *Cartography*, **24** (2), 1-15.

Robinson, A., Sale, R., and Morrison, J., *Elements of Cartography.* Fourth edition. (New York: John Wiley)

Tobler, W. R., 1973, Choropleth maps without class intervals? *Geographical Analysis* **5**(3), 262-265.

Tobler, W. R., 1975, Linear operators applied to areal data. In *Display and Analysis of Spatial Data*, Davis, J. C., and McCullagh, M. J., (eds), pp. 14-37. (London: John Wiley)

Tobler, W R., 1984, Application of image processing techniques to map processing. *Proc. Int. Symposium on Spatial Data Handling*, Zurich, Vol. 1, pp 140-144.

Tobler, W. R.,, and Kennedy, S., 1985, Smooth multidimensional interpolation. *Geographical Analysis* **17**(3) 251-257.

Tobler, W. R., 1987, Measuring spatial resolution. *Proc. Int. Workshop on GIS*, Beijing, pp 42-48.

Weibel, R., and Buttenfield, B. P., 1988, Map design for geographic information systems. *Proc. GIS/LIS '88*, San Antonio, pp 350-359.
Wilson, A. G., and Bennett, R. J., 1985, *Mathematical methods in human geography and planning*.(New York: John Wiley)
Yamane, T., 1973, *Statistics: An Introductory Analysis*. Third Edition.(New York: John Wiley)

Section VI

The last major section contains four papers which offer varying levels of solutions to the problems of inaccuracies in socioeconomic data attributable to spatial effects. None are solutions in themselves, but each develops a promising research direction.

Much traditional theory in human geography and spatial analysis is built at specific scales - the patterns of central place theory, for example, emerge at isolated scales in the hierarchy of service functions. Fractals offer an alternative basis, by being concerned with the extent to which phenomena are scale-free. Fotheringham's paper looks at the spatial structure of towns in relation to the fractal model. The implications of a good fit are powerful, because they suggest that scale effects might be predictable, suggesting that it might ultimately be possible to develop scale-independent models which would have some of the frame-independence envisioned earlier in Tobler's chapter.

Amrhein and Flowerdew analyze a dataset of Canadian migration statitstics, and find that despite expectations, the effects of aggregation on the outcomes of modeling are relatively minor, although aggregation clearly affects the statistics themselves. The results suggest that scale-dependence of certain widely-applied models might be a useful property to measure for such datasets.

Flowerdew and Green are concerned with cross-estimation, or the estimation of data values for one set of reporting zones based on known values for another set. In essence this is the applied perspective on the modifiable areal unit problem, as cross-estimation is vitally important to people working with demographic or socioeconomic data in health, political redistricting, regional economic modeling and a number of other practically significant areas. It is also an information-limited problem - the more one knows about the area, the better the estimates. Flowerdew and Green show how information on some underlying factor might be used to improve the target zone estimates over the results of more naive techniques.

The final paper in the section integrates many of these papers on socioeconomic data problems by proposing an integrating framework for all issues of estimating from one spatial basis to another. Arbia's paper describes a general class of problems which are within the capabilities if GIS on the one hand, but poorly understood statistically on the other. These include any change of basis or reporting zones, whether they be points, lines or areas. An integrated approach is attractive because it might be implemented within the constraints of a number of current vendor GIS products.

Chapter 19

Scale-independent spatial analysis

A. Stewart Fotheringham

Introduction

It seems likely that all spatial data and all types of spatial analysis contain some type of error. Therefore it is thus impossible to perform error-free spatial analysis and it becomes the task of the spatial analyst to reduce error to the point at which it does not interfere which the conclusions drawn from a particular analysis. Hence, any type of spatial analysis can be criticized in some way; what becomes important is to differentiate between substantive error, the presence of which is likely to affect the conclusions reached and trivial error, whose presence is unlikely to affect the conclusions of the study. In this paper several examples of substantive error are discussed and ways in which these errors can be minimized to the point where they are trivial are described.

Three main types of error are examined: the first concerns the effect of imposing a boundary on the system being investigated; the second concerns the estimation of spatial trends in data where the source of the trend is ambiguous, and the third concerns the effect of examining a relationship at different spatial scales. Each type of error is examined in the context of a specific type of analysis - that of estimating a density gradient - in which the local density of a spatial distribution at some point is related to the distance from the center of the distribution. While the objective of the discussion is to minimize error, bearing in mind the statements made in the first paragraph it is not expected that the solutions proposed are error-free; rather an attempt is made to trivialize error to the point at which it does not affect the substantive conclusions drawn from an analysis.

Before discussing ways of minimizing certain types of error that exist with spatial data, a brief review of what has become known as "*The Modifiable Areal Unit Problem*" (MAUP) is necessary because much of what follows is concerned with estimating the rate of change in a variable caused by changes in the scale at which that variable is measured.

The modifiable areal uint problem

The MAUP is the generic name given to two different, but related, problems with spatial data. Both problems refer to the situation where data and relationships between data are sensitive to the zoning system in which the data are reported. The scale problem relates to the situation where data values and inferences concerning these data are affected by the number of zones used to report the data. The aggregation problem relates to situations where this sensitivity is a function of the arrangement of the reporting zones.

There are many examples of the MAUP in geographical studies (*inter alia* Williams, 1977; Openshaw and Taylor, 1980; and Putman and Chung,1989) and its existence has been acknowledged for some time (Gehlke and Biehl, 1934; Neprash, 1934; and Robinson, 1956). Perhaps one of the more convincing examples of the effects of the reporting units on the conclusions reached through spatial analysis is that given by Openshaw (1978;1984). Openshaw examined the sensitivity of one of the simplest types of analysis - a correlation between two variables - to variations in the way the data for 99 reporting units (counties in Iowa) were aggregated into 12 analytical units. The data in question were the percentage of the population in each zone voting Republican and the percentage of the population who were greater than sixty years old. By varying the aggregation procedure, Openshaw demonstrated how it was possible in this instance to obtain a worst fit correlation between the two variables of -.058 and a best fit correlation of -.997. He also achieved a correlation coefficient of .993 for one zoning system showing that the range of possible relationships spanned almost the entire interval between +1 and -1. Clearly, the results reported for any one particular zoning system could be highly misleading if taken as representative of the 'true' underlying relationship between the two variables.

The above is a very simple example of the MAUP and variations in the correlation coefficient can be explained theoretically by reference to the underlying changes in variances and covariances. However, it serves to demonstrate a problem of spatial data and one that exists when much more complicated relationships are examined (for example in the Putman and Chung paper referenced above, a similar sensitivity to zoning system design is described for a six parameter spatial interaction model). It has led us to think of zoning systems as being similar to radio wave receivers in that different zoning systems "pick up" different signals and can lead us to very different conclusions about the same underlying relationship. In the terminology of the first paragraph this is clearly a substantive error rather than a trivial error and it is the task of the remainder of the paper to examine one way around the MAUP.

Possible ways forward given the MAUP

At first and even second glance the error in spatial data handling caused by the MAUP seems daunting. Given that space is continuous, is any discrete division of that space meaningful and will any such division yield data that are reliable for analysis? However, a number of ways around the MAUP are possible. These include:

(i) the derivation of "optimal" zoning systems;
(ii) the identification of basic entities;
(iii) sensitivity analysis;
(iv) abandonment of traditional statistical analysis;
(v) shifting the emphasis of spatial analysis towards relationships that focus on rates of change.

The derivation of an "optimal" zoning system has been discussed by several researchers at one time or another (*inter alia*, Moellering and Tobler, 1972; and Openshaw, 1984) but seems fraught with unobtainable aspirations. What constitutes 'optimal' is clearly subjective both in definition and in operation. Even if one can agree upon what is optimal, and in both the above papers there is agreement that an optimal zoning system should maximize interzonal variation and minimize intrazonal variation, there will be problems in measuring when such optimality has been achieved. Openshaw, for example, uses variance to model inter and intra zonal variation while Moellering and Tobler employ spectral analysis. Another immediate problem in attempting to define an optimal zoning scheme is answering the question "optimal for what?". What constitutes an optimal zoning scheme for one variable may not constitute an optimal zoning scheme for some other variable. In terms of calibrating a spatial interaction model, for example, the zoning system yielding the minimum interzonal variation in the relationship between interaction and destination attractiveness may not be the same as that yielding the minimum interzonal variation in the relationship between interaction and distance to the destination.

The identification of basic entities perhaps provides the clearest way out of the MAUP because the latter is a product of aggregation. If the analysis is performed with data on basic entities then no MAUP exists. For instance, the MAUP in spatial interaction modelling (Putman and Chung, 1989) only exists when the interaction data are reported for aggregated spatial units rather than at the level of individuals, the basic entities of the movement system. The trend in spatial interaction modelling in recent years from aggregate modelling to the modelling of individual choices of destination (see chapter 4 in Fotheringham and O'Kelly, 1989 for a fuller discussion) is thus a useful way of removing the MAUP. However, while it is possible to identify the basic entities of many relationships it is not always possible to collect data at this level. Moreover, there are relationships we wish to examine but where no basic entities can be identified. For instance, suppose we want to produce a map depicting population density throughout an urban area. Given that density is measured by dividing population within a unit area by that unit area, clearly our map will reflect an arbitrary decision regarding what this area is as well as the underlying structure of population density. In some cases, the former may outweigh the latter and a distorted pattern of population densities emerges. This is a situation where no clear measure of the area in which each population density should be reported is obvious. Many similar examples exist in physical geography and in land use studies. What are the basic units of measurement we should use when examining the relationship between a river's catchment area and its length or between the coverage of some agricultural system and a particular soil property? It would thus appear that while spatial analysis should be performed at the level of basic entities where possible, in many cases such entities cannot be defined.

A third way forward from the MAUP is not to try to correct it but to recognize its existence and to report the sensitivity of a particular relationship to variations in both scale and level of aggregation. For example, when providing the results of a model calibration with one set of data it would be useful to provide the extreme results that could have been reported if different zoning systems had been used. Presumably, we would have more faith in estimates that were relatively insensitive to changes in the arbitrary units in which the data were reported. The various types of Bootstrap methods and Computer Intensive significance testing procedures (Diaconis and Efron, 1983; Knudsen and Fotheringham, 1986; Knudsen, 1987) that have been proposed in the spatial analysis literature would be useful here. However, the use of these methods relies upon data being available at some more disaggregate level than that at which the analysis was performed which will never be the case if we are trying to minimize the MAUP. Two possibilities exist, though, that would still allow a sensitivity analysis to be performed. One is that the data used in the analysis could be aggregated in different ways and the sensitivity of the results reported for this higher level aggregation. It would have to be assumed here that the sensitivity to aggregation is roughly constant at all scales, a point which has not received much attention to date. The other possibility is that data at a more disaggregate level could be interpolated in order to provide a basis for examining the sensitivity of the reported results. Presumably, the inaccuracies in the spatial interpolation method used (see Lam, 1983 for a review of spatial interpolation methods) would make these data inappropriate for analysis but would allow them to be used to perform a sensitivity analysis.

A fourth solution to the MAUP is much more speculative in that it involves the abandonment of traditional statistics in favor of other ways of representing the inherent information in the data. What exactly these others ways are is still uncertain but increasing emphasis is being placed on the visualization of data rather than on statistical analysis (Tukey, 1977; Sibley, 1988) and it may be that these techniques are less susceptible to the MAUP. This point has also been argued by Batty and Longley (1986, 1989) who espouse the ability of a model to depict realistic patterns of spatial phenomena over some statistical reporting of performance. The "Geographical Analysis Machine" of Openshaw *et al.* (1987; 1988) similarly is based on the idea of representing the available data visually over a range of spatial scales in preference to performing a certain type of spatial analysis at only one level.

A fifth way forward from the MAUP on which this paper will concentrate proceeds as follows. Given that variables and relationships change with changes in spatial scale, can we provide information on the rate of such change? Do some variables and relationships generally show erratic fluctuations with scale changes while others are generally more stable? For example, if we take data on the area of a lake and a soil coverage from maps at different scales, it seems likely that our measure of the lake coverage will exhibit a greater degree of stability than our soil coverage which will tend to decrease with finer scale measurement. This leads to the measurement of the fractal dimension of spatial distributions (Goodchild and Mark, 1987; Batty, Fotheringham and Longley, 1989; De Cola, 1989) which is the topic of the remainder of this paper.

In what follows the measurement of the fractal dimension of a spatial distribution is discussed and several potential errors in this measurement are highlighted and solutions proposed. The context of the discussion is in terms of urban density gradients but the conclusions reached are expected to have more widespread application.

Fractal dimension, urban density gradient and allometric growth

Imagine a city in one point in time that has grown outwards from an initial point of settlement (the center) and where parcels of land have been developed through a contiguous process. Let N(r) be the number of parcels of land that are now developed within radius r of the center so that

$$N(r) \approx r^{D} \tag{1}$$

and

$$\frac{\partial N(r)}{\partial r} \approx r^{D-1} \tag{2}$$

where D is a parameter which scales developed area to distance from the center of the city. If all land were developed then N(r) in equation (1) could be replaced by A(r), the total amount of land within radius r, and clearly D would be equal to two. However, if there are parcels of land that are undeveloped and the density of such undeveloped parcels increases as distance from the center increases, then D will be less than two and will be defined as the fractal dimension of the structure.

It is a simple matter to relate D to urban density gradient and allometric growth. Define the density of developed land within radius r as

$$\rho(r) = \frac{N(r)}{A(r)} \approx r^{D-2} \approx r^{\beta} \tag{3}$$

so that the urban density gradient, β, will always be negative when D is less than 2. From equation (3) note that

$$\frac{\partial N(r)}{\partial A(r)} \approx r^{D-2} \tag{4}$$

In terms of allometry the ratio of total land area to developed land area is given by A(r):N(r). Since total area has dimension 2 and developed area has dimension D,

$$A(r) \approx N(r)^{\gamma} \tag{5}$$

where, $\gamma = 2 - D$. The total area grows at a faster rate than the area of developed land implying that the margins are increasingly sparse as the structure grows.

The statistical dimension of fractal dimension

Fractal surfaces can be of two types: those with regular geometric shapes where fractal dimension can be measured exactly, such as the Sierpinski triangle, and those which are irregular where the fractal dimension has to be estimated statistically. Most spatial distributions fall into the latter category and Batty *et al.* (1989) and Fotheringham *et al.* (1989) have shown that there are at least three issues that need to be resolved in order to improve the accuracy of the statistical estimation of the fractal dimension of such structures. The first concerns the equation used to estimate D; the second concerns the type of measurement used for each variable in the equation to be estimated (the options are one-point and two-point measurements); and the third concerns the distance from the center of the structure over which measurements are taken.

The appropriate estimation equation

Any one of the first five equations presented in this paper could be used to estimate D. In operationalizing these equations it is necessary to divide the surface into discrete rings around the central point. The relevant data are taken from within these rings with the independent variable being a function of the radius of each ring's outer edge. Equations (1), (3) and (5) are structurally identical and can therefore be treated as one expression. They are recommended in preference to equations (2) and (4) which will yield less accurate estimates of D due to the inaccuracy introduced through the approximation of a continuous change variable with a discrete one.

One and two point correlation measures

Critical to the estimation of D in the above equations is the distance from the central point of the distribution. In the calibration of equation (1), for instance, data would be measured on the number of developed parcels within a distance, r, from the central point. Such a methodology is known as one-point correlation and it has been the technique used in all studies of urban density to date. An alternative approach is to take measurements around each developed parcel in the system and compute the average number of developed parcels within radius r of each developed parcel. This methodology is known as two-point correlation and has been suggested as being more appropriate for the examination of complex structures since it is more likely to capture the effects of self-similarity (Witten and Sander 1981, 1983; Meakin 1983a, 1983b). It also removes the error in the usual one-point correlation measures associated with identifying the central point of a distribution from which the distance to every other point is measured.

The two-point equivalent of equation (1) is obtained by first computing the total number of parcels of developed land T(r) within a distance r from all developed land parcels as

$$T(r) = \sum_{k=1}^{K} N_k(r) \tag{6}$$

where k denotes a developed land parcel and K denotes the total number of such parcels. The average over all developed parcels is then given by

$$\overline{T}(r) = \sum_{k} N_k(r) / K \tag{7}$$

which is related to r^D in the same manner as in equation (1). Just as the number of developed land parcels within a band around the central point can be computed as,

$$\Delta N(r) = N(r) - N(r-1) \tag{8}$$

the two-point equivalent can be computed as

$$\Delta\overline{T}(r) = \overline{T}(r) - \overline{T}(r-1) \tag{9}$$

As Batty *et al.* (1989) demonstrate, two-point relationships contain much less random noise than their one-point equivalents.

Identifying a region of stability

Probably the most critical decision to be made in the estimation of the fractal dimension of a spatial distribution concerns the identification of what Fotheringham *et al*. (1989) term the "completed portion" of the structure which in essence removes boundary effects from the estimation procedure. The completed portion of a spatial distribution is defined as being that part of the structure within which the probability of growth as the whole structure grows is so low that as the process continues the statistical relationship between N(r) and r is unaltered. Rings outside the completed structure are subject to further infilling as the growth process continues. The limit of the complete structure is defined to be at radius R. This definition of R is critical to the accurate estimation of D.

Fortunately, the separation of the complete and incomplete portions of the structure is relatively easy to achieve and in a broader context the procedure may provide a useful mechanism for removing boundary effects in empirical analyses of spatial distributions. The critical radius of the structure, R, occurs when the gradient of the relationship in question, in this case $\partial \ln\Delta N(r)/\partial \ln r$, changes sign. Up to R, $\ln\Delta N(r)$ increases as r increases whereas beyond R where the structure is not yet complete, $\ln\Delta N(r)$ decreases as r increases. Fotheringham *et al.* (1989) demonstrate the "escarpment" shape of the relationship between $\ln\Delta N(r)$ and $\ln r$ which may be characteristic of all structures undergoing growth at the periphery.

Conclusions

A basic problem that spatial data analysis faces is that measurements of variables and relationships between variables are affected by the scale at which these measurements take place. This is commonly referred to as the Modifiable Areal Unit Problem. One solution to this problem that is examined here is to recognize that a sensitivity to scale does exist and to measure rates of change with scale. This leads to the concept of the fractal dimension of a spatial distribution, and possibly of a spatial relationship although the latter is not examined here. It would be useful to examine which variables and which relationships are most sensitive to scale changes. Are there any generalizations to be drawn? Presumably there are because entities such as a lake change relatively little with scale relative to a soil coverage or the definition of an urban area. In the latter case, which was the focus of the second half of the paper, it is quite easy to appreciate how the definition of what is urban changes with scale variations. On a map of the US, Buffalo may appear as a dot; on a New York state map the same city may appear as a solid area; and on more detailed maps the city loses its uniform appearance and becomes a complex intertwining of developed and undeveloped land.

Given the premise that the concept of measuring rates of change is a valid one, this paper examines three types of error that can be present in measuring rate of change accurately. The first two concern the equation used to estimate rate of change and the third concerns the identification of a region of stability wherein relationships are constant (or nearly so).

While no procedures for the analysis of spatial data are likely to be error-free, it seems likely that the measurement of rates of change through the concept of a fractal dimension contains only trivial errors at the cost of yielding only limited information. Other statistical analyses have the appearance of yielding more information but are liable to contain substantive errors making the results misleading or, at best, highly context-specific.

References

Batty, J. M.,and Longley, P. A., 1986, The fractal measurement of urban structure. *Environment and Planning A* **18**, 1143-1179.

Batty, J. M., Fotheringham, A. S., and Longley, P. A., 1989, Urban growth and form: scaling, fractal geometry and diffusion-limited aggregation. *Environment and Planning A* at press.

Diaconis, P., and Efron, B., 1983, Computer-intensive methods in statistics. *Scientific American* **5**: 116-130.

De Cola, L., 1989, Fractal analysis of a classified landsat scene. Unpublished manuscript.

Fotheringham, A. S., Batty, J. M., and Longley, P. A., 1989, Diffusion-limited aggregation and the fractal nature of urban growth. *Papers of the Regional Science Association* at press.

Fotheringham, A. S., and O'Kelly, M. E., 1989, *Spatial Interaction Models: Formulations and Applications*. (Norwell, MA: Kluwer Academic)

Gehlke, C. E., and Biehl, K., 1934, Certain effects of grouping upon the size of the correlation coefficient in census tract material, *Journal of the American Statistical Association* Supplement **29**, 169-170.

Goodchild, M. F., and Mark, D. M., 1987, The fractal nature of geographic phenomena. *Annals of the Association of American Geographers* **72**, 265-278.

Knudsen, D. C., 1987, Computer-intensive significance - testing procedures. *Professional Geographer* **39**, 208-214.

Knudsen, D. C., and Fotheringham, A. S., 1986, Matrix comparison, goodness-of-fit and spatial interaction modelling. *International Regional Science Review* **10**, 127-147.

Lam, N. S. N., 1983, Spatial interpolation methods: A review. *The American Cartographer* **10**, 129-149.

Longley, P. A., and Batty, J. M., 1989, On the fractal measurement of geographical boundaries. *Geographical Analysis* **21**, 47-67.

Meakin, P.,1983a, Diffusion-controlled cluster formation in two, three and four dimensions *Physical Review A* **27**, 604-607.

Meakin, P., 1983b, Diffusion-controlled cluster formation in 2-6 Dimensional Space. *Physical Review A* **27**, 1495-1507.

Moellering, H., and Tobler, W., 1972, Geographical variances. *Geographical Analysis* **4**, 34-50.

Neprash, J. A., 1934, Some problems in the correlation of spatially distributed variables. *Journal of the American Statistical Association*, Supplement **29**, 167-168.

Openshaw, S., 1978, An empirical study of some zone design criteria. *Environment and Planning A* **10**, 781-794.

Openshaw, S , 1984, *The Modifiable Areal Unit Problem.* CATMOG **38**. (Norwich, England: Geo Books)

Openshaw, S., and Taylor, P. J., 1980, A million or so correlation coefficients.in three experiments on the modifiable areal unit problem. In *Statistical Applications in the Spatial Sciences* , Wrigley, N., (ed.), pp. 127-144. (London: Pion)

Openshaw, S., Charlton, M., Wymer, C., and Craft, A., 1987, A Mark I geographical analysis machine for the automated analysis of point data sets. *International Journal of Geographic Information Systems* **1**, 35-51.

Openshaw, S., Charlton, M., and Craft, A., 1988, Searching for Leukaemia clusters using a geographical analysis machine. *Papers of the Regional Science Association* **64**, 95-106.

Putman, S. H., and Chung, S.-H., 1989, Effects of spatial system design on spatial interaction models l: the spatial system definition problem. *Environment and Planning A* **21**, 27-46.

Robinson, A. H., 1956, The necessity of weighting values in correlation of areal data. *Annals of the Association of American Geographers* **46**, 233-236.

Sibley, D., 1988, *Spatial Applications of Exploratory Data Analysis.* CATMOG **49**. (Norwich, England: Geo Books)

Tukey, J. W., 1977, *Exploratory Data Analysis* . (Reading, MA.: Addison-Wesley)

Williams, I. N., 1977, Some implications of the use of spatially grouped data. In *Dynamic Models for Urban and Regional Systems*, Bennett, R. J., Thrift, N. J., and Martin, R. L., (eds.), pp 53-64. (London: Pion)

Witten, T. A., and Sander, L. M., 1981, Diffusion-limited aggregation: s kinetic critical phenomenon. *Physical Review Letters* **47**, 1400-1403.

Witten, T. A., and Sander, L. M., 1983, Diffusion-limited aggregation. *Physical Review B* **27**, 5686-5697.

Chapter 20

The effect of data aggregation on a Poisson regression model of Canadian migration

Carl G. Amrhein and Robin Flowerdew

Introduction

It is generally assumed that the process by which data describing a spatial process are organized into mutually exclusive and collectively exhaustive areal units affects analytical results. It is further assumed that these analytical results will vary in response to a change in either the number of units, or the organization of smaller units into larger units. This apparent indeterminacy has been called variously the aggregation problem, the ecological fallacy, or the modifiable areal unit problem. While the literature dealing with this issue is growing, it is surprisingly small given the rather fundamental nature of the problem as it pertains to spatial analysis (Openshaw 1984).

The most extensive treatment of these issues, and most relevant to the current effort, is contained in a series of papers by Openshaw (1977a, b, c, 1978) [much of this material is summarized in Openshaw 1984]. Openshaw first demonstrates that a wide range of correlation coefficients can be obtained by recombining a given set of areal units into the same total number (the zonation problem), or by combining a set of smaller areal units into a smaller number of larger areal units (the scale problem). As a possible solution strategy to manage the variablility of results, Openshaw (1977c) proposes a method for designing an optimal zoning system based on a set of four gravity models.

In each of these studies the criterion used to judge model performance is critical. The nature and degree of the observed aggregation bias changes with the statistic chosen to describe the effect (this will be demonstrated below). In the early studies, Openshaw uses a standard correlation coefficient. Later, the performance of the gravity models is judged using a standardized residual. In addition, the parameters of the models from different scales are compared.

In another set of experiments (Goodchild and Grandfield 1984), the set of US counties is aggregated to produce "states" that minimize the intransitivity of migration flows. The authors were unable to produce a set of aggregate "states" that eliminated the intransitive flows. The conclusion of these simulations, therefore, indicates that intransitivity is fundamental to the process of migration, and not an artifact of the original aggregation of counties into states. In a second study Goodchild and Hosage (1983) aggregate regions to form election districts. In these experiments, population constraints are imposed to generate a distribution of all possible election districts. Finally, also using US county level migration flows, Slater (1985) fits a spatial interaction model to "macrozones" constructed from the county-level data.

Research purpose

In the context of an interaction model of Canadian migration, this study examines the validity of the assumptions stated above. In other words, this study examines whether the national migration system of Canada described by a Poisson regression model exhibits any aggregation effects (see Lovett and Flowerdew 1989 for a description of the Poisson regression model). The results discussed below are preliminary findings from the second stage of a larger study. The first stage results, the calibration of several Poisson models using the complete data set, are reported in Flowerdew and Amrhein (1989).

The original data set consists of a 266 by 266 matrix of 1986 migration flows for Canada derived from income tax returns. The 266 regions represent the census divisions. The study uses 260 of these regions, excluding the divisions from the territories due to lack of comparable socio-economic data. The Poisson model used is essentially an unconstrained gravity model in which migration flows between origin region i and destination region j are predicted with an exponential function of the origin population, the destination population, and the separating distance as follows:

$$y_{ij} = exp\ (b_1 + b_2 ln\, p_i + b_3 ln\, p_j + b_4\ ln\, d_{ij}) \qquad (1)$$

where P_i is the population at origin location *i*, P_j is the population at destination location *j*, and d_{ij} is the distance between origin *i* and destination *j*. Distance is measured between regional centroids. In the original (260 region) data, centroids are provided by the census department. Some of these original centroids were changed to coincide with the dominant population center in the division (especially in the northern divisions of Montreal and Ontario). In the regions created by aggregating census divisions, the new centroid is a population-weighted average of the original centroid locations.

For the original data (260 regions), the model is calibrated as:

$$y_{ij} = exp\ (-10.10 + 0.805\ ln\, p_i + 0.821\ ln\, p_j - 0.893\ ln\, d_{ij}) \qquad (2)$$

(See Flowerdew and Amrhein 1989) for details. This model accounts for about 76% of the total deviance between the simplest Poisson model (the grand mean model) and the original data.

The purpose of this study is to test the degree to which these data exhibit an aggregation bias when the same model is calibrated using (1). fewer regions than 260, and (2). given a number of regions (less than 260), different zonings. The first part of this study will thus be testing scale effects when compared to the original model. The second part will test the stability of the results obtained in (1) under different combinations of 260 regions into fewer regions. Two different scales (130 and 65 regions) and three different combination rules will be examined.

There are a number of practical reasons for conducting such a set of experiments. While Openshaw (1984) and others have demonstrated the existence of some type of aggregation effect, the data sets used in these studies have been relatively small and spatially well-behaved. Goodchild and Grandfield (1984) used a very large data set, but the purpose of the study was to examine one specific quality of the data set, not to conduct a comprehensive study of the effect of aggregating the counties of the US into alternative states. As a result, the nature of aggregation effects is still rather poorly understood. This study certainly does not answer all questions, but it does contribute another case-study thus increasing the understanding of aggregation effects. Secondly, from a practical perspective, it may be possible at the conclusion of this stage of the research to identify an efficient number of regions for studies of Canadian migration patterns. Using the smallest number of regions necessary to preserve the essential characteristics of the migration process saves computational and research effort while maintaining the quality of the results. If using a model with 65 regions (with 4160 flows) produces results statistically as good as those produced using a model with 260 regions (with 67340 flows), the smaller problem is clearly preferred despite the loss of spatial resolution. Finally, from a modelling perspective it is important to determine whether the observation of an aggregation bias is constrained by the choice of the statistics used in the analysis.

Research design

There are two basic dimensions to the results reported below. First, for each set of experiments, a scale had to be defined. For the original data there are 260 regions. There are two sets of results, one based on 130 regions and the other on 65 regions, that are compared to the original data. Within each set of results, three zonation rules are tested (producing six groups: two levels of scale for each of the three rules). These rules are described below as Models II, III, and IV. Model I is described by equation (2) above.

The four models might be named as follows. Model I is the original model. Model II is a 'random-constrained random' model in which the region around which the aggregation is to be built (the seed region) is chosen randomly, and the neighbors to be added to this seed region are then chosen randomly from the regions contiguous to this seed. Model III is a 'random-random' model. The seed region is chosen as in Model II, but the region to be added must be contiguous to the evolving aggregation, not necessarily contiguous to the seed region as in II. Model IV is a 'random-maximum flow' model. In Model IV, the seed is chosen as in II, but the regions added to the seed are chosen from the contiguous regions in order of decreasing interaction with the original seed.

The primary difference between Model II and Model III concerns the shape of the resulting aggregations. Both models require contiguity of regions to be combined. Model II however, attempts to achieve relatively compact aggregations by adding regions contiguous to the original seed region. Model III requires only that the region to be added to a growing aggregation be contiguous to the growing region, not necessarily the original seed region. As a result, Model III can produce long, linear aggregations whereas Model II produces more compact aggregations. Secondly, Model II targets the number of regions to be combined into an aggregation. For example, combining 260 regions into 130 regions produces a target value of two. There is no such constraint in Model III. Again, as a result Model III produces a set of aggregates containing variable numbers of original regions. Model II produces a majority of aggregates composed of the targetted number of regions, with a few containing more regions, and a few less. However, the distribution is very tightly clustered around the target. While the preliminary results have not yet indicated any noticeable difference in the performance of the two models, there is an implicit spatial difference. Model II implicitly creates aggregations focussed on a core region, Model III does not. In the case of Canada, the relatively sparse distribution of higher order central places might argue in favor of an aggregation that is focussed on these few, widely dispersed places. Model III ignores this geographical fact by allowing aggregations to spread over much greater distances than Model II. However, in other situations, for example the East Coast of the US or Southeast England, the type of aggregation produced by Model III might appear more appropriate. In either case, since the seed region is chosen randomly, the appropriateness of the aggregations is left to future work.

Model IV is a more constrained version of Model II. Since ultimately it makes sense to build aggregations based on the context of the problem, Model IV is an example of other rules currently under development. This will be discussed further in the conclusions. As in the case of Models II and III, the preliminary results presented below do not indicate any substantial difference in the performance of the three rules when compared among themselves.

In rough outline form, the four models can be described as follows:

Model I:
1. Model reported in Flowerdew and Amrhein (1988).
2. 260 regions corresponding to the Canadian Census Divisions.
3. Migration Matrix derived from tax return data for 1986 (adjusted as required to fit the 1981 census divisions).
4. Census divisions as of 1981.

Model II: Algorithm I
1. Randomly choose an initial seed region.
2. Randomly choose as many contiguous neighbors as necessary to create a new region from original census divisions.
3. Target the number of "new" regions to be produced (at 130 or 65 in this case).
4. Identify the islands. Create as many aggregations as possible with determined number of regions, then identify the remaining unassigned regions. Produce the targetted number of regions by then adding any remaining unassigned regions to existing aggregations to ensure every original region is assigned.

Model III: Algorithm II (modified Openshaw (1977b) algorithm):
1. Randomly choose an initial seed region.
2. Set the number of aggregations desired (130 or 65).
3. Randomly choose the number of neighbors (less than some maximum number).
4. Randomly choose a region contiguous with the existing aggregation (the original seed region in the first instance) to create a new aggregation one region larger.
5. Repeat steps 3. and 4. until the desired number of aggregations is produced, or until all of the original regions are assigned.

Model IV: Algorithm III:
1. The same as Algorithm I except that neighbors are chosen for aggregation as follows:
2. Choose the contiguous neighbor of a seed region for aggregation that has the greatest interaction in terms of migration out of the seed region. Repeat this procedure until the desired number of neighbors is chosen.

Results

The outcomes of six sets of simulation experiments comprise these preliminary results. These results are displayed in the following tables. Each of the three alogorithms was used to combine the original 260 census divisions into two predetermined scales, 130 and 65 regions. There is a wide range of statistics that can be reported for each set of experiments. In this first instance, two key statistics are used. The proportion of deviance explained is used to assess the performance of each model as compared to the original model (260 regions), as well as every other model for each scale, and each

algorithm. The square of the correlation coefficient is used to compare these results with the results obtained by Openshaw.
A second set of tables includes the regression coefficients from the various models. These results are used to compare the nature of the model calibrated with the aggregated data.

Model	Mean Proportion of Deviance Explained (d)[2]	Standard Deviation of d	Min Value of d	Max Value of d
I[1]	.7634			
II[3]	.7621	0.007	0.740	0.780
III[3]	.7751	0.011	0.740	0.800
IV[3]	.7612	0.006	0.750	0.780

1 Original set of 260 census divisions
2 $d = 2\sum_{ij} n_{ij} \ln\left(n_{ij} / \hat{n}_{ij}\right)$
3 100 experiments each with 130 regions in each experiment

Table 1 Proportion of deviance explained (130 region model).

Model	Mean Proportion of Deviance Explained (d)[2]	Standard Deviation of d	Min Value of d	Max Value of d
II[1]	.7442	0.013	0.713	0.774
III[2]	.7641	0.016	0.726	0.792
IV[3]	.7419	0.012	0.720	0.773

1 68 experiments
2 24 experiments
3 58 experiments

Table 2 Proportion of deviance explained (65 region model).

Model	Mean r-square	St.Dev.	Min	Max
I[1]	0.456			
II[2]	0.675	0.056	0.494	0.763
III[3]	0.638	0.067	0.546	0.762
IV[2]	0.668	0.617	0.470	0.764

1 Original model with 260 census divisions.
2 100 experiments
3 10 experiments

Table 3 r-square statistic for the 130 region models.

Model	Mean r-square	St. Dev.	Min.	Max.
II[1]	0.680	0.059	0.501	0.800
III[2]	0.680	0.064	0.525	0.800
IV[3]	0.668	0.052	0.548	0.768

1 68 experiments
2 24 experiments
3 58 experiments

Table 4 r-square statistic for the 65 region models.

Model		Coeff.	St. Dev.	Min.	Max.
	I	-10.10			
130 regions	I	-8.97	0.364	-9.91	-8.24
	III	-9.84	0.670	-10.91	-8.87
	IV	-9.33	0.370	-10.40	-8.67
65 regions	I	-8.46	0.787	-10.13	-6.75
	III	-9.11	1.320	-11.56	-6.91
	IV	-8.68	0.748	-10.96	-6.72

Table 5 Mean regression coefficients: constant.

Model		Coeff.	St. Dev.	Min.	Max.
	I	0.805			
	II	0.761	0.017	0.729	0.796
130 regions	III	0.781	0.022	0.741	0.806
	IV	0.772	0.017	0.756	0.813
	II	0.719	0.029	0.658	0.787
65 regions	III	0.743	0.054	0.651	0.877
	IV	0.722	0.029	0.652	0.806

Table 6 Mean regression coefficients: origin population.

Model		Coeff.	St. Dev.	Min.	Max.
	I	0.821			
	II	0.805	0.012	0.782	0.831
30 regions	III	0.825	0.018	0.800	0.856
	IV	0.817	0.010	0.797	0.843
	II	0.800	0.024	0.743	0.854
65 regions	III	0.813	0.037	0.750	0.877
	IV	0.807	0.024	0.743	0.887

Table 7 Mean regression coefficients: destination population.

Model		Coeff.	St. Dev.	Min.	Max.
	I	-0.893			
	II	-0.927	0.020	-0.960	-0.835
130 regions	III	-0.864	0.044	-0.915	-0.799
	IV	-0.917	0.022	-0.943	-0.831
	II	-0.871	0.027	-0.915	-0.725
65 regions	III	-0.843	0.031	-0.905	-0.775
	IV	-0.861	0.017	-0.908	-0.817

Table 8 Mean regression coefficients: distance term.

Note that there are a different number of experiments for each model due to computing and time constraints. Any impression about the significance of the differences reported should recognize this limitation.

One result is immediately apparent in a comparison of Tables One and Two. There is no obvious difference in the aggregate performance of the models as captured by the proportion of the deviance explained. The deviance statistic is lower for models II and IV in the 65 region case, but the difference, given the smaller number of experiments, is not likely to prove significant. The deviance values generated appear to be more variable. The 65 region models generate a greater range in each direction. In this case, the smaller number of experiments indicates that the full range may not yet be represented.

The first outcome is rather surprising and runs counter to *a priori* expectations. As the number of observations is reduced, the ability of the model to fit the data should improve. However, this reasoning reflects a linear model in which every observation is treated equally. In this case, the presence of many zeros in the flow matrix would pose problems which should diminish with aggregation. This outcome is observed below when the r-squares are discussed. With the Poisson model, the large values in the flow matrix are more important. Aggregation reduces the number of smaller values, but their presence had no effect on the power of the model in the first place, hence, aggregation shows no change in the model's ability to explain the variance. The existence of a greater range is also expected in the context of the Poisson model. As the number of new regions decreases, the sensitivity of the model to a particular zonation increases since 260 original regions must be recombined. With 65 regions, the variability in the zonations increases, hence the ability of the model to fit the aggregated data. Note that these results are based on random zonation rules. Under deterministic rules, the variability in the results may decline due a certain stability in the aggregation process.

A comparison of Tables Three and Four reveal that when the squared correlation coefficient is used to assess model performance, results similar to those in the literature are obtained. There is a range of coefficients obtained in response to different scales as well as zonation rules. Furthermore, consistent with the expectation based on a linear regression model, the range of results obtained with 65 regions is smaller than that obtained with 130 regions. Since aggregation essentially reduces the range of entries in the flow matrix (removes zero cells especially), the linear correlation coefficient stabilizes. Presumably, as the number of regions is reduced further, the range of results should narrow.

The actual value of the correlation coefficient also performs according to expectation. As the number of observations is reduced (the number of observations is N*N-N where N is the number of regions) due to aggregation, the ability of the linear model to fit the data improves.

The question arising from these two results concerns the determination of the appropriate statistic with which to analyze model results. In this case, the deviance statistic provides an impression exactly opposite to that obtained based on the correlation coefficient. The deviance statistic finds no aggregation effect, while the correlation statistic does. How does one choose? In this case, since the Poisson model theoretically is more appropriate given the data, one would be tempted to accept the conclusion of no effect. However, might there be an equally appropriate statistic that conflicts with this finding? An examination of the performance of the model's parameters might help.

Compare the results in Tables Five through Eight, especially Six through Eight. Consistent with the suggestion from Tables One and Two, there does not appear to be any identifiable trend in the parameter values. The difference in the values appears in the second decimal place. The lack of movement in the population parameters may be explained as follows: the large flows in the flow matrix are most important to the Poisson model. The large flows are associated with the large population centers. After the aggregation process is completed, the regions to which the large population centers are assigned also have large populations. The aggregation process results largely in the removal of zero, or small cells from the table. As a result, with 130 or 65 regions, the structure of the large flows in the matrix still resembles the urban structure captured by large population centers and the parameter values are therefore stable.

The expectation in the case of the distance parameter is different. As the aggregation process proceeds, the distance matrix among the regions will change. As these distances change the associated parameter value should also change. However, since the centroid of each region is calculated as a population weighted average of the centroids of the original regions, the distances associated with the large flows remain relatively constant, giving rise to stability in the parameter value.

Conclusions

Since these results are incomplete (experiments with 30 and then 10 regions representing the actual provincial structure of the country are underway), any conclusions are speculative and perhaps premature. We suspect that there will be some critical number of regions below which (that is, with fewer regions) the deviance and the model parameters begin to change dramatically. This anticipated number might indicate the appropriate scale for the analysis for national migration studies. At this stage, it is clear that the choice of the statistic with which to examine the model's results is critical. Since the statistics used in this case are closely associated with specific models, the implication is the obvious one that choosing the appropriate model is critical as well. The results obtained so far support the Poisson model in this case.

The questions concerning the nature and degree of the aggregation effect on the Canadian migration data remain. If further experiments fail to reveal any aggregation effect at any level, then either there is no such thing as an aggregation bias (a rather remote possibility we think), the Poisson fails to capture the effect (other models would have to be tried to show this), Canadian migration is peculiar in having no scale dependency (other data using the same model would have to be tried), or some combination of these conditions prevail in the current situation. Final results await further study.

Acknowledgements: Data for this study were purchased with a grant from the Institute for Market and Social Analysis, Toronto, Ontario. Computing time was provided by a research grant from the Research Council of the University of Toronto. Computer time was subsidized by a grant from the Province of Ontario to the Ontario Centre for Large Scale Computation. Algorithm II was programmed by Felipe Calderon of the Department of Geography, University of Toronto. This research was conducted in part while Dr. Flowerdew was a research fellow at the Institute for Market and Social Analysis.

References

Flowerdew, R., and Amrhein, C., 1988, A Poisson regression model of Canadian census division migration flows based on tax file data. Presented to the *35th North American Meetings of the Regional Science Association,* November 11-13, 1988, Toronto, Ontario, Canada.

Goodchild, M., and Grandfield, A., 1984. Spatial aggregation and intransitivity in U.S. migration streams. *Modeling and Simulation 15*, Proceedings of the 15th Annual Pittsburgh Conference on Modelling and Simulation, 501-05.

Goodchild, M., and Hosage, C., 1983, On enumerating all feasible solutions to polygon aggregation problems. *Modeling and Simulation* **14,** Proceedings of the 14th Annual Pittsburgh Conference on Modelling and Simulation, 591-95.

Lovett, A., and Flowerdew, R., 1989, Analysis of count data using Poisson regression. *Professional Geographer* **41**, forthcoming.

Openshaw, S., 1977a, A geographical solution to scale and aggregation problems in region-building, partitioning, and spatial modelling. *Transactions of the Institute of British Geographers* New Series, **2**, 459-72.

Openshaw, S., 1977b, Algorithm 3: A procedure to generate pseudo-random aggregations of N zones into M zones, where M is less than N. *Environment and Planning A* **9**, 1423-28.

Openshaw, S., 1977c, Optimal zoning systems for spatial interaction models. *Environment and Planning A* **9**, 169-84.

Openshaw, S., 1978, An empirical study of some zone design criteria. *Environment and Planning A* **10**, 781-94.

Openshaw, S., 1984, *The Modifiable Areal Unit Problem.* CATMOG **38**. (Norwich, England:Geo Abstracts)

Slater, P., 1985, Point-to-point migration functions and gravity model renormalization: approaches to aggregation in spatial interaction modelling. *Environment and Planning A* **17**, 1025-44.

Chapter 21

Statistical methods for inference between incompatible zonal systems

Robin Flowerdew and Mick Green

All collectors and users of geographical data, whether they are in commerce, administration or research, face a major problem in combining data collected for different zonal systems. Data of many kinds in the social and environmental sciences are collected for areal units, but the systems of units used differ between different sets and are subject to change over time. This is a perennial source of difficulty in comparing data originating from two different sources, in studying change over time, and in studying relationships between variables of all types. This paper introduces a project intended to develop improved methods for tackling this problem.

If data for a variable of interest are available for one zonal system but are required for a different system, a zone in the former can be described as a source unit, and one in the latter as a target unit. The problem is then one of estimating the value for that variable for the target units given its values for the source units. A variety of approaches have been developed to this question; the one of most interest here is known as areal interpolation. The standard form of areal interpolation is based on weighting by area. Essentially, if the variable is a ratio or proportion, the value for a target unit is estimated as the weighted average of the values for source units which overlap it. The weights are derived from the proportion of the target unit overlapping each source unit. If the variable is a count or aggregate, a different but analogous method is used.

This method may give the best estimate if there is no additional information about how the values recorded are distributed within the source units, but in many cases common sense and elementary geography suggest that the even distributions this method assumes are most unlikely to be found in reality. Often other information is available for target units which we know is likely to be linked to the variable of interest. We intend to develop methods for using such information, so that estimates from areal interpolation techniques can take account of our knowledge of the areas concerned. Such methods would help reduce the many practical problems stemming from the incompatibility of zonal systems in use today.

The aim of the project is to develop a set of methods for *intelligent* areal interpolation, in which our knowledge of other variables can be used to predict target unit values for a variable of interest. A set of methods, not just one, will be needed because the statistical methods necessary will vary according to the measurement scale of the variables concerned. Thus a regression estimate for a binary variable will require a different error specification from that for a count or for a continuous variable. There will also be differences in the methods required depending on the measurement scale of the predictor variables.

Our project will consist of several components. Methods will be developed for the various subproblems defined by the measurement scales of the variables involved; we estimate there may be some eight to ten subproblems sufficiently distinct to require different treatment. Second, they will be operationalised using the statistical package

GLIM, writing new procedures where necessary. Third, they will be tried out using real-world examples. Finally, they will be written up and disseminated to potential users.

Background

The problem of comparing data collected for incompatible zonal systems (hereafter referred to as the areal interpolation problem) arises in many different situations. Openshaw, Wymer and Charlton (1986) report that the Domesday project involved twenty-three different sets of areal units. In addition to the fact that different data collection agencies use different boundaries, most of the zonal systems used are liable to change over time. The simplest type of problem arising from this situation is that of estimating the value of a variable for an areal unit different from that for which it is measured. To clarify the discussion, the areal units for which data exist will be referred to as source units, and those for which data are desired as target units (Markoff and Shapiro 1973). An extension of this problem arises when it is desired to study changes over time in a variable for an areal unit which has been subject to boundary changes. A further problem is that of studying the relationship between two (or more) variables which are defined for different sets of areal units.

The appropriate methods for areal interpolation depend on the exact nature of the problem, including the scale of measurement of the variable of interest and the nature of the areal units concerned. The major cases to be considered are those where the variable of interest is measured as

a) dichotomous
b) polytomous
c) count
d) interval extensive
e) interval intensive

Cases d) and e) can be distinguished by considering the relationship between the value for an areal unit and the values for its component parts (following Goodchild and Lam 1980). If the former is equal to the sum of the latter, the variable is extensive; if it is a weighted average of the latter, as with ratio or percentage variables, the variable is intensive. Different methods for areal interpolation are appropriate in each case.

Two main approaches to areal interpolation have been employed, which can be referred to as polygon overlay and smoothing methods (see Lam 1983). Polygon overlay methods (Markoff and Shapiro 1973) consist essentially of computing averages weighted by area. These are the methods used in those geographical information systems which offer facilities for areal interpolation. The main problem with such methods is that they assume the variable concerned is distributed evenly in the source zone - in practice, there may be good evidence from the distribution of other variables that this is not so.

In contrast, smoothing methods (e.g. Tobler 1979) work by fitting a surface to data for the source units and using that surface to interpolate values for the target zones. This is likely, at least for some types of data, to give a more realistic distribution for the target units, but in many cases there is no reason to suppose that the distribution is smooth and the method can perhaps be criticised for giving a spurious impression of precision.

Aims

The principal aim of the project is to develop methods of areal interpolation that differ from earlier methods in two important respects. First, they will take into account the values of other (predictor) variables to which the variable of interest may be related, information not systematically used in existing methods but often highly relevant to our intuitive feelings about the data. Second, they will be statistical methods, which will give maximum likelihood estimates of values for target zones with respect to the models constructed. In addition, unlike existing methods, they will be accompanied where possible by measures of goodness of fit, so that the reliability of the estimates can be assessed.

Different models will be necessary for making these estimates, depending on the measurement scales involved. Not only are there differences according to the measurement scale of the variable of interest; there will be differences in method according to the scale of the predictor variable. There will also be differences in method if more than one predictor variable is used. Further complications may arise if the estimating model is to take spatial autocorrelation effects into account.

Progress to date

Our interest in these problems stems from our involvement in the North West Regional Research Laboratory, which has been established at Lancaster with funding from the British Economic and Social Research Council (ESRC). One aspect of our work has been the development of spatial analysis techniques for use in regional research, in which workers at Lancaster have co-operated with colleagues at Newcastle, especially Stan Openshaw. Our interest in areal interpolation has grown out of this work, together with the recognition of the importance of the topic for our applied work.

We have completed a review paper (Flowerdew and Openshaw 1987) which includes a classification of the different types of areal interpolation problem. We have already developed a method for one case, that where the variable of interest is a count and the predictor variable a dichotomy. This might apply, for example, where population figures are available for local government units (wards) but are needed for land-use categories (e.g. grassland and forest). If the area of each ward in each land-use category is known (information available from a geographical information system), ward population can be regressed on area under grassland and area under forest. Because population is a count variable, this can be done using a form of Poisson regression (Lovett and Flowerdew 1989); in GLIM terminology, a Poisson error term and identity link are appropriate. The parameters of this model can be used to estimate the populations of parts of wards with each land-use, which are then adjusted to fit known ward totals, and combined to get target unit estimates. Clearly the reliability of such a procedure will depend on how good the regression estimates are. However, experience so far suggests that even a poorly fitting model can produce estimates much better than those derived from areal weighting alone. The procedure is explained in more detail in the next sections.

Estimating count data from a binary variable

In the polygon overlay method, values for target units are derived by adding values for the areas of intersection of the target unit with each source unit. The areal interpolation problem therefore becomes one of estimating values for these areas of intersection or, equivalently, of dividing the source unit value among the subareas defined by the target unit boundaries. In the standard polygon overlay approach, this is done on the basis of area; in the approach to be demonstrated here, the value of a binary variable defined for the target units is used as well.

Figure 1 shows a hypothetical example. Assume we have population data available for grid squares; this might well be the case in a rural area. The objective is to estimate population for target units defined according to the value of a binary variable (these units are represented by the areas with irregular boundaries in Figure 1). Assume this represents patches of land which are grassland (category 1) and patches which are woodland (category 2). The binary variable concerned could alternatively measure any of several aspects of land cover, land use, soils or geology. The target units are natural units in Unwin's terminology (1981), and hence we are dealing with Type I data in Flowerdew and Openshaw's classification (1987).

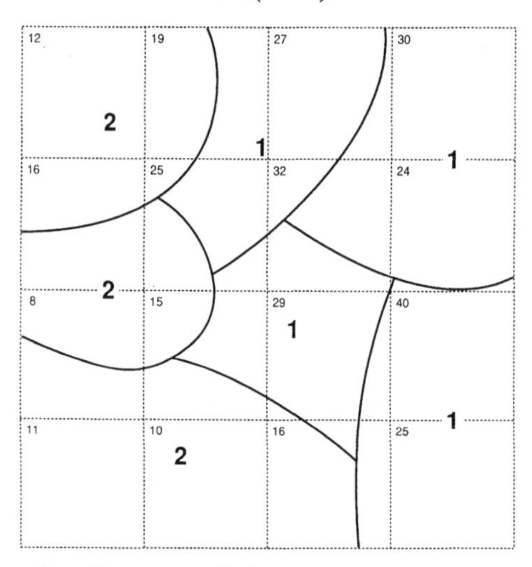

Figure 1 Hypothetical example to illustrate method.

If the binary variable bears any relationship to population, knowledge of its value will be relevant to prediction of population distribution for the subunits. Even if the relationship is weak, estimates produced using this information should be an improvement over estimates based solely on area. If the expected population in a subunit is thought of as being generated by a probability distribution, then the mean of this distribution will have a relationship to the binary variable. If people are more likely to live in grassland than woodland, then the grassland probability distribution should have a higher mean than the woodland probability distribution. As a count variable, population must follow a discrete distribution; if it has a Poisson distribution, then the population living in a unit area of grassland can be regarded as a realization of a Poisson process with mean λ_1 and the population in a unit area of woodland as a realization of a Poisson process with mean λ_2, with $\lambda_1 > \lambda_2$.

It is well known that the sum of two Poisson processes is itself Poisson. It therefore seems natural to regard the source unit population as a realization of a Poisson process whose mean λ is equal to the area of the source unit in category 1 (grassland) multiplied by λ_1 plus the area in category 2 (woodland) times λ_2. Writing A_i for the area in category i gives the formula:

$$\lambda = \lambda_1 A_1 + \lambda_2 A_2 \quad (1)$$

The areas can be calculated by a GIS system (or measured in any other way), and the equation can be regarded as a regression equation where the (known) populations of source units can be used as input data to calculate the coefficients λ_1 and λ_2.

It may be argued that the Poisson assumption is invalid for population (and for many other count variables). In particular, it is based on the assumption that individuals behave independently of each other; in this case, that an individual's location in a zone is independent of all other individuals. This is not true, if only because of people's tendency to live together in families, and it has been argued in the context of migration modelling (Flowerdew and Lovett 1989) that this basic fact makes a pure Poisson model inappropriate, and a compound or generalised Poisson model should be used.

However, in this context our main concern is to improve our estimates of population, rather than to produce a model to explain population distribution, and Davies and Guy (1987) have shown that the estimated coefficients (though not their standard errors or overall goodness of fit) from a compound or generalised Poisson model are the same as those from an ordinary Poisson model. As it is these coefficient estimates that are used in areal interpolation, the computationally easier Poisson model can validly be used despite the non-Poisson nature of the underlying probability process.

Poisson regression analysis can be used to fit the regression model (1); Flowerdew and Aitkin (1982) describe the technique and Lovett and Flowerdew (1989) discuss its implementation in the computer package GLIM. There is an important difference from standard Poisson regression. Usually the Poisson-distributed response variable is assumed to be logarithmically linked to the explanatory variables. In this case, however, the response variable in equation (1) is simply a linear combination of the two area variables; there is no logarithmic term and the link is therefore the identity function. It should also be noted that the regression has no constant term.

In the case of the data set depicted in Figure 1, input data for the regression should be organised by source unit, with each record containing the population for that unit, the area under grassland (category 1) and the area under woodland (category 2). Source units should be included in the data set even if they are wholly within one of the two categories (and even if they have no population). The necessary GLIM commands would be as follows (assuming input data are on file 1):

```
$UNITS 16 $
$DATA POP A1 A2 $
$DINPUT 1 $
$YVAR POP $
$ERROR POISSON $
$LINK IDENTITY $
$FIT A1 + A2 - 1 $
$DISPLAY E R $
```

The '-1' part of the $FIT directive instructs GLIM that there should be no constant term in the regression, and the last line asks for parameter estimates and residuals to be displayed (the residual display includes fitted values for each case).

The parameter estimate for A_1 (the area under grassland) gives the value for λ_1 (29.00 for the data in Figure 1), expected population for a unit area of grassland, and that for A_2 gives the value for λ_2 (11.34 in this case), expected population per unit area of woodland. The standard regression output indicates model goodness of fit with the scaled deviance statistic. If the data could have been generated by the Poisson model being fitted, scaled deviance should not be greatly in excess of the degrees of freedom. In this case, scaled deviance is 10.38 with 14 degrees of freedom; as the numbers were assigned subjectively to show a difference between woodland and grassland population, the apparent good fit is not surprising.

The next stage is to calculate expected population for the subunits into which the source units are divided by the target unit boundaries. This can be estimated simply by multiplying the parameter estimate for each category by the area covered by that category. This method, however, does not have the pycnophylactic property. In other words, the estimates for the subunits do not sum to the known population for the source unit. Instead they sum to the fitted value produced by the regression. The residual display indicates that the fit is not all that good for some source units: grid squares 12 and 5 are under-predicted and 9, 2 and 8 over-predicted. Estimates can be improved by taking the source unit population into account through scaling them up or down to fit the source unit total. For each source unit, this involves multiplying by the observed value divided by the fitted value. The estimated populations in the category 1 and category 2 subareas can be calculated with the following GLIM commands:

```
$CALC EST1 = (29.00 * A1) * POP / %FV $
$CALC EST2 = (11.34 * A2) * POP / %FV $
$PRINT POP EST1 EST2 $
```

Fitted values from a regression are automatically stored in the total population, the estimated population in category 1 subunits and the estimated population in category 2 subunits. It is then a simple matter to assemble the estimates for the subunits to reach the desired estimates for the target units.

If the parameter estimates for A_1 and A_2 are substantially different, the estimates derived by this method will also be substantially different from those derived by an ordinary polygon overlay technique. The quality of the estimates will depend on the closeness of the relationship between the count and the binary variable.

The example described above uses hypothetical data, but the method has also been applied to an example using Lancashire census data (see Figure 2a and 2b). In this case, the units concerned were parliamentary constituencies and local government districts, units of comparable size which sometimes but not always have coincident boundaries. The districts were regarded as source units and the parliamentary constituencies were regarded as target units. Two interval-scale variables were considered: first, total population, and second, population born in the New Commonwealth and Pakistan (the usual surrogate for race in the British census, where an explicit question on race is not asked). The political representation of the constituency was used as a binary predictor variable (Conservative or Labour). Clearly political representation is not responsible for the distribution of population, and regressing the latter on the former may be thought absurd, but there is a relationship between population density and political party (rural people are more likely to vote Conservative and inner-city residents to vote Labour). In the absence of any better information, political representation can be used to improve our estimates of population density.

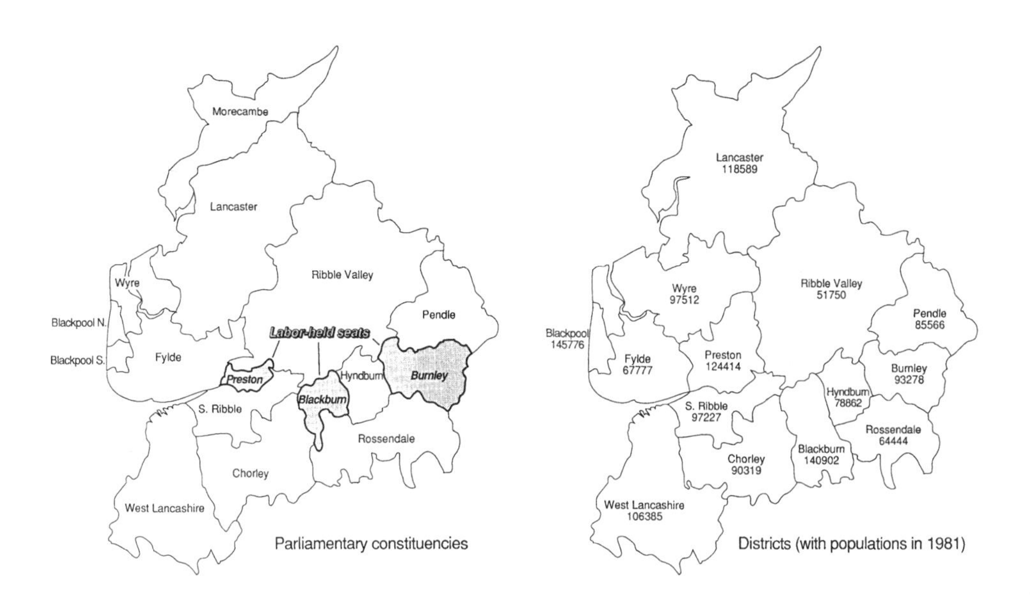

Figure 2 Districts and parliamentary constituencies in Lancashire, 1983.

The standard polygon overlay approach, based on area, was used first to estimate the population of the target units. This method proved very unsuccessful. The new method was then used and produced population estimates which were far closer to the true values. Repeating the procedure for New Commonwealth and Pakistan born population produced an even more dramatic improvement. Details of the results are available in Flowerdew (1988).

Generalizing the approach

As stated earlier, this approach is restricted to the case where a count variable is to be predicted on the basis of a binary variable. In order to apply the ideas underlying the method to a wider set of situations, it must be generalized. Use of the EM algorithm (Dempster, Laird and Rubin 1977) may be an appropriate way of doing this.

The EM algorithm is a general statistical technique which unifies a collection of specific methods designed to cope with problems of imperfect observations. Examples include:

a) Missing values. We may wish to use an estimation procedure that requires a complete data set, but some observations may be missing due to accidental loss of the information or as the result of an invalid experiment. The EM solution is to impute the values of the missing data and apply the estimation procedure to the "complete" data set.

b) Censored observations. In many cases observation of some data values may be incomplete. For example in medical studies the survival times of patients still alive at the end of the study are not known: we simply know that they survived beyond the study period. The EM solution is to impute survival times for these patients and use these in a model as if they were observed survival times.

The EM algorithm is an iterative procedure in which each iteration consists of two steps:

the E-step:	compute the conditional expectation of the missing data given the model and observed data;
the M-step:	fit the model by maximum likelihood to the 'complete' data set treating the imputed values from the E-step as real observations.

To apply the technique to the areal interpolation problem, consider a table whose rows are the source units for which we have an observation on a count variable y and whose columns are the target units for which we require estimates. Each cell st of the table represents the intersection of source unit s and target unit t and should contain the value of y for this intersection. In most cases these values will be zero because s and t are disjoint. If we now imagine that this table had existed and that someone had calculated the row totals y but had then lost the original table, we can consider this as a missing data problem. If the value of an ancillary variable x is available for the target units, we can use this information and the areas of the zones of intersection to estimate what the values of y should be for each zone of intersection. The estimation can be done using the EM algorithm. If x is a binary variable, the method is equivalent to that outlined above.

This approach has the advantage that it can be generalized to use target unit data of other measurement scales, including continuous variables. It can also be extended to incorporate more than one target unit variable. Future papers will give details and examples of how this can be done.

Conclusion

This paper has described a new method for areal interpolation. The standard procedure currently in use is based on the assumption that the variable of interest is distributed evenly over each source unit. In contrast, this method allows the analyst to take into account information available for target units in a formal and statistically appropriate way. This information is assumed to be available for a binary variable which has some relationship, not necessarily causal, to the variable of interest, which is assumed to be a count.

The method consists of regressing this variable on two explanatory variables, the areas within each source unit having each value of the binary variable. A Poisson model with identity link, with no constant term, is the appropriate form of regression. The regression coefficients are then used to form estimates for the zones of intersection between source and target units; these estimates can be aggregated to the target units to achieve a solution to the areal interpolation problem. The quality of the solution is likely to depend on the degree to which the variable of interest is correlated with area and with the binary variable used in estimation.

The approach discussed is highly restrictive in that it is usable only when the variable of interest is a count and the predictor variable is binary. The EM algorithm offers a method for extending it to deal with other kinds of predictor variable, and the project described above will be concerned with implementing the method and extending it to other situations. In some cases, extension of the methods should be relatively straightforward; in other cases, it is likely to prove difficult and perhaps impossible.

Acknowledgements . We have had much useful discussion with Tony Gatrell and Peter Vincent (Geography, Lancaster), Stan Openshaw (Geography, Newcastle), Richard Davies (Town Planning, Cardiff) and others, to whom we are indebted. Support of the

ESRC through the Regional Research Laboratory initiative is also acknowledged. Parts of the paper appeared in Research Report 16 of the Northern Regional Research Laboratory.

References

Davies, R. B., and Guy, C. M., 1987, The statistical modeling of flow data when the Poisson assumption is violated. *Geographical Analysis* **19**, 300-314.

Dempster, A. P., Laird, N. M., and Rubin, D. B., 1977, Maximum likelihood for incomplete data via the EM algorithm. *Journal, Royal Statistical Society B* **39**, 1-38.

Flowerdew, R., 1988, Statistical methods for areal interpolation: predicting count data from a binary variable. *Research Report* **16**, Northern Regional Research Laboratory, Lancaster and Newcastle.

Flowerdew, R., and Aitkin, M., 1982, A method of fitting the gravity model based on the Poisson distribution. *Journal of Regional Science* **22**, 191-202.

Flowerdew, R., and Lovett, A., 1989, Compound and generalised Poisson models for inter-urban migration. In *Advances in Regional Demographic Analysis*, Congdon P., and Batey, P. (eds.), pp. 246-256. (London: Belhaven)

Flowerdew, R., and Openshaw, S., 1987, A review of the problems of transferring data from one set of areal units to another incompatible set. *Research Report* **4**, Northern Regional Research Laboratory, Lancaster and Newcastle.

Gatrell, A. C., and Charlton, M., 1987, An introduction to ODYSSEY, a geographical information system. *Research Report* **5**, Northern Regional Research Laboratory, Lancaster and Newcastle.

Goodchild, M. F., and Lam, N. S. -N., 1980, Areal interpolation: a variant of the traditional spatial problem. *Geo-Processing* **1**, 297-312.

Lam, N. S. -N., 1983, Spatial interpolation methods: a review. *American Cartographer* **10**, 129-149.

Lovett, A., and Flowerdew, R., 1989, Analysis of count data using Poisson regression. *Professional Geographer* **41**, 190-198.

Markoff, J., and Shapiro, G., 1973, The linkage of data describing overlapping geographical units. *Historical Methods Newsletter* **7**, 34-46.

Openshaw, S., Wymer, C., and Charlton, M., 1986, A geographical information and mapping system for the BBC Domesday optical discs. *Transactions, Institute of British Geographers* **11**, 296-304

Tobler, W. R., 1979, Smooth pycnophylactic interpolation for geographical regions. *Journal of the American Statistical Association* **74**, 519-530.

Unwin, D., 1981, *Introductory Spatial Analysis* . (London: Methuen)

Chapter 22

Statistical effect of spatial data transformations: a proposed general framework

Giuseppe Arbia

Introduction

Spatial data can be observed in various forms. A first distinction can be drawn between continuous areal data and discontinuous data. A similar distinction is usually made when dealing with time series data. For example a time series of the stock market prices is essentially a continuous one which can be recorded at any point in time. In contrast, the total level of production in one country can be defined only over intervals of time, as the sum of units produced in that interval.

Continuous areal data are fairly common in meteoreology, in remote sensing and in ecology. For example annual rainfall totals can be, in principle, observed at all points over the entire study area. In the case when data are intrinsically continuous we can consider the observations at some points in space as a sample from a continuous random process. It is, however extremely difficult to find examples of such data when we deal with socio-economic analysis where phenomena occur which often have sharp discontinuities between areas.

While in time series analysis we have only one way in which data can be discrete, in spatial analysis we can distinguish between three different kinds of discontinuities, namely points, lines and areas. A point refers to a single place and is usually considered as having no dimension or having a dimension which is negligible if compared with the study area. There is a large number of examples of point data such as, for instance, the location of industrial plants or houses or the distribution of trees in a forest. Furthermore it may happen that areal data are reduced to point form as, for instance, when studying central-place theories of settlement structure (Wilson and Bennett, 1985) or in order to interpolate a trend surface (Haggett, 1967). Point data lead to studies of the *spatial pattern* of geographical entities. Line data or networks arise in geography to specify features like travel-to-work flows or transportation networks. Flow data lead to studies on *spatial interaction.* Areal data have a paramount relevance especially in studies concerned with socio-economic phenomena. In fact most empirical studies in human geography are based upon Census information collected over discrete areal units such as counties or regions. Locational data, such as regional population or regional income, lead to studies on the *spatial structure* of geographical phenomena which are concerned with the distribution of entities in areal units and with their associated properties.

When we deal with areal data a second distinction can be made between data laid on a regular Cartesian lattice and data located on an irregular grid. Examples of the first kind of data in human geography are very rare. Some Census information for the U.K. are grid-referenced (e.g. in the 1971 Census). Similarly other countries like Sweden and the U.S.A. record some population data on regular grid. Finally some examples exist dealing with geography at a micro scale. The far more common typology of spatial data in human geography, however, is when we have our information recorded on territorial units, often administrative units, which, in general, are of variable size and shape. It is usual to refer to this kind as data from irregular collecting areas (Haggett, 1967).

It sometimes happens that, for various reasons, data are not available in the form required by the phenomenon under study but, instead, in a different form. A typical example is the case in which we want to study the point pattern of a particular geographical entity and we can only access data aggregated into areal units. A second example can be the case in which we study the rainfall level which is a continuous phenomenon, but is only observable as a point process. In these cases, and in many others, the spatial data base we have in hand turns out to be the result of a transformation of what we can call the *original* geographical process, which is, instead, unobservable.

The major aims of this paper are:

(i) to present some examples of spatial data transformations (hereafter SDT)
(ii) to define formally the problem of SDT and
(iii) to propose a general framework to study the effects of these transformations on the statistical analysis of spatial data.

The setup of the paper is as follows. In Section 2 we will review briefly the major results obtained in previous work (Arbia, 1988) on a particular kind of SDT, namely the transformations from areal data to areal data. These transformations are the basis of the *modifiable areal unit problem.* In Section 3 we will introduce some possible extensions of the methodological framework exploited in Arbia (1988) to a wider set of problems generated by SDT, such as, for instance, those originated by point-to-point transformations, or by transformations from surface-to-areal data. Finally in Section 4 we will summarize the content of the paper and draw some tentative conclusions.

Area-to-area transformations: the modifiable areal unit problem

A first kind of spatial data transformation refers to the case in which we have in hand areal data and we wish to study a geographical phenomenon at a given level of spatial resolution, but we have at our disposal data collected at a different spatial scale or following different criteria.

The problem is very well known and is usually referred to as the *modifiable areal unit problem* (Openshaw, 1984). The problem is particularly serious in regional economics where data are constituted, as a rule, by aggregation (or accumulation) of individuals' characteristics in portions of space, either in the form of sums or in the form of averages. For example, population in a county is the sum of the individuals living in that county, the total income of a region is the sum of the individual income of the population in that region, the per-head consumption in an area is the mean of the individual consumption in that area and so on. Furthermore, for various reasons it sometimes happens that aggregate spatial units are further aggregated into larger portions of space e.g. by shifting from a town to a county level, or from one zoning system to another. In each of these cases we start by considering a certain generating process and we end up, after aggregation, with a different process. We will from now on refer to the original generating process as the individual-process, whether it is constituted by actual individuals or by small areas, and we will refer to the process after aggregation as the group-process.

In the quoted work (Arbia, 1988) our aim was to derive the group-process probability distribution in terms of the individual-process probability distribution. We followed three different approaches.

The first approach consisted in developing a general scheme to derive the joint probability density function of a group-process given a distribution of any kind underlying the individual-process. This is only possible if the form of the individual-process is completely specified.

The second, less ambitious, approach consisted in deriving only some significant moments of the group-process in terms of the moments of the individual-process. The study was restricted to the lower order univariate moments (mean, variance, skewness and kurtosis), and to the spatial autocorrelation and cross-correlation between the two processes.

Thirdly we considered that for the special case where we deal with Gaussian processes the study of only the first and the second moments exhausts the problem because these are enough to determine the whole distribution of the process. Furthermore we have the convenient property analogous to the central limit theorem, that for a large family of well-behaved stationary processes, linear combinations of observations tend to normality even when they are constituted by mutually dependent components. For this reason a third section is devoted to transformations of Gaussian processes.
We will now review some of the results obtained with the three approaches outlined above.

The method based on density functions

Following the first of the approaches we seek a relationship between the joint density function of the generating process before aggregation and the same after aggregation. Let $\{X_i\}$, $i=1,...,n$ be a stochastic process with a joint density function $f_{1,...n}$ $(x_1,...,x_n)$ and consider n real-valued functions of the n observed values

$$x_i^* = x_i^* (x_1,...,x_n) \quad i = 1,...,n \qquad (1)$$

The inverse transformation is

$$x_i = x_i(x_1^*,....,x_n^*) \qquad (2)$$

Let the random variables $X^*_1,...,X^*_n$ be similarly defined as

$$X_i^* = x_i^*(X_1...X_n) \qquad i = 1 ,..., n \qquad (3)$$

Then the joint probability density function of $\{X^*_i\}$ is (Anderson, 1958, p.11)

$$g_{1,...,n} (x_1^*,...,x_n^*) =$$

$$= f[x_1(x_1^*...x_n^*);...;x_n(x_1^*...x_n^*)]J(x_1^*...x_n^*) \qquad (4)$$

where $J(x_1^*,...,x_n^*)$ is the Jacobian of the transformation. In the case of linear transformations we have, for instance, that

$$x_i^* = \Sigma_j \, g_{ij} \, x_j \qquad i = 1,...,n \qquad (5)$$

or, in matrix notation

$$x^* = G\,x \qquad (6)$$

The previous formula (6) assumes that a one-to-one transformation occurs from the x-space to the x*-space. A more common situation in geographical analysis is, however, the case in which, through linear transformations, we make a reduction from the original space of n random variables to a subspace of m (m < n) random variables. In this case the matrix of derivatives becomes singular and the Jacobian cannot be computed. A method to overcome this problem has been proposed (Arbia, 1988) and consists of introducing a number of identical transformations so as to have a non singular matrix of derivatives, and then integrating out the redundant variables.

Formally, suppose we are given a process $\{X_j\}$ j = 1,...,n with joint probability density function $f_{1,...,n}(x_1,...,x_n)$ and then consider m real-valued functions (m < n)

$$x_i^* = x_i^*(x_1...x_n) \quad i = 1,...,m \qquad (7)$$

and (n-m) real-valued functions

$$x_l = x_l \qquad l = 1,...,(n-m) \qquad (8)$$

In the linear case this can be summarized in the matrix notation

$$X^* = G\,X \qquad (9)$$

with $X^*=(X^*_1,...,X^*_m \mid X_1,...,X_{(n-m)})$, $X = (X_1,...,X_n)$, and G an appropriate grouping matrix. The inverse of the transform is

$$x_j = x_j\,(x_1^*...x_m^*\; x_1...x\;\;),\quad j = 1,...,n \qquad (10)$$

and the joint probability density function of the group-process is given by

$$\int\!\!\int f_{1...n}\left[x_1\left(x_1^*...x_m^*\; x_1...x_{(n-m)}\right)...\; x_n\left(x_1^*...x_m^*\; x_1...x_{(n-m)}\right) J\left(x_1^*...x_m^*\; x_1...x_{(n-m)}\right)\right]$$

$$dx_1...dx_{(n-m)} =$$

$$g\left(x_1^*\,...x_m^*\right) \qquad (11)$$

This kind of approach was only outlined in the quoted work and no application was attempted. In fact for Gaussian processes, on which most of the quoted work was concentrated, the situation is much more simple as we will see in a later section.

The method of moments

When it is not possible to specify fully the density function of the individual-process, a practical way of attacking the problem could be to confine our attention to obtaining the lower moments of the group-process given the moments of the individual-process and, if possible, to try to fit one of the most common tabulated distributions (Ord, 1972). This approach will now be summarized.

Suppose we are given a random process {X } which is in the strict sense stationary. Consider then the m real-valued functions, defined for short:

$$x^*_i = x^*_i (X_1, X_2,..., X_n) = g (X) \qquad (12)$$

If the function g has finite derivatives we have

$$g_j (\mu) = dg (X) / dX_j \qquad (13)$$

evaluated at the point μ If we now expand g(X) around the mean using the Taylor's series expansion we have (Kendall & Stuart, 1976. p.231)

$$g(X) = g(\mu) + \Sigma_j g'_j(\mu) (X_j - \mu) + \Sigma_j g''_j (\mu) (X_j - \mu)^2 + ...+ O(n^{-1}) \quad (14)$$

where $O(n^{-1})$ is a term of order n^{-1}. In the case of linear aggregation we have g'= g"=...= 0 and the previous formula approximately yields

$$g(X) = g(\mu) + \Sigma_j g_j'(\mu) (X_j - \mu) \qquad (15)$$

This formula allows us to derive all the univariate moments of the group-process in terms of the moments of the individual-process simply by observing that

$$E\{x^{*r}\} = E \{[g(X)^r]\} = E \{[g(\mu) + \Sigma_j g'_j (\mu) (X_j - \mu)]^r\} \qquad (16)$$

Furthermore an analogous formula can be applied for the joint moments (See Kendall and Stuart, 1976. p.232). This approach has been exploited in Arbia (1988) to obtain the lower order moments of the group-process given the values of the moments of the individual-process.

A solution for Gaussian processes

If it is possible to assume that the individual-process is Gaussian, a third approach is available. In rather formal terms we want to describe the situation in which we are given n sites with a given spatial sequence indexed by j, and a vector $\mathbf{x} = (x_1, x_2,..., x_n; y_1, y_2,....y_n)$ of observations on the variables X and Y which is considered a finite realization of an individual process $\{X_jY_j\}$ of dependent random variables $X_1, X_2,...,X_n, Y_1, Y_2,...,Y_n$.
The spatial process is now assumed to be Gaussian and stationary with vector of means $\mu = \mathbf{0}$ and dispersion matrix

$$V = \begin{bmatrix} xV & xyV \\ yxV & YV \end{bmatrix} \qquad (17)$$

The submatrices $x\mathbf{V}$ and $y\mathbf{V}$ are the matrices of variance and covariance of the process $\{X_j\}$ and, of the process $\{Y_j\}$ respectively, and the submatrix $xy\mathbf{V} = yx\mathbf{V}$ is the matrix of the cross-covariances between the process $\{X_j\}$ and the process $\{Y_j\}$.

We now wish to introduce some hypotheses about the form of the matrix **V** which embody a plausible pattern of spatial dependence. The "first law of geography" (Tobler, 1970) holds that: "Everything is related to everything else, but near things are more related than distant things". This is the basis of the local dependence hypothesis and assumes a "distance decay" or "neighborhood" effect. Formally this is expressed as the conditional distribution of X at site j depending only upon the values of those sites which are, somehow, in the proximity of site j (Besag,1974).

Let us define N(k) as the set of neighbors of site k, and ${}_x\delta_{jk}$ the typical element of xV; then we have the condition that a Gaussian process $\{X_j\}$ is locally dependent (or locally covariant) if:

$$
\begin{aligned}
{}_x\delta_{jk} &= E[(X_j - {}_x\mu_j)(X_k - {}_x\mu_k)] = {}_x\delta\, {}_x\sigma^2 \text{ when } j \in N(k) \\
&= 0 \text{ otherwise} \qquad (18)
\end{aligned}
$$

where ${}_x\sigma^2$ and ${}_x\delta$ are, respectively, the variance and the spatial autocorrelation (Cliff and Ord, 1981) of the process $\{X_j\}$. Analogous definitions hold for the process $\{Y_j\}$. Furthermore, defining ${}_{xy}\delta_{jk}$ as the typical element of xyV we have that a bivariate Gaussian process $\{X_jY_j\}$ is locally dependent (or locally covariant) if :

$$
\begin{aligned}
{}_{xy}\delta_{jk} &= E[(X_j - {}_x\mu_j)(Y_k - {}_y\mu_k)] = {}_{xy}\delta'\, {}_x\sigma {}_y\sigma \text{ when } j \in N(k) \\
&= {}_{xy}\delta\, {}_x\sigma {}_y\sigma \text{ when } j = k \\
&= 0 \text{ otherwise} \qquad (19)
\end{aligned}
$$

The term ${}_{xy}\delta$ is usually referred to in the literature as the cross-covariance between the process $\{X_j\}$ and the process $\{Y_j\}$ (Anderson, 1958). In contrast, we will refer to the term ${}_{xy}\delta'$ as the lagged cross-covariance between the two processes with a term borrowed from time series analysis. This term is sometimes referred to in the spatial literature as the *spill over* effect (Paelinck and Nijkamp, 1975) and describes the effect that a variable at site k produces on a second variable in the neighbourhood of k.

Let us now consider a group-process $\{X_i^*\}$; i = 1,...,m (m<n) derived from the individual-process through the linear transformation

$$X^* = G X \qquad (20)$$

where G is an m by n averaging matrix with elements g_{ij} which assume the value

$$
g_{ij} = \begin{cases} r_i^{-1} \text{ if } j \in g(i) & r_i \text{ times in the i-th row} \\ 0 \text{ otherwise} & n - r_i \text{ times in the i-th row} \end{cases} \qquad (21)
$$

In the previous formula G(i) is the set of random components of the individual-process included in the i-th random component of the group-process and r_i is the cardinality of G(i).

Let us now recall a famous result (Anderson, 1958, p.25) which states that, if $\{X_j\} = X_1,...,X_n$ is distributed according to the law $N({}_x\mu, {}_x\mathbf{V})$ then $\{X^*_i\} = X^*_1,...,X^*_m$,(m < n) such that $\mathbf{X^*} = \mathbf{GX}$ is distributed according to the law $N(\mathbf{G}\, {}_x\mu, \mathbf{G}\, {}_x\mathbf{VG'})$, where **G** is an m by n grouping matrix of rank m < n. An immediate extension to a bivariate process states that if $\{X_jY_j\}$, j = 1,..,n, is a bivariate process distributed according to the Gaussian law with vector of means $({}_x\mu\ {}_y\mu)$ and dispersion matrix as in formula (17) then the bivariate process $\{X^*_i Y^*_i\}$, $i = 1,\ldots,m$, $(m<n)$ such that $\mathbf{X^*} = \mathbf{GX}$ and $\mathbf{Y^*} = \mathbf{GY}$, is distributed according to the Gaussian law with vector of means $(G_x\mu, G_y\mu)$ and dispersion matrix

$$\begin{bmatrix} G_x VG' & G_{xy} VG' \\ G_{yx} VG' & G_y VG' \end{bmatrix} \qquad (22)$$

This result allows us to find the formal relationship we sought to link the moments of the group-process to the moments of the individual-process.

A fuller discussion of the results that can be derived through the foregoing discussion can be found in Arbia (1988). We wish to review here only some results concerning the variance. Assuming that the individual-process is stationary with dispersion matrix given in (17), the variance of the i-th group is given by

$$ {}_{x^*}\sigma_i^2 = {}_x\sigma_i^2\, r_i^{-1}\left(1 + A_i r_i^{-1}\, {}_x\delta\right) \qquad (23)$$

where, in addition to the previous notation, A_i is the number of non-zero cross-covariances at an individual-process level between X_j and Y_j, that is the connectedness (Unwin, 1981) within the i-th group.

A first consequence of formula (23) is that, even if the individual-process is stationary, the group-process is not stationary unless $A_i r_i^{-1}$ and r_i are both constants in the different groups. This is a condition only rarely found, the only remarkable exception being when the observations are laid on a regular lattice grid.

Secondly, formula (23) provides an indirect measure of spatial autocorrelation. In fact when $\delta=0$ the variance of the group-process is simply damped by a constant term r_i^{-1} being ${}_{x^*}\sigma^2_i = {}_x\sigma_i^2 r_i^{-1}$. The decrease in variance is less sharp when we introduce a value of $\delta>0$. It is, instead, more dramatic when $\delta<0$. If we have at our disposal the variance of the process at different levels of aggregation, we can plot the value of the group-process variance against the group size. If the observed value of the variance falls above the line corresponding to $\delta=0$, then we detect a positive spatial autocorrelation; when, in contrast, it falls below this line, we detect a negative spatial autocorrelation.

The third consequence of the expression in formula (23) is that we can use it to derive measures of the loss in accuracy of a spatial database when we aggregate the information.

This measure could be simply the ratio

$$I = {}_{x^*}\sigma^2_i\ {}_x\sigma^{-2}_i = r_i^{-1}\left(1 + A_i r_i^{-1}\, {}_x\delta\right) \qquad (24)$$

For instance, in the case of data laid on a regular square lattice grid if we aggregate following a quadtree structure with group of equal size 4 we have $r_i=4$ and $A_i r_i^{-1} = 1$. As a consequence formula (23) simplifies into

$$I = 0.25\,(1 + {}_x\delta) \qquad (25)$$

where $0< I< 0.5$. This measure shows that the loss in accuracy depends on the grouping criterion and on the level of spatial autocorrelation.

Through the methodological framework described in this section we are therefore able to study how any moment of the bivariate process is affected by a transformation (aggregation or averaging) of the original areal data.

Spatial data transformation

As we stated in the Introduction, the aim of this paper is to propose a general framework to study the effects of transformations of spatial data on statistical analysis. The work contained in Arbia (1988), and reported in Section 2 considers only a particular kind of SDT, namely a transformation leading from an areal distribution to another areal distribution. However spatial data may occur in four different guises (points, lines, areas and surfaces) each of which can, in principle, be transformed to any of the others.

It is perhaps useful to distinguish at this point between two kinds of transformations that may occur in practical situations. Some transformations are, in fact, made by the researcher in order to perform some kind of statistical procedure. This is the case, for instance, in the reduction of areal data to points (e.g. centroids) to interpolate trend surfaces or when we have information about the individual location of geographical entities, but we superimpose a regular square lattice grid in order to study the pattern of the phenomenon. Other transformations, in contrast, have to do with the criteria of data collection and with the nature of the phenomenon under study in the sense that the true phenomenon is unobservable and the only data available are presented in a transformed form. This is the case for studies on personal income when we have at our disposal only regional aggregates or the case in which we do not have detailed information about the location of industrial plants, but only the number of plants in each zone.

The nature of the process is	But we observe/ analyse instead	Examples
Points	Points	- Sampling from a point process
Points	Areal data	- Quadrat count - Dirichlet tesselation (or Thiessen polygon) Count of individual geographical entities in a study area - Ecological Fallacy - Triangulation
Points	Surfaces	- Trend surface analysis
Areal data	Points	- Centroids - Spatial median
Areal data	Areal data	- Modifiable areal unit problem - Spatial sampling - Missing data - Boundary problem
Flows	Areal data	- Gravity model
Flows	Flows	- Modifiable areal unit problem and analysis of flows
Surfaces	Points	- Observing continuous surfaces at random points
Surfaces	Areal data	- Remotely sensed data - Image analysis

Table 1 Some examples of spatial data transformations.

Notice that the above distinction implicitly defines a taxonomy of errors occurring in spatial databases. The first typology of transformations leads to an error in the representation of the geographical information or to a *model error*. In contrast, the second kind of transformations derive from a *measurement error*.

Some examples of the most frequently occurring spatial data transformations are shown in Table 1 and are fairly common in geography. In each of these cases we start by considering a certain generating process and we end up, for various reasons, with a different process. We may, therefore, ask, analogously to the case presented in Section 2, what are the properties of the observed process given some hypotheses about the distributional form of the original process. This can hopefully help us to draw inferences about the generating process .

It is certainly not the aim of this paper to discuss thoroughly all the examples listed in Table 1, but only to indicate a general approach and priorities for research. It can be argued that geographical phenomena occur in practice as points, lines or continuous surfaces, but the statistical observation of them is almost invariably within areal units. My personal view is that the most interesting research area is the study of SDT leading to areal distributions and in particular the cases of point-to-area and surface-to-area transformations.

Point-to-area transformations

The problem of point-to-area data transformations can be tackled with three different strategies, namely, (i) a theoretical study of the properties of the process which generates the areal distribution in terms of the properties of the point process; (ii) the analysis of computer simulated data; and (iii) the analysis of case studies to test the validity of the theoretical derivations and, also to explore their practical relevance.

Dealing with the theoretical derivation of the areal distribution which results from a certain point pattern, we can again exploit the method based on the process density function and the method based on moments whereas, the method based on the assumption of normality is no longer applicable, since we now have to assume a discrete distribution at the individual-process level. Furthermore, we might base our study on two different assumptions.

First of all we could assume the number of individuals in each areal unit as given and attach to each of them a certain discrete distribution incorporating a contagion effect (see Neyman, 1939) to indicate, for instance the presence or the absence of the characteristic under study. In this way, using the framework briefly outlined in the previous section, it would be possible to derive the areal distribution given the individuals' distribution. For example we could derive the areal distribution of employed within administrative regions given the probability of being employed for the single individual and given the spatial distribution of individuals within the study area.

Alternatively, we could assume that the number of individuals or objects in each cell is itself a random variable, so that the areal distribution of a certain variable would be the outcome of two compounded random effects: (1) a first effect which embodies the uncertainty we have about the number of individuals in each areal unit; this number should follow some discrete contagious distribution, such as the negative binomial (see Feller, 1943; Ord, 1972); (2) a second effect which embodies the uncertainty we have about the value assumed by the variable under study by each of the individuals. In this case, we could again exploit the framework of Section 2, but with the further complexity involved in the fact that the summation index follows a probability distribution.

Concerning the problem of analyzing artificial data, we could simulate with the computer a point pattern obeying certain distributional rules. Once a series of artificial maps of geographical objects is obtained, we can then superimpose on them a grid and study the resulting spatial structure of the phenomenon.

This procedure should not create particular problems since there is now a large amount of software available to simulate observations drawn from various probability distributions. Finally the analysis of real data is required to test the validity and the practical relevance of the theoretical analysis.

Surface-to-area transformations

In the case of surface-to-area transformations we have the interesting case of the discretization of a continuous phenomenon. This is the case, for instance, of the analysis of remotely sensed images in which the underlying process is intrinsically a continuous one, but we observe it in a rectangular n by m array of (n m) pixels.

Formally we start by considering a continuous stationary stochastic process $\{X(r,s)\}$, with r and s real numbers. For instance, the process could be Gaussian with mean

$$E[X(r,s)] = \mu \tag{26}$$

and with a continuous autocorrelation function obeying the simple negative power-law (Arbia, 1989)

$$E[X(r,s);X(r',s')] = \delta(s) = s^{-v} \tag{27}$$

in which $s \perp 0$ is the distance between the point (r,s) and the point (r',s') and v is a parameter. Given these definitions, the resulting areal distribution after aggregation into pixels in a rectangular array can be derived through the integral

$$x_{ij} = {}_{i-1}\int^{i} {}_{j-1}\int^{j} x(r,s)\, dr\, ds \tag{28}$$

and the new process $\{X_{ij}\}$: i = 1,...,n; j= 1,...,m, is now the observed discrete stochastic process.

Concluding remarks

The aim of this paper was to suggest that it is possible to treat within the same methodological framework, a set of problems frequently encountered in the statistical analysis of spatial databases. In all the cases in which spatial data are not available in the form required by the problem under study we need to be able to specify what are the consequences on the statistical analysis, in order to be aware of possible distortions and, also, in order to introduce corrections. Some of the most common examples of spatial data transformations have been considered and two remarkable cases have been discussed in some detail. The case of area-to-area transformations, (as those occurring in the case of the scale problem or, more generally, when we encounter the modifiable areal unit problem) has already been analysed in a previous work (Arbia, 1988) and I have done my best to summarize the results obtained there in Section 2 of this paper. The case in which we shift from a set of point data to a set of areal data occurs fairly often in geographical analysis. Examples are the quadrat count, the Dirichlet tesselation, the count of individual geographical entities in a study area, and the ecological fallacy problem. A second frequent problem is that of transformations from a continuous surface (such as agricultural or geological phenomena) to a discrete set of observations (such as the pixels of a remotely sensed image). This paper presents some suggestion in the direction of the analysis of these point-to-area and surface-to-area transformations.

References

Anderson, T.W., 1958, *An introduction to multivariate statistical analysis.* (New York:: John Wiley)

Arbia, G., 1988, *Spatial Data Configuration in the Statistical Analysis of Regional Economics and Related problems.* (Kluwer, Dordrecht)

Arbia, G., 1989, A note on a lower bound to the negative correlation of stationary processes on a two dimensional lattice, (To appear in *Stochastic Processes and Their Application).*

Besag, J., 1974, Spatial interaction and the statistical analysis of lattice systems (with discussion), *Journal of the Royal Satistical Society* **36**, 192-235.

Cliff, A. D., and Ord, J. K, 1981, *Spatial Processes.* (London: Pion)

Feller, W., 1943, On a class of contagious distributions. *Annals of Mathematical Statistics* **14**, 389-400.

Haggett, P., 1967, *Locational Analysis in Human Geography.* (Edward Arnold: London)

Kendall, M. G., and Stuart, A., 1976, *The Advanced Theory of Statistics.* Second edition, (London: Griffin)

Neyman, J., 1939, On a class of contagious distribution applicable in entomology and bacteriology. *Annals of Mathematical Statistics* **10**, 35-57.

Openshaw, S., 1984, *The Modifiable Areal Unit Problem.* Catmog **38**, (Norwich, England: Geo Abstracts)

Ord, J. K., 1972, *Families of Frequency Distributions.* (London: Griffin)

Paelinck, J. H., and Nijkamp, P., 1975, *Operational Theory and Method in Regional Analysis.* (Farnborough, England: Lexington Books)

Ripley, B. D., 1988, *Statistical Inference for Spatial Processes.* (London: Cambridge University Press)

Tobler, W. R., 1970, A computer movie simulating urban growth in the Detroit region. *Economic Geography*, supplement **46**, 234-40.

Unwin, D., (1981), *Introductory Spatial Analysis.* (London: Methuen)

Wilson, A.G., and Bennett, R.J., 1985, *Mathematical Methods in Human Geography and Planning.* (New York: Wiley)

Section VII

It seems obvious from the content of many of the preceding chapters that errors in spatial databases are here to stay. All spatial data are inaccurate to some extent (within the meaning of accuracy adopted in this book): the important questions have to do with measurement of inaccuracy, and the development of strategies to minimize its effects.

With this in mind, it seems appropriate to end with a paper which addresses strategies for coping. Openshaw has perhaps contributed more than any other researcher in spatial analysis to the identification of spatial accuracy problem, particularly in social data. In this final chapter he discusses a number of techniques, some traditional and some novel, for survival in a world of inaccurate data.

Chapter 23

Learning to live with errors in spatial databases

Stan Openshaw

Introduction

Error and uncertainty have always been a feature of cartographic information so it hardly surprising that these aspects are also present in digital versions of analogue maps. Neither should it be imagined that any map-related spatial data exist which are error-free. Errors and uncertainty are facts of life in all information systems. There are many different causes of uncertainty and those explicitly due to GIS-based manipulations of mappable information are merely a more recent problem. However, it is also obvious that GIS has the potential to dramatically increase both the magnitude and importance of errors in spatial databases. Urgent action is needed to minimise the potential worst case consequences and also to provide a methodology for coping with the effects of processing uncertain data. Yet as Burrough (1986) points out "It is remarkable that there have been so few studies on the whole problem of residual variation and how errors arise, or are created and propagated in geographical information processing, and what the effects of these errors might be on the results of the studies made" (page 103). Indeed, there is a remarkable lack of information about the level of errors in map and remotely sensed data and, there are seemingly no available tools for measuring error in the outputs, and no methodology for assessing their significance.

Error transmission is very important because although GIS provides an accurate and virtually error-free system for manipulating map data, the data being processed are often of variable precision. The effects of combining data characterised by differing levels of error and uncertainty need to be identifiable in the final outputs. The problems result from the power of GIS to routinely perform operations on cartographic data which traditionally would not have been done, or else performed only under special circumstances, because of the problems of scale, complexity, and feature generalisation that might be involved. With manual cartographic methods many of the problems are visible and the highly skilled operator makes the necessary adjustments and knows how far the information can be relied upon. With GIS the equivalent operations are transparent, the operators are no longer so knowledgable in or aware of all the limitations of the data, and the problems are more or less completely invisible. The apparent ease by which data from different scales and qualities of map document, with different levels of innate accuracy, can be mixed, integrated, and manipulated totally disguises the likely reality of the situation. The inter-mixture of digital map and associated attribute data with different error and uncertainty characteristics, and the largely unknown and possibly data specific error reducing or error amplifying properties of a particular sequence of GIS operations all help create a spatial database of completely unknown qualities of accuracy. Its reliability is probably also spatially varying.

The resulting uncertain data may then be cherished and saved for use in subsequent operations, or else mapped, or input into a spatial analytical method, or used with a complex model as part of a spatial decision support system. Yet all these operations typically assume error free inputs. The results may well look nice. They may be used for decision making and planning despite possessing levels of uncertainty that are completely unknown and usually cannot even be guessed. Whether the spatial data errors matter depends to a large extent on the application. Near map operations based on single source data are going to be the least error prone and also the most self-correcting. The more the full power of GIS is used to build multi-source, multi-scale, integrated databases the greater the potential levels of uncertainty but whether it matters again depends on what it is used for and how it is used.

The problem is not that uncertainty exists in GIS but that the traditional response of both researchers and users has too often been to simply ignore it because the methods needed to handle it in a more explicit fashion simply do not exist. This neglect may not always matter. Many applications are probably fairly error tolerant but it would be nice to have some means of quantifying and evaluating its significance in terms of a particular application. Serendipity is no justification for the neglect of basic scientific research. Yet too much emphasis on the possible effects of errors will make new applications of GIS far more difficult than at present and may cast possibly unjustified doubts over the validity of current and past applications. Users tend therefore to be in a difficult position. If they acknowledge the presence of errors that they cannot take into account, due to the lack of relevant tools for handling it in the standard GISs, then they run the risk of being responsible for potentially unusable projects and are susceptible to the accusation that they are neglecting a possibly important source of noise and confounding factors. On the other hand, it may be that if an error audit could have been performed, then the worst fears might not be realised and the application could stand unchanged. It is very easy therefore to understand the widespread adoption of a conservative approach.

It is possible to adopt such a naive approach towards spatial data error for a number of reasons: (1) current practice is a continuation of the past albeit using new and more precise tools, it was not previously a problem so why now; (2) there are no widely used methods for measuring the uncertainty properties of the various types of spatial data and of the outputs from GIS; (3) it is not known how serious the problem really is due to a lack of empirical research; (4) there is no consensus that it is a problem that matters because its magnitude is both data and application specific; (5) there are no established algorithms and procedures for dealing with error in many of the GIS functions; (6) there are no standard methods for tracking error propagation through spatial databases; and (7) there is little knowledge about what effects the various GIS operations have on error, whether amplification, maintenance, or removal.

Despite this neglect, the major sources of spatial data error are reasonably well understood; see Burrough (1986). Goodchild (1988) lists the following: errors in the positioning of objects, errors in the attributes associated with objects, and errors in modelling spatial variation over or between objects for instance by assuming spatial homogeneity. To these may be added: errors resulting from GIS operations on spatial data (viz. transformation and interpolation), the effects of generalisation operations (viz. aggregation), the effects of model error in predictions stored as spatial data, errors due to differences of a temporal nature, and representational errors (viz. referencing area objects as point objects). A GIS working with several different data layers, from different sources and scales, might be expected to propagate errors in a highly complex fashion. There is clearly an urgent need for basic research to resolve many of the issues. This research can take many forms. It is important to measure and document the extent to which there is a problem, to develop a better understanding of error propagation through spatial databases, to identify and then classify those application areas and operations most sensitive to error, and to provide basic tools that can be added to existing GIS to handle error in various operational situations. It might well be imagined that the GIS of the future will have "error buttons" which once enabled will automatically perform various types of standard error audit and a sensitivity analysis of the results. The problem is essentially how to design and build a sufficiently general purpose system that can handle the full range of data and GIS operations.

The real problem therefore is not the elimination or the reduction of errors, instead the more immediate and seemingly simpler task is purely that of learning how to live with error and uncertainty in the spatial databases being manipulated by GIS. This task is seen as involving two major components; firstly, to develop an adequate means of representing and modelling the uncertainty and error characteristics of spatial data; and secondly, to develop GIS related methods and techniques that can explicitly take error into account during their operations with spatial data. Both tasks are clearly important but it is argued here that it is the latter which is most pressing and urgent, and can be most readily achieved within the constraints of current technology and knowledge. It is suggested that in many applications it is only necessary to be reasonable about the error assumptions rather than seek high degrees of accuracy, on the grounds that accurate error information may never be available. Furthermore, what many applications seem to need is not precise estimates of error but some confidence that the error and uncertainty levels are not so high as to render in doubt the validity of the results in a particular data specific situation. This second error handling task may proceed therefore without waiting for accurate error models to be developed and it is possible to go at least part of the way to avoiding the worst excesses of the problem.

This paper focuses on some of the methods that are available, or could be made available on a short timescale, as a pragmatic response to the broad spectrum of problems that are perceived to exist. Section 2 describes a general approach to the problem. Section 3 provides an example of how it is possible to incorporate an error handler in a point pattern analysis method and also in spatial database operations in geodemographics. Section 4 examines a different type of error that occurs when spatial data are aggregated. Finally,, Section 5 outlines a possible research agenda.

A general method for handling errors in GIS

In principle the problems of error transmission through GIS operations on spatial data can be tackled by using the existing statistical theory of error propagation (Burrough, 1986). In practice, the problems are far too complex for such a simple minded approach. The general lack of single continuous differentiable functions renders the use of explicit equations for error propagation either impossible or at least very difficult to apply. Instead it is much simpler and also far more general to seek a universal solution based around a Monte Carlo simulation approach. Openshaw (1979) describes a method for using urban and regional models in planning when the data inputs are assumed to contain errors. It has been used in a small number of modelling applications. It is suggested that similar technology can also be used here and is applicable to virtually any GIS operation or sequence of operations. The only cost is one or two orders of magnitude increase in computing times although these can be greatly reduced by a little ingenuity. Additionally, computer time factors are not always a major constraint in GIS and are becoming increasingly less so.

The basic algorithm as might be applied in GIS is as follows:

Step 1. Decide what levels and types of error characterise each data set as input to a GIS.

Step 2. Replace the observed data by a set of random variables drawn from appropriate probability distributions designed to represent the uncertainty in the data inputs. This can apply to both the geographic references as well as to the attributes.

Step 3. Apply a sequence of GIS operations to the Step 2 data. Any errors or uncertainties in exogenously supplied models, equations, and parameters also need to be simulated by randomisation.
Step 4. Save the results. The result may either be a single value or a set of values for geographic objects (e.g. coverages) or rasterised.
Step 5. Repeat Steps 2 to Step 4, M times. In practice the actual value of M may be considerably smaller than might be expected.
Step 6. Compute summary statistics or apply a Monte Carlo significance test

The distribution of results contains information of the end-state effects of simulating uncertainty in the data inputs to the GIS. These distributions may be summarised by a few statistics (e.g. median, inter-quartile range as a proxy for 50 per cent confidence limits, etc.). Alternatively, if the output coverages are rasterised, then frequency distributions can be displayed directly on top of the deterministic and error free results. The use of a Monte Carlo significance test is very important because it provides a means of identifying parts of the resulting map where the effects of simulating data uncertainty really matter.

This Monte Carlo approach is also completely general, makes very few assumptions, and is computationally tractable because in practice surprising few simulations are required. To summarise, the method involves the simulation of the uncertainty in the geographic locations of the data, by replacing the observed coordinates with random numbers drawn from a probability distribution centered on the observed values. The probability distributions would be selected to be either an accurate representation of actual levels of error or a plausible estimate of the possible degree of uncertainty that may exist if no accurate information is available. The method is independent of the actual nature of the error model. If no details of applicable error models are known then a normal distribution with a conservative standard deviation might be good enough in many applications. Remember the purpose is to offer an indication of the possible extent of output uncertainty given a reasonably plausible set of data error assumptions. This data uncertainty simulation is easily applied to point data. Line segments require more careful handling to avoid nodes being shifted and a more complex serial error model might be more useful than an error independency assumption. Similar principles can be applied to other spatial data. For example, errors or uncertainty in a pixel classification can be simulated by generating from the classification probabilities a sample of alternative categorisations.

The sequence of GIS operations in Step 3 can be as lengthy or as complex as necessary for a particular application. The availability of a macro language for many GISs greatly facilitates re-running a fixed set of GIS operations a large number of times. However, the randomisation process should use a number of different random number generator seeds to avoid possible correlation effects in the x-y coordinates.

The key question is the appropriate size of M in Step 5. Typically, it can be fairly small since precise estimates of output uncertainty are seldom required and the central limits theorem can be invoked. If the results are to be summarised by a mean statistic then values as small as 20 or 30 might be usable. A little numerical experimentation can be used to determine whether further sampling would be worthwhile. If mean values are being summarised for areas or for a grid then larger values might be needed because of spatial autocorrelation effects. In the first instance, as few as 10 simulations might be a useful starting point that will at least give some feel for what is going on.

A very useful feature is the application of a Monte Carlo significance test to the results of the sensitivity analyses. If a normal degree of confidence is to be applied (viz. a type I error of 5 per cent) then M=99 should be sufficient. This test is particularly useful for two reasons: first, to check whether the presence of error has caused the simulations to be statistically different in a consistent direction from the observed assumed to be error-free data; and second, to allow the simulations to be compared with a target value that may well have some substantive importance (viz. a threshold) and it is useful to know whether the simulation of data uncertainty causes the results to wander across the specified threshold. The significance procedure provides a quantification of the effects of the sensitivity analyses. However, too much emphasis should not be placed on the significance levels that are obtained. Classical inference is not really appropriate here because of multiple significance testing problems and the absence of a formal experimental design. However, there is no reason why this testing procedure cannot be used as a form of descriptive inference, as a guide to action rather than as a precise test of a hypothesis.

A value of M=99 might be a useful maximum initial limit but it is emphasised that it is worth checking the results for intermediate values as once the target result is ranked higher than 5th (assuming a one tail test at the 5 percent level) then the process can be stopped. So it is only necessary to run the full number of simulations if it appears that the null hypothesis is not going to be rejected. This is useful and ensures that full length computer runs are only needed if the results are likely to be of greatest interest.

Consider a classic polygon overlay as an example. Suppose that a composite multi-map overlay weighting procedure is being used to define areas less than a fixed threshold. A run is made using the assumed error-free data and the results stored in raster form. Conservative estimates are made of the likely magnitudes of map error and probability models selected. The GIS process is now repeated 99 times and the numerical results stored in raster form. It is now possible to rank the results for each raster. If the observed data occurs in either tail of the distribution then there is evidence that the input of error has caused a bias or shift in the distribution of results. It is also possible to test for each raster how many times the composite score exceeds a given threshold and a comparison made with the assumed error-free data. The ranks provide a measure of significance. For instance, if a given raster is declared to be in category 1 based on the assumed error-free data, then a count of how many times the simulation results are assigned to non-category 1 classes gives an approximate measure of significance (if M=999 then divide this count by 10). The results can be mapped. Care may need to be taken if ties are possible in which case the results should be expressed as a range of values. Those rasters where the category 1 result is found in less than say 95 per cent of the runs can be shaded and displayed on top of the deterministic results as a means of visualising the effects of data uncertainty. If the data perturbations cause a large variation in the outputs, then the user will at least know about it. The stability of the results under different error model assumptions can also be investigated.

Simpler GIS based queries can be dealt with in an analogous manner. If more complex GIS based models are used with exogenously determined parameters and error distributions, then the appropriate amounts of uncertainty need to be added to the outputs and carried forward. However, if a model is run within the GIS and any parameters estimated on the uncertain data then only the error estimates of the results would have to be added to the data if they can be computed; for example, an interpolation operation. However, there are now possible technical and logical complications. If the errors in the model(s) can be computed then their effect cannot really be judged by a single randomisation but will require a large sample. This large sample could be provided by increasing M to a larger value or by simulating the model error effect a further M times so that all subsequent GIS operations would then need to be performed M*M times instead of M. The alternative is to ignore the errors contributed by the models on the grounds that their variability will typically be dependent on the uncertain data that was input and that additionally simulation would not only greatly increase computer times but it would probably also result in double counting. These problems might be clarified by some numerical experimentation.

In all cases it would be possible to investigate the effects of error in an individual data source by holding all the other inputs constant. The error assumptions could be reduced to investigate the potential benefits of greater data precision. They could also be increased in order to determine how robust the final results really are and the extent to which they might survive much poorer quality data inputs.

The weaknesses of this approach are threefold: (1) additional computer processing times, (2) the subjective nature of the error assumptions that are used; (3) whether the users can be educated sufficiently to use this technology which might well be infeasible in a low powered PC environment; and (4) it provides only a form of sensitivity analysis which might serve to mislead. On the other hand, it is generally applicable, the technology is available now, it is well understood, it is fairly easy to apply without too much damage to existing GISs, and the methodology can be upgraded as better knowledge about the nature of spatial errors is provided by basic research.

Incorporating a spatial error handler in a sophisticated spatial analytical technique

The simulation procedure in Section 2 is generally applicable to standard GIS style of operations. However, it may not be so convenient or a satisfactory solution for more specialised analytical operations, especially where the GIS operations are embedded within a spatial analytical technique. Under these circumstances there may be no easy way to operationalise the Section 2 error simulation approach and it is necessary to consider an alternative strategy whereby spatial error handlers are explicitly built into the analytical technique itself. This is more complicated but it does offer some opportunities for inventing new error tolerant spatial analysis procedures. The possibilities and problems of this approach to spatial data error handling are discussed with respect to a novel spatial analysis technique and by reference to the design of a fuzzy geodemographic system.

Introducing a spatial data error handler into a spatial analysis technique

The Geographical Analysis Machine (GAM) of Openshaw *et al*. (1987) is used as an example of a complex spatial analytical technique in which a GIS has been embedded. It provides a framework within which to examine the types of error handlers that could be developed to deal with these special cases but which, with a little ingenuity, may also be seen as representing procedures that could be adapted to many other kinds of spatial analytical technique.

The GAM is one of the few spatial analysis techniques that tries to take spatial data error into account. The original GAM mark 1 (GAM/1) provides a form of automatic point pattern analysis that will systematically search a spatial region for evidence of localised deviations from an underlying Poisson process model. It attempts to achieve this goal by examining circles drawn around a finely spaced lattice of points in order to provide a good discrete approximation to examining all possible point locations on a plane subject to the scale of resolution of the data. The circles overlap both to ensure coverage and to take into account small amounts of locational error in the point distributions. The costs of building this analysis machine are increased computing time such that a supercomputer had to be used for the development work and the adoption of a Monte Carlo simulation approach to handle the problems of multiple significance testing and non-independence of the circle specific tests for non-randomness.

The GAM/1 strategy can be further developed to increase the opportunities for handling error. The basic GIS operation in GAM is to retrieve for millions of circles of varying sizes cancer and population at risk counts, both items are represented by point data sets accurate to 10 metres resolution but with possible positional and representational errors. These unknown errors can be considered to be area size dependent; the cancer counts related to unit post code areas which typically contain between 1 and 16 households, the population at risk data relates to census enumeration districts which typically contain about 20 households. In neither case is digital boundary information available and no details of area size are recorded although both exhibit large and systematic size variation.

One error handling strategy involves the following procedure. The search process retrieves point data for a square approximation of the circular search region. From these data for the rectangular search region (in this case a slightly larger region of which the search circle is part) it is possible to estimate average area size (number of data points divided by area of search region). One technique would be to use this information to classify the larger census enumeration district (ED) based points in the region that contains the search circle) as: within and away from the edge, within but near the edge, and outside but near the edge; see Blakemore (1984). The usual GAM/1 statistical test can be applied to data for all three regions and if significant departures from Poisson are recorded for all three then the result is obviously robust to spatial data error. Another method would be to use the average size information to generate M random versions of the data (with uniformly distributed random coordinates within 0..5*SQRT(AREA) of each coordinate within the region of the search circle and then apply the Monte Carlo test procedure discussed in Section 2. The circle would be considered to contain an excess if no more than say 5 of the 100 results were greater than the specified threshold. This route would also be possible by saving this point size information and using the fact that GAM/1 can deal with 500 or more different random data sets simultaneously (this is normally used for handling type 1 error problems) and input instead 500 different randomised point data sets to investigate the effects of positional errors.

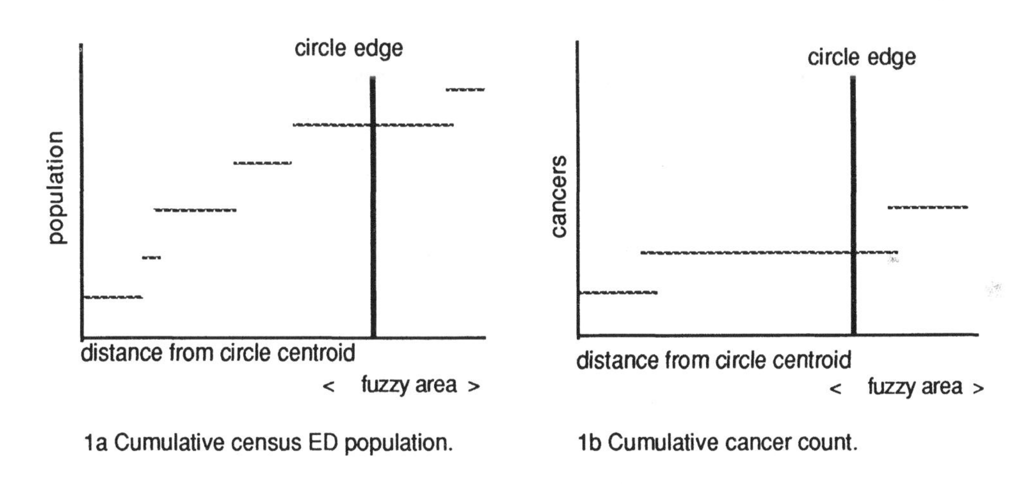

Figure 1 Discreteness in point data distribution.

A simpler alternative involves exploring the discreteness in the data near the edges of the circular search regions; see Figure 1. Rank all the nearby points by distance from the centre of the search circle. Compute cumulative population and cancer counts. Look for and examine the breaks in these distributions resulting from the lumpiness of the data near the edges of the circles. If these near edge plateau regions are still characterised by an excess cancer count then flag the circle in the normal GAM fashion for plotting. It would also be possible to estimate how much in either direction from the circle edge it is possible to go before either a significant result becomes insignificant or vice versa; if these distances are thought to be within the fuzzy region then appropriate action needs to be taken. The population and cancer counts in these fuzzy regions can be expressed as ratios of that for the complete circle and a measure of result robustness obtained.

A final suggestion is to use an adaptive circle size. Determine the range of circle sizes of interest and re-apply the test for significance after each increment in either the cumulative population at risk or cancer counts. The overall test statistic could either be a circle size adaptive measure, for example, Stone(1988)'s Poisson maximum statistic, or a standard Poisson probability statistic with an arbitrary robustness rule (viz. the circle is significant if there is a sequence of at least K contiguous distance bands which are themselves individually significant either near the target radius or over a range of radii). This greatly de-emphasises the importance of assuming that the point data are precise, or that the associated attribute data are correct, and allows flexible spatial data error handling to be incorporated into what is a fairly complex analytical procedure. In all cases the objective is that of robustness. At the end of the process there has to be some confidence that "real" results are found that are robust to spatial data uncertainty.

It is argued that these heuristic devices developed for GAM are illustrative of what can also be done for many other complex analytical tools into which a GIS is embedded rather than as is usually the case with the analytical tool being part of the GIS. In many cases this may well involve the use of Monte Carlo significance testing for statistics that previously had an analytical expression for the error-free case. Computer processing costs have fallen rapidly and look set to continue to fall for a number of years yet, so the previously largely unthinkable computationally intensive numerical approaches become practicable and in this area are probably essential.

Fuzzy geodemographics

Spatial data error is not the unique preserve of the explicitly GIS domain but occurs in both implicit and explicit forms wherever and whenever spatial data are being used. An example of such an area is geodemographics. In both Britain and the USA a large market research industry has been established that uses what might be considered a GIS product. Typically, it consists of a spatial classification based on multivariate small area census data and a geographic link to postal address files. In Britain, this involves a classification of 130,000 census EDs which have been linked to the 1.6 million unit postcodes by performing nearest neighbour matches between point references for both the EDs and unit postcodes. The resulting geodemographic system contains the following sources of error: (1) the grid-references for the postcodes are of variable precision; (2) the postcodes and EDs do not nest into each other so that a completely accurate assignment is impossible; and (3) there are ties in 10.7 per cent of the nearest neighbour linkage of postcodes to EDs. In addition, there are other sources of uncertainty that are currently excluded but which need to be taken into account, in particular: (4) the census classification is deterministic so that many EDs may be quite close in the taxonomic space to more than one neighbourhood type; (5) many of the postcodes are near to more than one ED which may belong to different neighbourhoods; (6) there is no means of incorporating genuine neighbourhood effects in the spatial sense; and (7) there are errors in the neighbourhood classification either resulting from the nature of census data or ecological inference errors due to the varying but unknown representativeness of an area profile of the people who live there. In practice these various errors and uncertainties interact to produce a situation in which the typical client has no way of knowing whether the best areas are those that were correctly targetted or else were the result of errors.

Some of the geographical linkage errors can be removed by improving the postcode-to-census ED directories or by basing small area census statistics on unit postcodes. But neither is sufficient to cope with the nature of the problem. The former can never be completely accurate and the use of smaller census reporting areas (deemed impossible for 1991) merely causes problems due to the unreliability of ratio data based on small numbers. The use of larger postcode based census aggregates removes the linkage problems but worsens precision by increasing aggregation and ecological inference errors. There is also a feeling that focusing on the possible errors in the linkage between the neighbourhood classification and postal addresses is in fact the wrong problem to tackle. An alternative approach is to see whether it is possible to handle and put to good use the uncertainty in the geo-part of geodemographics rather than attempt the currently impossible task of eliminating all such errors. Perhaps, target marketing systems are trying to be too precise and by over-targetting under conditions of considerable uncertainty important market segments are being missed.

Consider an example. Suppose you wish to mail all the households in neighbourhood type 42 areas. The traditional approach would be to identify as accurately as possible all the postcodes assigned to type 42 areas and then mail the associated addresses. This is not really very ingenious. There will be in all the cluster systems other clusters which have minority type 42 features but which are excluded from this mailing; this results from the all or nothing deterministic nature of the classifications (each postcode can only belong to one neighbourhood type). Is it not also important to be mailing some of these "nearly" type 42 areas as well? Furthermore, because of the unavoidable misassignment of some postcodes to the wrong census enumeration districts, some "incorrect" postcodes will have been included and others that should have been included left out. In addition to these geographic linkage and classification uncertainties, there is also the problem of whether to try and make some allowance for genuine neighbourhood effects. A key target area may well be the neighbours of the type 42 areas. These areas may have non-type 42 characteristics, but possess either have some minority type 42 features or aspire to the same lifestyle as their type 42 neighbours. At present these neighbourhood effects are completely missed. Should geodemographics not be trying to handle all these areas of uncertainty? Indeed, it is possible that in many applications it is these fuzzy areas, largely missed by conventional geodemographic targetting, that hold the greatest promise for new prospects and business. Perhaps the real challenge is, therefore, how to make best use of this fuzziness factor within geodemographics.

One solution is currently being tested. The object is to allow for simultaneous variation in the fuzziness of both the geo- and the demo- parts of geodemographics. The geo- part contains error and uncertainty because of linkage problems and also the need to at least consider the existence of possible spatial neighbourhood effects. The demo- part contains error and uncertainty because of the possibility of missclassification and ecological inference errors in the spatial classification. By categorising the levels of both sources of uncertainty into a small number of classes and then classifying a mailing response by both variables, it is possible to obtain a more reasonable picture of what is actually happening.

Figure 2 provides an illustration of this fuzzy targetting system. Cell 0-0 contains no error and is equivalent to current practice. Cell 1-0 allows for a small amount of error in the geographic linkage and will include in any targetted neighbourhood cases which would now be assigned to the target neighbourhood although originally their nearest neighbour assignment allocated them to a different neighbourhood type. Cell 2-0 increases this process and cell 3-0 even more so. Going down the columns, cell 0-1 consists of those cases which would now be assigned to the target neighbourhood with only a small amount of error in the classification process (for instance, the loss in classification efficiency would be small in making such a slightly suboptimal assignment). With cell 0-2 the loss in taxonomic performance would still be small but slightly greater, and so on. The really interesting cells are those in between these two simple cases. Cell 1-1 defines those cases which are in the target area only because of a small amount of fuzziness in both the geographic and the taxonomic assignments; and so on. From such a table, it is but a small step to determine the optimal subset that provides the best results for target marketing a particular client's data, with an optimal degree of geo- and demo- fuzziness. The tabulation is adaptive and will adjust according to the types of residential areas that respond best. This smart target marketing system operates by seeking to exploit (rather than remove) the areas of uncertainty in geodemographics. It

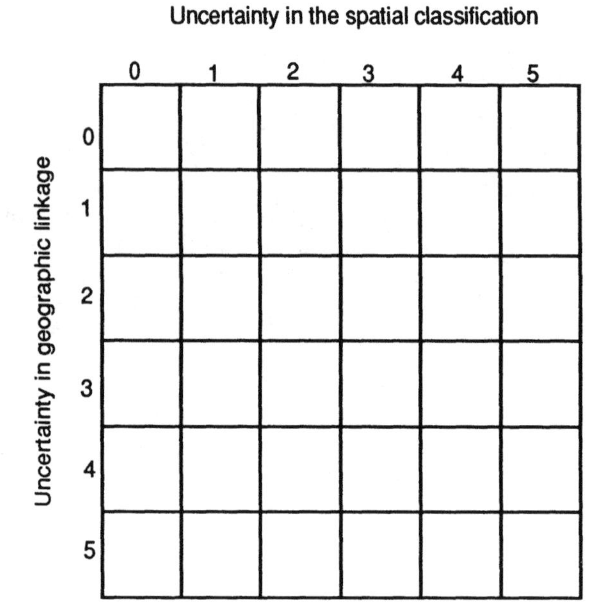

Figure 2 A fuzzy geodemographic system.

no longer matters that the geo- and the demo- parts of the technology are not perfect, the imperfections are taken into account and handled to maximum benefit. It is expected that systems based on this technology will be available in 1989 and that by the mid 1990's most good geodemographic systems could be using similar concepts.

The message is again quite clear. There is a considerable amount that can be done to take into account, in a realistic manner, all the sources of uncertainty that impinge on spatial data as viewed from a particular application. Some of the spatial error handling procedures will by their nature tend to be universal in applicability, others specific to particular techniques and data. In all cases there is a cost, due to the greater complexity of the process. In this example, the cost is not processing power; instead it comes in the form of a relational database whereas previously all that was required was a random access lookup in a fixed directory. Now each client record has to be tagged with the various classification and geography fuzziness factors. There is also the added complexity of deciding how best to explain all this hidden technology to the end user who more than anything else expects any improved method to be no more difficult to understand or explain than that which it seeks to replace. It is possible by careful packaging but doubtlessly similar end-user acceptability problems will recur in many other areas.

The modifiable areal unit problem

Some forms of spatial data uncertainty are not properly described as being errors but are of considerable practical significance and behave as if they are errors. The aggregation problem is a major source of potential uncertainty in spatial data. The spatial aggregation of geographic data is error inducing only if there is some notion of what the "true" or "correct" result should be. One definition of truth might be based on the micro-data that becomes areal data only after aggregation but much of the spatial data that are aggregated have no obvious pre-spatial form with which meaningful comparisons can be made or the micro-data are not released for analysis. Additionally, once micro data (for example, individual census data) are spatially aggregated (for example into census tracts) the measurement scales change and the nature of the object (viz. person compared with the small area in which the person lives) also changes in a rather fundamental way. As a result with spatially aggregated data there is seldom any simple definition of truth against which to measure the errors or uncertainty induced by the generalisation inherent in the aggregation process.

The aggregation problem is now of much greater operational importance and is no longer purely an academic curiousity because GIS frees the user from having to use a limited number of fixed, usually "official", sets of areas as objects for geographic analysis; areas which reflect the whims and fancies of whoever designed the zoning systems. This is important because as Kendall and Yule (1950) so clearly recognised ".. we must emphasise the necessity in this type of work, of not losing sight of the fact that our results depend on our units" (p 313). That is, the results of studying zonal data are dependent to some degree on the definition of the zones and these definitions are arbitrary, modifiable, and usually represent a particular arbitrary choice out of a tremendously large number of alternatives. Suddenly, it is possible to design bespoke spatial reporting units and to consider using data aggregations that are most appropriate for a particular application. The problem is the absence of guidance as to how best to perform this design process given the immense degrees of freedom that exist and in the knowledge that whatever zones are selected will influence the results to some unknown but maybe large degree.

This is the very essence of the modifiable areal unit problem. The same base data can be aggregated in a immensely large number of different ways producing a distribution of results. Currently there is no technology for determining what might be considered the "best" aggregation for any particular purpose. It is not surprising that different zoning systems may yield what are considered to be different results because the zoning system interacts with whatever model or method of analysis is being used, and acts as a variable performance spatial pattern detector. The clearest example is with interaction data where micro-level trips are only detected by a zoning system when the trip crosses one or more zonal boundaries. As far as the zoning system is concerned, intra-zonal trips are non-detected interaction, whereas the inter-zonal trips are of considerable interest but whether they can be well represented by a spatial interaction model depends on their nature. The patterns of interaction and the fit of the model can often, it seems, be manipulated within wide limits by changing the zoning system. However, beware! The range of results obtained in real world examples also depends on the metric of the test statistic being examined and the nature of the aggregation generator. Randomly generated zoning systems seem to produce a distribution of results centred on the values obtained from analysing the data at the finest spatial scale. The range of results obtained in this way have limited applicability to the real world where the zoning systems are not generated in a random fashion.

Patterns can also be made to vary by changing the scale of the analysis. The existence of scale effects is well known but their importance is often over-emphasised. The scale problem can be viewed as deciding how many large regions (M) the original base data of (N) small areas are to be aggregated to. It is over-rated because there are only relatively few choices (N-1) and the selection of data for any particular scale (value of M) also requires a decision as to which of the tremendously large number of alternative aggregations of N small areas into M larger areas is to be used. There is a combinatorial explosion once N exceeds a very small number and there is no longer any basis for a rational choice. One useful aspect here is that spatial data aggregation might be expected to diminish the effects of error in the data being aggregated, because of the averaging and smoothing operations. However, if the importance of spatial data errors is diminished, it is more than compensated for by a massive increase in uncertainty due to scale and aggregation effects. One problem is that there are no rules for designing zoning systems for use as pattern detectors, as viewed by a particular analytical procedure. There are also strong and conflicting arguments about whether it is safe from a scientific point of view to be seen to be explicitly engineering zoning systems. However, the conventional scientific view is hardly helpful here, in that it would insist that any zoning systems are defined prior to any data analysis or knowledge of the data and that any rules should be completely independent of any intended analysis. This is of course not particularly useful in a spatial context because it is equivalent to the current practice of using virtually any zoning system and then expecting to obtain useful results that have a substantive interpretation. It should not be forgotten that modifiable data will result in modifiable results! It is not possible to overcome this problem by ignoring it and by pretending that it does not exist. The only real solution is to stop analysing zonal data and that is hardly helpful advice.

Openshaw (1984) has commented that perhaps the modifiable areal unit problem is not so much a massive problem than a potentially very powerful pattern analysis tool. If the patterns as detected by a particular method vary in their nature and intensity when the zoning system is changed, then one obvious answer is to try and manipulate the zoning system to maximise a particular method's pattern detection properties, perhaps as measured by its goodness of fit. The analogy with parameter estimation might be helpful here. The configuration of the resulting optimal zoning system might well be considered as a visualisation of the best fitting data as seen from the point of view of a particular model or technique. In some ways this view of the modifiable areal unit problem is analogous to a cartogram, so perhaps it might be important one day to try and interpret what the resulting zonal map patterns mean.

This optimal zoning approach is not really all that different from a regionalisation procedure. It is seemingly quite acceptable to aggregate data to minimise either a within cluster sum of squares or to optimise an intramax function. The resulting classifications are often regarded as useful descriptions of the data and afforded a geographic interpretation. So why not extend the basic principle and develop an aggregation process that no longer optimises a taxonomic function but uses instead some measure of the performance of a particular model of spatial pattern?

One solution to handing aggregational uncertainty is therefore to develop not aggregation invariant methods (which might well be impossible) but to invest in formalised zone design procedures. The criteria used can be left to the investigator to determine. A general zone design procedure has been developed (Openshaw, 1977, 1984, 1987). The zone design task is viewed as an optimisation problem in which the principal unknown variable is the zoning system. The basic optimisation problem can be specified as:

$$\textit{optimise } F(W^*, P^* Z, C) \qquad (1)$$

where F(..) is some nonlinear function defined on the spatial data that result from the aggregation in W*; W* represents an aggregation of data for N small original zones or objects into M larger zones subject to the contiguity restrictions contained in C; Z is the original N object spatial data set; and P* is optionally a set of unknown parameters dependent on W* that may also require estimation. The W* defines the approximately optimal aggregation which is of interest. Implicit also are linear constraints on W* to ensure that each of the N small zones are included exactly once and that each region is composed of contiguous small zones. Equation (1) might be regarded as a kind of impossible to solve integer nonlinear optimisation problem. At present it can only be attacked via stochastic optimisation heuristics and there is no proof of optimality; indeed, global optima will not exist for this class of problem. Two procedures have been developed that do seem able to identify what appear to be approximately optimal solutions; one is based on a single and multiple edge swapping heuristic, and another works by embedding the former within a simulated annealing framework. Cellular automata alternatives would also appear to be feasible. Finally, it is noted that a range of additional constraints can be added. These may be zone specific (viz. approximate equality of size) or data related (viz. that the aggregated zonal data are approximately normal) by using a penalty function approach; see Openshaw (1978).

This automated zoning procedure (AZP) allows a unified approach to be adopted. A range of different objective functions could be used to allow the GIS user to at least try and bring the modifiable areal unit problem under control. Whether the AZP will ever be successful as a means of visualising spatial pattern does depend on further research but it should not be dismissed out of hand merely because it is unorthodox in relation to what is currently considered the orthodoxy. GIS is a revolution in the ways of handling spatial data and it is quite likely that it will also occasion new styles and methods of spatial data analysis. Meanwhile, it is important to provide the GIS user with at least a tool by which to manipulate spatial aggregation in the same way as it is now possible to manipulate other aspects of spatial map data and by so doing provide a means of handling the errors and uncertainty caused by the aggregation process.

A research agenda

The real challenge is to develop useful tools for handling the effects of errors in spatial data based on currently available technology. It may be possible under certain conditions to adopt an analytical approach but this is unlikely to be sufficiently general to meet most needs. The design goal is to devise methods that can be applied to all types of geographic data, that incorporate any error model, can handle any sequence of GIS operations no matter how complex, and provide some indication of confidence in the output results. A class of error handlers is needed that can be applied to the whole of GIS and to any of its parts. It is suggested that the Monte Carlo simulation approach outlined in Section 2 does meet this need, whilst the procedures discussed elsewhere are indicative of other methods that have been developed to handle more specialised analytical methods.

It is argued, therefore, that the task of developing spatial data error handlers, based on arbitrary but reasonable assumptions about the nature and magnitude of errors in spatial databases, is feasible with current technology. Maybe the problem cannot currently be completely or accurately solved but it is possible now to develop methods and procedures that offer operational and pragmatic solutions that at least seem to be heading in the right direction. The research strategy advocated here has three components to it: (1) improve user awareness and education as to the existence and importance of errors in spatial databases by performing empirical research into the magnitude of the problem for a broad cross-section of databases and applications; (2) make available GIS-portable software for the principal types of error handlers and error simulation along the lines of those described here and persuade the principal vendors to include them into their systems; and (3) develop a similarly portable set of zone design procedures. The additional computer resources needed to drive some of these computationally intensive procedures may in the short-term result in the need to offer a limited spatial database error audit service as part of the user awareness campaign and as a means of investigating the extent of the problem in real applications. However, given the expected trends in workstation performance there are no major problems in operationalising any of these suggestions in the longer term (within 3 years).

Acknowledgements. The author gratefully acknowledges the support of the ESRC in funding the NorthEast Regional Research Laboratory in Newcastle.

References

Blakemore, M., 1984, Generalisation and error in spatial data bases. *Cartographica* **21**, 131-139.

Burrough, P. A., 1986, *Principles of Geographical Information Systems for Land Resources Assessment*, Oxford University Press, Oxford

Goodchild, M. F., 1988, The issue of accuracy in global databases. In *Building Databases for Global Science*, Mounsey, H., and Tomlinson, R., (eds.), pp. 31-48. (London: Taylor and Francis)

Kendall, M. G., Yule, G. U., 1950, *An Introduction to the Theory of Statistics*. (London: Griffith)

Openshaw, S., 1977, A geographical solution to scale and aggregation problems in region-building, partitioning, and spatial modelling, *Transactions of the Institute of British Geographers*, New Series **2**, 459-472.

Openshaw, S., 1978, An optimal zoning approach to the study of spatially aggregated data. In *Spatial Representation and Spatial Interaction*, Masser, I., and Brown, P. J. B., (eds), pp. 95-113. (Leiden: Martinus Nijhoff)

Openshaw, S., 1979, A methodology for using models for planning purposes. *Environment and Planning A* **11**, 879-896.

Openshaw, S., 1984, *The Modifiable Areal Unit Problem*, CATMOG **38**. (Norwich: Geo Abstracts)

Openshaw, S., 1987, The aggregation problem in the statistical analysis of spatial data. In *Informazione ed analisi statistica per aree regionalie subregionali*, pp 73-93, Societa Italiana di Statistica, Perugia

Openshaw, S., Charlton, M., Wymer, C., Craft, A. W., 1987, A Mark 1 geographical analysis machine for the automated analysis of point data sets. *International Journal of GIS* **1**, 335-358.

Stone, R.A., 1988, Investigations of excess environmental risks around putative sources: statistical problems and a proposed test. *Statistics in Medicine* **7**, 649-660.

Subject index

Page numbers in italics refer to Figures and Tables

Author index

Page numbers in italics refer to Figures and Tables